Viral Nanotechnology

Viral Nanotechnology

Edited by Jake Mason

SYRAWOOD
PUBLISHING HOUSE
New York

Published by Syrawood Publishing House,
750 Third Avenue, 9th Floor,
New York, NY 10017, USA
www.syrawoodpublishinghouse.com

Viral Nanotechnology
Edited by Jake Mason

International Standard Book Number: 978-1-64740-366-9 (Hardback)

Cataloging-in-publication Data

Viral nanotechnology / edited by Jake Mason.
p. cm.
Includes bibliographical references and index.
ISBN 978-1-64740-366-9
1. Nanotechnology. 2. Virology. 3. Viruses. 4. Biotechnology. 5. Nanomedicine. I. Mason, Jake.
T174.7 .V57 2023
620.5--dc23

TABLE OF CONTENTS

PREFACE

Viruses are submicroscopic, subcellular agents that cannot replicate outside of a host cell. They can infect all living things, including birds, bacteria, reptiles, mammals, protozoa, insects, plants, fungi, and fish. Virology refers to the scientific study of the biology of viruses and viral diseases, comprising their evolution, physiology, biochemistry, clinical aspects, molecular biology, distribution, and ecology. Due to their small size, viruses are considered to be naturally occurring nanoparticles. The usage of viruses for biomedical purposes is studied under the field of viral nanotechnology. Viruses are used within this field as vectors for gene delivery, immunotherapy and vaccination. Bacteriophages are a type of viruses which are used extensively for medical purposes. These viruses infect and kill bacteria without harming human cells. This book unravels the recent studies in the field of viral nanotechnology. It is a valuable compilation of topics, ranging from the basic to the most complex advancements in this field. This book will prove to be immensely beneficial to students and researchers in the field of viral nanotechnology.

Significant researches are present in this book. Intensive efforts have been employed by authors to make this book an outstanding discourse. This book contains the enlightening chapters which have been written on the basis of significant researches done by the experts.

Finally, I would also like to thank all the members involved in this book for being a team and meeting all the deadlines for the submission of their respective works. I would also like to thank my friends and family for being supportive in my efforts.

Editor

1

Research Progress of M13 Bacteriophage-Based Biosensors

Jong-Sik Moon [1,*,†], Eun Jung Choi [2,†], Na-Na Jeong [3,†], Jong-Ryeul Sohn [3], Dong-Wook Han [4] and Jin-Woo Oh [2,5,*]

1 National Core Research Center for Hybrid Materials Solution, Pusan National University, Busan 46241, Korea
2 Department of BIT Fusion Technology Center, Pusan National University, Busan 46241, Korea; eunjung721203@gmail.com
3 Department of Public Health Sciences, Graduate School, Korea University, Seoul 02708, Korea; nana8931@naver.com (N.-N.J.); sohn1956@korea.ac.kr (J.-R.S.)
4 Department of Cogno-Mechatronics Engineering, College of Nanoscience & Nanotechnology, Pusan National University, Busan 46241, Korea; nanohan@pusan.ac.kr
5 Department of Nanofusion Technology, College of Nanoscience & Nanotechnology, Pusan National University, Busan 46241, Korea
* Correspondence: jsmoon@pusan.ac.kr (J.-S.M.); ojw@pusan.ac.kr (J.-W.O.)
† These authors contributed equally to this work.

Abstract: Recently, new virus-based sensor systems that operate on M13 bacteriophage infrastructure have attracted considerable attention. These systems can detect a range of chemicals with excellent sensitivity and selectivity. Filaments consistent with M13 bacteriophages can be ordered by highly established forms of self-assembly. This allows M13 bacteriophages to build a homogeneous distribution and infiltrate the network structure of nanostructures under mild conditions. Phage display, involving the genetic engineering of M13 bacteriophages, is another strong feature of the M13 bacteriophage as a functional building block. The numerous genetic modification possibilities of M13 bacteriophages are clearly the key features, and far more applications are envisaged. This paper reviews the recent progress in the application of the M13 bacteriophage self-assembly structures through to sensor systems and discusses future M13 bacteriophage technology.

Keywords: virus; biosensor; M13 bacteriophage; color sensor; phage display

1. Introduction

1.1. M13 Bacteriophage

Distinguishing infinitesimal amounts of chemicals or microbials in a precise and simple manner is one of the most important techniques in academia and industry. Classical sensor technology is categorized as electromagnetic (gamma radiation, optics, microwave, radio wave, and Eddy current), mechanical (sound, MEMS, and fluid), magnetic, chemical (affinity, catalytic reactions, electrochemistry, and biochemistry), nuclear (nuclear magnetic resonance), and a combination of them (opto-acoustics and membrane technology) [1,2]. Generally, the key features of sensor technology are the low cost, small size, robustness, selectivity, and sensitivity [1,3].

Recently, the use of ecofriendly materials has been one of the main points in research and industrial manufacturing. The M13 bacteriophage, which can be categorized as a natural polymer, has been a major research focus owing to its target-specific response in chemical reactions with outstanding sensitivity [4]. The bacteriophage is a human-safe viral material that only infects certain criteria of bacteria, such as *Escherichia coli.* [5].

The filamentous type of M13 consists of approximately 2700 copies of the spirally arranged pVIII major coat protein on the body, with 5 to 7 copies of pIII, pVI, pIX, and pVII minor coat proteins at each end. M13 has a regular form, with a length and diameter of ~880 nm and ~6.6 nm, respectively. The genome structure and protein sequences of the M13 bacteriophage are fully understood. Therefore, M13 bacteriophages are very simple to genetically engineer to achieve the desired biochemical properties [6–9]. Figure 1 shows the basic protein structures and genetic engineering pathways of M13 bacteriophages. For example, genetically modifying the pVIII protein of M13 bacteriophage to possess the tryptophan–histidine–tryptophan (WHW) peptide sequence results in it showing specific binding affinity to nitrotoluene derivatives [4,10–14]. Consequently, the specific binding affinity to target materials, including chemical and biological materials, by simple genetic engineering provides great possibilities to use M13 bacteriophages as the core material of sensors.

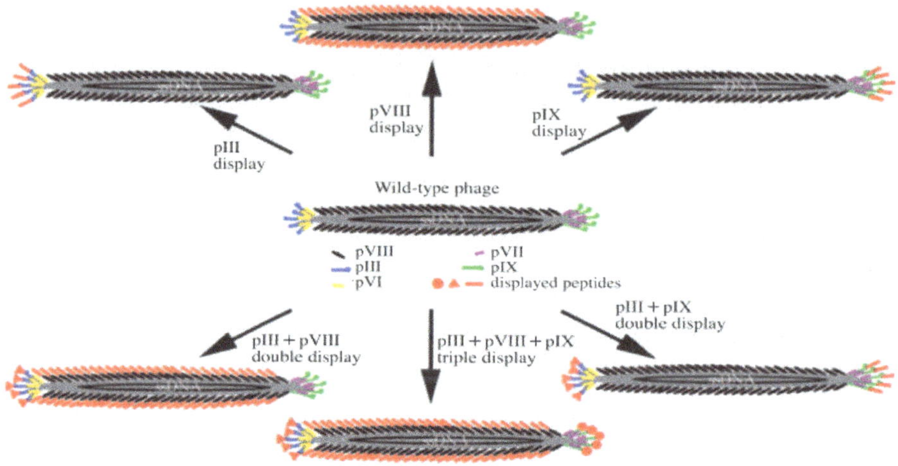

Figure 1. Basic structure of the M13 bacteriophage and the possible pathway of generic engineering. Reproduced from [9], with permission from WILEY-VCH, 2009.

M13 bacteriophages have a filamentous shape and different protein configurations at each end. One end has the pVI and pIII protein and the other has the pVII and pIX protein. By the well-defined, anisotropic structure and monodispersity caused by the natural character of M13, the material can be fabricated easily in hierarchical self-assembled structures using appropriate techniques [4,14]. The structure and behavior of M13 bacteriophages during fabrication are similar to those of liquid crystals. The bacteriophage can also be modified chemically and genetically to reveal a range of features, such as desired functionality and target-specific affiliation [11–13].

The target-specific reactivity character of the M13 bacteriophage makes it advantageous for sensor applications. The other strong point of the M13 bacteriophage is its ability to form higher-order dimensional structures in a specified order over self-assembly technology [13]. High-dimensional structures using natural building blocks, such as the M13 bacteriophage, were inspired by natural systems, such as the collagen bundle structure [4]. Although the imitation of a natural assembly structure on the micro scale using artificial building blocks, such as polymers or artificial nanomaterials, is progressing slowly, due to uncertainty and complexity, using natural building blocks such as M13 bacteriophage can be a solution. The filament structure and short dispersion of the M13 bacteriophage allow liquid crystal like behavior in suspension under controlled conditions [15,16]. This liquid crystal behavior of the M13 bacteriophage can be managed mostly by the concentration of the M13 bacteriophage aqueous suspension. The crystal structure of self-assembled M13 bacteriophages shows nematic-, cholesteric-, and smectic-phase at low, medium, and high concentrations, respectively [16].

M13 is also attractive owing to its easy growth and handling characteristics. The use of natural substances as a building block makes a synthetic process possible at low cost under a non-toxic environment. The fabrication process associated with M13 bacteriophages using aqueous self-assembly technology allows easy processing without the need for organic solvents and additional processing. By using the pulling technique, concentration of M13 bacteriophage aqueous solution and polling speed of substrate are main factors in the process [17,18].

An additional outstanding benefit of M13 bacteriophages is the easy genetic modification for the specific binding affinity to definite target substances. Each coat protein can be modified genetically using phage display technology. Random phage libraries containing more than 1.0×10^{11} random peptide sequences of M13 bacteriophages are screened for specific target materials [19,20]. The binding phage on the target material is then selected and the bacteria are infected and amplified. After repeating this process to isolate the target-specific binding phage, the target-specific peptide can be recognized by DNA analysis [21,22]. Recently, the technology has developed a M13 bacteriophage that binds specifically to several inorganic resources, e.g., GaAs, GaN, Ag, Pt, Au, Pd, Ge, Ti, SiO_2, quartz, $CaCO_3$, ZnS, CdS, Co, TiO_2, ZnO, CoPt, FePt, $BaTiO_3$, $CaMoO_4$, and hydroxyapatite [21–25]. The M13 bacteriophage can be made to have specific functionality by genetically engineering to carbon-based nanomaterials, such as C_{60}, graphene, and carbon nanotubes [24,25]. Moreover, these specific binding capacities of the M13 bacteriophage can be used as templates to produce various nanostructures, such as highly monodisperse nanostructures [26–28]. This review introduces the current advances of M13 bacteriophage-based biosensors.

1.2. M13 Bacteriophage-Based Biosensor

The M13 bacteriophage has been used to improve existing sensor systems. Förster resonance energy transfer (FRET) is based on long-term bipolar interactions between excited state donors and ground state recipients. The distance of each N-terminal to the end of the peptide on the surface of a M13 bacteriophage is approximately 3.2 nm (oa) and approximately 2.3 nm (ob), respectively. Wang et al. recently reported the fabrication of FRET-based lattice probes using the M13 bacteriophage [29,30]. Owing to the regular organization of the M13 bacteriophage on the substance, a thin coating film of M13 bacteriophage can be used for surface plasmon resonance (SPR) measurements. Genetically engineered M13 bacteriophages, which possess the Arg-Gly-Asp (RGD) peptide sequence, have been used to detect the SPR signal of the cell proliferation rate and the morphology of cells [31]. The signal strength of surface-enhanced Raman spectroscopy (SERS) depends strongly on the distance between the target molecules and novel metal surface. Therefore, the controlled organization of metal nanoparticles and Raman active dye is the critical point. Cha et al. introduced an Au@Ag-core shell nanoparticle using an M13 bacteriophage as the template. In this approach, nanoparticles were functionalized by DNA-conjugated M13 bacteriophage, and showed a 75 times stronger Raman signal than DNA-functionalized nanoparticles without the M13 bacteriophage. This was mainly caused by the large number of functional moieties on pVIII protein of the M13 bacteriophage [32]. The M13 bacteriophage could also be used medically because of the strong specific reactivity to a particular reactant. Mao et al. fabricated M13–liposome–ZnPc (zinc phthalocyanine) for a more stable and upgraded cancer drug delivery system [33–35]. Simple genetic engineering could reveal target-specific affinity to the M13 bacteriophage. Oh et al. engineered the pVIII protein of M13 bacteriophage to have the AXXXWHWQXXDP (WHW) sequence and showed excellent binding affinity to trinitrotoluene (TNT), as shown in Figure 2a,b [36,37]. The modified M13 bacteriophage was then fabricated as a color film structure through a simple pulling technique. As shown in Figure 2c,d, the M13 bacteriophage-based color sensor could detect the gas phase of TNT down to the 300 p.p.b level and showed superior selectivity among TNT, Dinitrotoluene (DNT), and Nitrotoluene (MNT) [4].

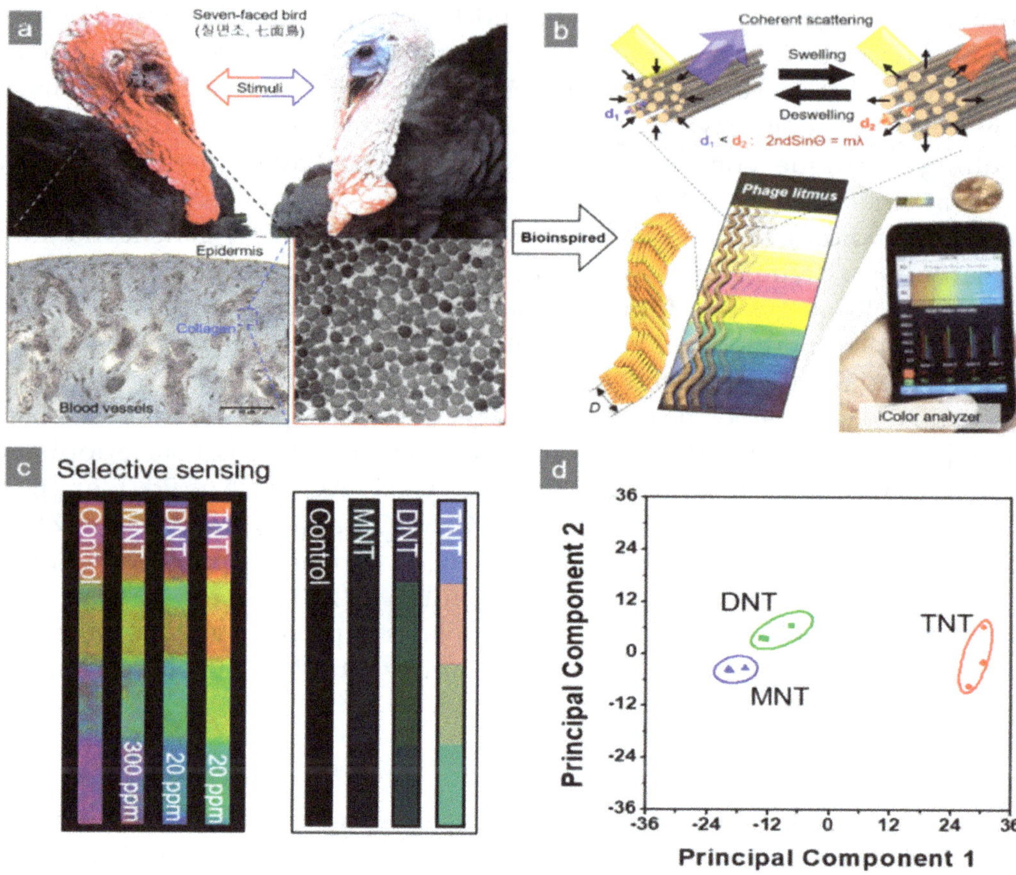

Figure 2. Nitrotoluene derivative detection using a M13 bacteriophage-based color sensor. (**a**) Bio-mimic structure of turkey collagen, (**b**) reversible color responding of a color sensor, (**c**) sensitivity of M13 bacteriophage-based color sensor upon exposure to nitrotoluene derivatives, and (**d**) principal component analysis (PCA) plot of the color changes. Reproduced from [4], with permission from Springer Nature, 2014.

2. M13 Bacteriophage-Based Protein and Microorganism Sensing

Indeed, using a natural M13 bacteriophage without any genetic engineering provides a great opportunity because of the abundant negative charge on the surface protein. The α-helical major coat protein (pVIII) has glutamate amino acid on the end of its sequence. Therefore, the dipole moment of the whole pVIII protein is directed from the outside (*N*-terminus) to the core (*C*-terminus). The charge distribution of the *C*-terminus (positive) and *N*-terminus (negative) induces a strong dipole to M13 bacteriophages possessing a natural negative charge [38]. This property allows the M13 bacteriophage to mix easily with positively charged materials, such as carbon nanofiber (CNF). Recently, Niedziòłka-Jönsson et al. reported a M13 bacteriophage and CNF complex structure for cysteine detection. The M13 bacteriophage and CNF were mixed in an aqueous solution, and it was assumed that the electrostatic interaction and π-π interaction between the M13 bacteriophage and CNF promoted an even distribution of CNF without aggregation, which occurred constantly when CNF was used alone in solution [39–41]. As a result, the addition of M13 bacteriophage to CNF increased the capacitive current with the growth of the faradaic current. An overall improvement of the electrochemical properties of the glassy carbon electrode (GCE)–CNF–M13 bacteriophage complex electrode applied to the electrocatalytic oxidation of the cysteine was observed and enhanced results were obtained, as shown in Figure 3 [39]. The fabrication in this study simply dropped CNF solution on the substrate (indium tin oxide (ITO) glass) then M13 bacteriophage dropped on the dried CNF. Although the process is not yet optimized, the study shows promising results. Using wild type M13 bacteriophage for this experiment was another strong point due to the simple preparation.

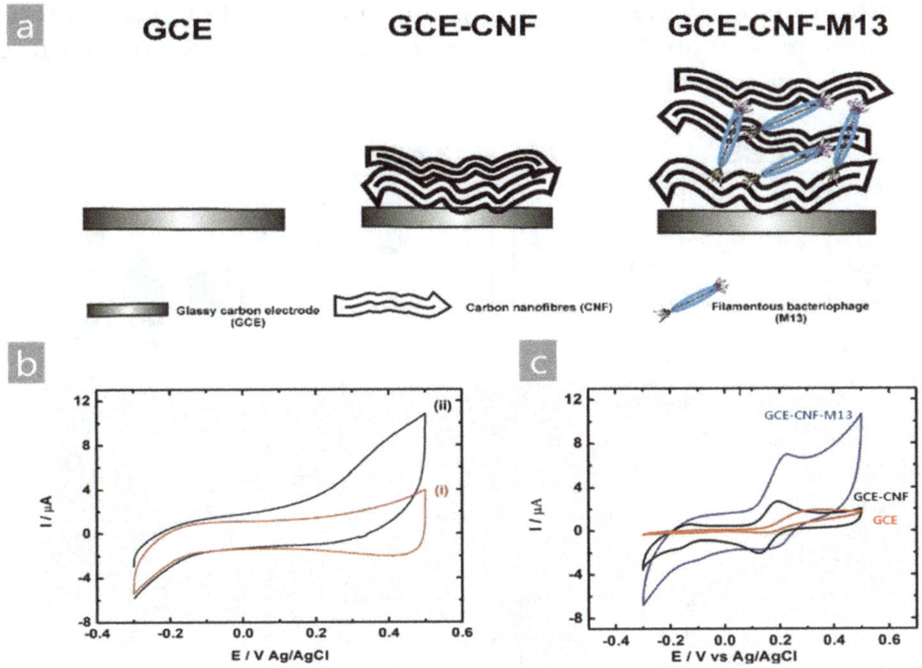

Figure 3. (**a**) Schematic diagram of the device fabrication of the CNF and M13 bacteriophage complex electrode, (**b**) cyclic voltammograms (3rd scan) of (i) glassy carbon electrode (GCE)–CNF and (ii) GCE–CNF–M13 bacteriophage treated in 1 mM cysteine solution in PBS, and (**c**) cyclic voltammograms (2nd scan) of the bare GCE, GCE–CNF, and GCE–CNF–M13 bacteriophage. Reproduced from [39], with permission from Elsevier, 2019.

As mentioned above, simple genetic engineering provides a specific interaction to a designated target material. Lladser et al. genetically engineered colorectal cancer (CRC)-specific M13 bacteriophages. The M13 bacteriophage was modified to have carcinoembryonic antigen (CEA)-specific moiety. CEA exists abundantly in colorectal cancer and is known to support the malignant features of colorectal cancer cells such as cell adhesion, cell-to-cell interaction, and signal transduction [42,43]. The CEA level clearly has metastatic potential, cancer progression, differentiation, and apoptosis of CRC cells [44–46]. In this study, CEA-specific (αCEA) M13 bacteriophages were applied to tumor cells for the tumor infiltration of neutrophils, macrophages, and the maturing of dendritic cells in tumor-draining lymph nodes [44]. For M13 bacteriophage genetic engineering, they transformed *E. coli* with a pSEX81 surface expression phagemid vector, which possesses CEA-specific single-chain fragment variable (scFv). The scFv was inserted at the NcoI and BamHI restriction sites upstream of the pIII protein sequence. *E. coli* were infected with a multiplicity of infection by a hyperphage that contains all genes of structural proteins of M13 bacteriophage except for the pIII protein [47]. The αCEA-M13 bacteriophage exhibited strong binding affinity to both CEA and CEA-expressing tumor cells (CT26) in vitro. The applied αCEA-M13 bacteriophage effectively suppressed a mouse cancer model compared to the phosphate-buffered saline (PBS)-control and wild type M13 bacteriophage experiment by successful intratumoral and systemic administration. Here, they confirmed that the tumor protection provided by αCEA-M13 bacteriophage occurred via $CD8^+$ T cells because the reduction of circulating $CD8^+$ T cells completely eliminated the antitumor protection. Figure 4 shows that the macrophage ($F4/80^+$) increased from approximately 24% to 54%; the neutrophils increased from approximately 6% to approximately 42%, and the tumor-infiltrating dendritic cells decreased from approximately 6% to approximately 2%. In other words, the application of αCEA-M13 bacteriophages to tumor cells resulted in an increase in the tumor infiltration of innate immune cells and maturing of dendritic cells at lymph nodes. Figure 5 shows decreased tumor volume and increased survival rate of mice by treating with αCEA-M13 bacteriophage. Their studies can be regarded as a successful suggestion of potential immunotherapy against CRC [44]. Antigen–antibody reaction is one of the most simple and widely used treatment methods in the medical field. However, it is still the most useful clinical method. Using highly concentrated functional moiety on the M13 bacteriophage, these types of treatments can be more effective as immunotherapy against CRC.

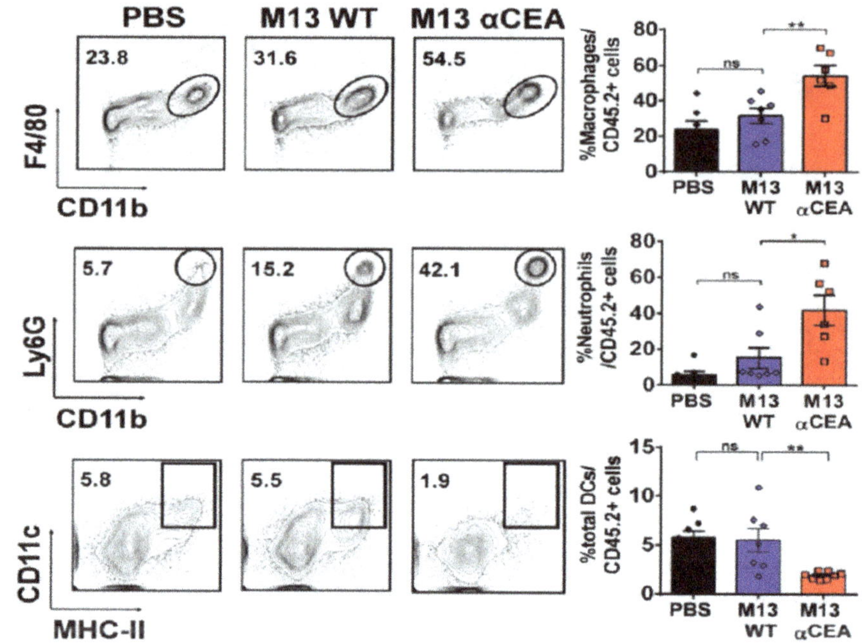

Figure 4. Treating the tumor microenvironment using PBS, wild type M13 bacteriophage, and αCEA-M13 bacteriophages for 24 h. Contour plots shows tumor infiltrating macrophages ($F4/80^+CD11b^+$), (**upper left**), neutrophils ($CD11b^+Ly6G^{high}$), (**middle left**), and dendritic cells ($CD11c^+MHC\text{-}II^+$), (**lower left**). The percentage of each plot (**right-hand panel**). Reproduced from [44], with permission from Springer Nature, 2017.

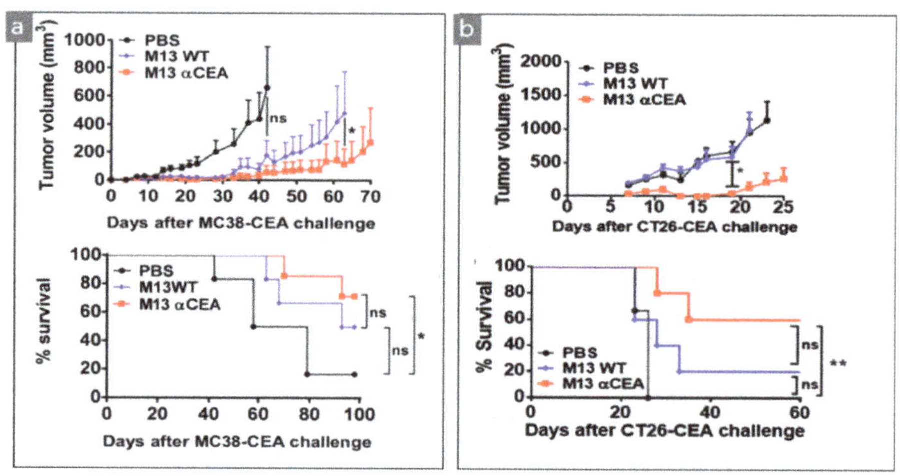

Figure 5. Growth of different CEA tumors ((**a**) MC38-CEA, (**b**) CT26-CEA) and survival of mice treated with M13 bacteriophage with or without αCEA. Reproduced from [44], with permission from Springer Nature, 2017.

Salmonella detection using a M13 bacteriophage is one of the largest fields of M13 bacteriophage research and could feasibly replace conventional methods, such as the polymerase chain reaction (PCR) and immunology-based assays [48–50]. Among the many types of pathogenic bacteria, *Salmonella* detection has received more attention because of its prevalence [51]. Recently, Thavarungkul et al. described a capacitive flow injection system for *Salmonella* spp. detection using a *Salmonella*-specific M13 bacteriophage in a working electrode [48]. They genetically modified the wild pVIII protein to have the NRPDSAQFWLHHGG sequence based on the phage display technique. They simply placed the *Salmonella*-specific M13 bacteriophage on a polytyramine (Pty)-coated gold electrode, and the amount of injected target *Salmonella* was determined by the capacitance of *Salmonella* over the total capacitance. Figure 6a shows the overall sensing procedure of this sensor device. The M13 bacteriophage-based electrode sensor was located in the running flow buffer. When the target *Salmonella* spp. was injected, the capacitance rolled up and down, and the new capacitance could be obtained. This capacitance was lower than that without a reaction to *Salmonella*. The difference in capacitance is denoted as ΔC and the total capacitance could be calculated as follows:

$$\frac{1}{C_{total}} = \frac{1}{C_{Pty}} + \frac{1}{C_{phage}} + \frac{1}{C_{Slamonella}} \tag{1}$$

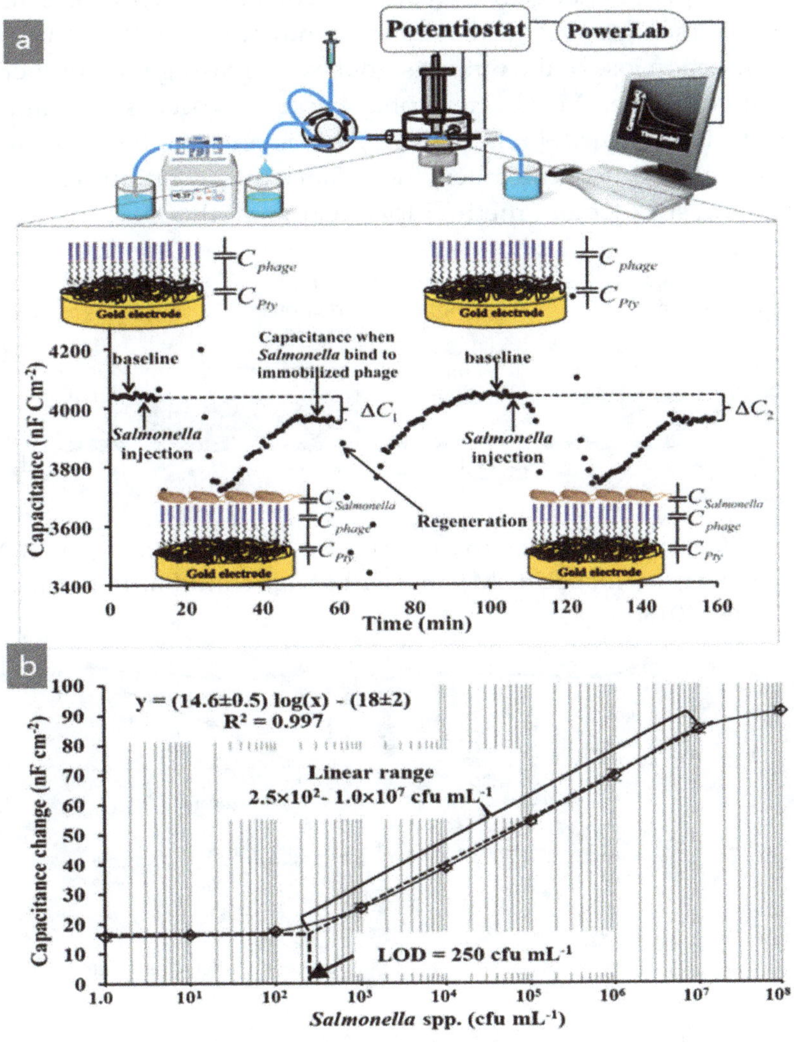

Figure 6. (**a**) Schematic procedure of the capacitive flow injection system for sensing *Salmonella* spp. using a M13 bacteriophage-based electrode sensor and (**b**) linear response of the sensor by number of *Salmonella* spp. Reproduced from [48], with permission from Elsevier, 2018.

The M13 bacteriophage-based electrode sensor was then washed with a recovery solution (5.0 mM of NaOH solution) to break the binding between the M13 bacteriophage and *Salmonella*. The M13 bacteriophage-based *Salmonella* sensor showed great selectivity to the *Salmonella* samples and exhibited an excellent recovery ratio, almost 100%. As a result, in Figure 5b, the sensor showed good sensing capability to detect *Salmonella* in chicken samples with linear range from 1.0×10^3 to

1.0×10^7 cfu/mL [48]. *Salmonella* detection is one of the main M13 bacteriophage biosensor research areas. Meanwhile, this research was performed using magnetoelastic, SPR, and microcantilever technique. The capacitive flow system was introduced very recently, and showed great results.

The target-specific binding affinity of the M13 bacteriophage could alter the macro-structure or morphology as well as the charge density of the M13 bacteriophage suspension. Therefore, the acoustic signal that passes through the M13 bacteriophage suspension can indicate the target-specific reaction both quantitatively and qualitatively [52–54]. Formin et al. reported an M13 bacteriophage-based biosensor for the detection of microbial cells with antibodies on the M13 bacteriophage [52,53]. They reacted initially with the specific antibody on the M13 bacteriophage to have a specific interaction with the target cell line (*Azospirllum basilense* strain Sp245). The cell suspension was located directly on the thin piezoelectric plate, which propagates the piezoactive acoustic wave by the input signal. Figure 7 presents a schematic diagram. After a specific interaction of M13 bacteriophages with the target microbial, they could measure significant insertion loss and phase change by a reaction. The parameters were also changed sensitively by different microbial numbers. Figure 8 shows increase conductivity of the solution caused a decrease in change of the phase and insertion loss of the output signal by increasing the number of cells. By increasing cell numbers, the interaction between M13 bacteriophage and cells occurs. Resulting clear signal changes were observed from 10^4 to 10^7 cells/mL [52]. Thus far, this is one of a few results using electroacoustic techniques to detect microbial target-specific interaction. Since this technique can be performed directly in suspension, it can be a useful supplement method for existing methods.

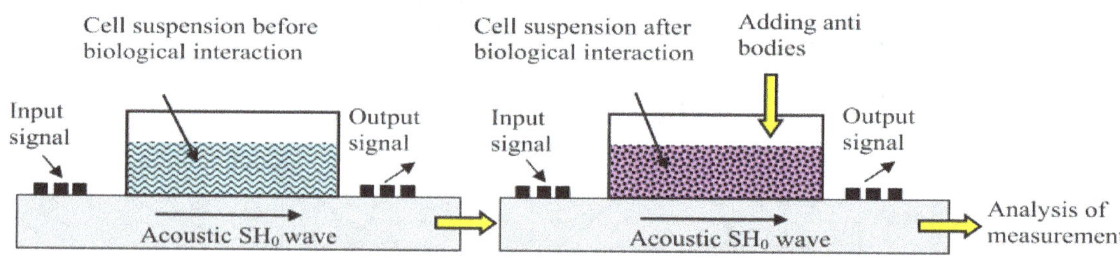

Figure 7. Schematic diagram of the electroacoustic measurements using an X–Y lithium niobate piezoelectric plate and target cell-specific M13 bacteriophage solution. Reproduced from [52], with permission from Elsevier, 2018.

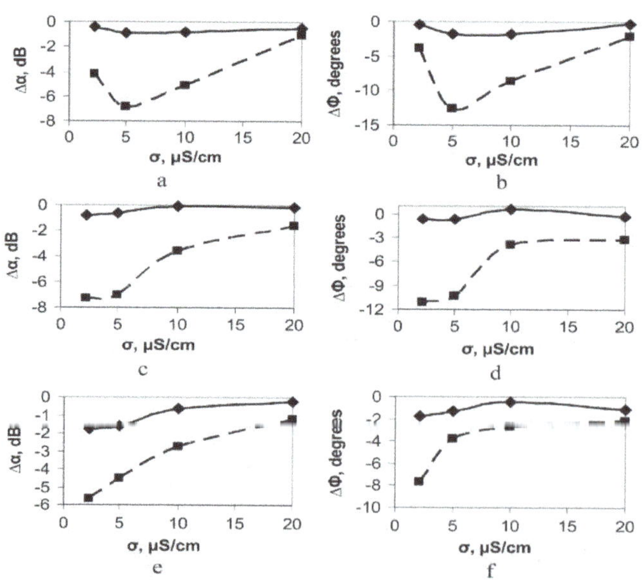

Figure 8. The change in the insertion loss (Δα, left) and phase of the output signal (ΔΦ) of the conductivity of the buffer solution. (**a**,**b**) 10^4 cells/mL, (**c**,**d**) 10^6 cells/mL, and (**e**,**f**) 10^8 cells/mL. The solid and dotted lines correspond to the electrically open and shored channels, respectively. Reproduced from [52], with permission from Elsevier, 2018.

3. M13 Bacteriophage-Based Chemical Sensing

The M13 bacteriophage has a systemically regular shape because it is a natural substance. The highly certain shape and specific reactivity to the target material provided to the M13 bacteriophage have great potential to become a building template [6]. Haberer et al. discussed the gold nanopeapod structure covered with polypyrrole (PPy) located directly between platinum electrodes. In this particular experiment, an M13 bacteriophage was genetically engineered to have the VSGSSPDS protein sequence on its pVIII major coat protein. A gold-specific M13 bacteriophage was covered and formed a bridge between two pre-fabricated platinum electrodes, and then reacted with a gold solution. A commercial gold electroless deposition reagent was used to increase the size of the gold nanoparticle.

Finally, electrodeposition of PPy was conducted to complete the gold PPy nanopeapod structure between the electrodes [55,56]. The product was applied as an NH_3 gas sensor due to nucleophilic NH_3 attack of the polymer backbone, decreasing the conjugation length, and irreversibly increasing the resistance. In other words, NH_3 molecules in this reaction act as an electron donor. Consequently, decreasing the electron hole concentration of the polymer results in a reversible de-doping process [57–59]. In this experiment, the irreversible reaction has competitive advantage over the reversible reaction, and partial recovery after gas sensing was achieved. This study showed an excellent detection limit of 0.005 ppm by NH_3 exposure to a M13 bacteriophage template gas sensor [55]. This technique implies resistance of one-dimensional nanowires which were fabricated using M13 bacteriophage as gas sensor. Due to the extremely thin layer of PPy and the catalytic effect of Au NPs, this technique showed great sensitivity. This structure highly depends on random connection between electrodes. Better performance can be expected from this structure by optimized structure such as nano pattern between two electrodes.

SERS technology based on the nanowire structure can be improved significantly by controlling the density of "hot spots", where Raman signal amplification occurs [60,61] For more effective signal amplification, the sensor surface needs to be highly functionalized, particularly around hot spots [62,63]. Jung et al. recently reported an M13 bacteriophage-covered silver nanowire SERS sensor to detect a certain pesticide, paraquat (PQ), which is a widely used herbicide. The M13 bacteriophage was functionalized genetically to have the WHW peptide sequence for the PQ specific binding affinity on its pVIII major coat protein. When the sensor was compared with the wild type M13 bacteriophage sensor and a sensor without M13 bacteriophage, the genetically functionalized sensor showed superior selectivity to the PQ sample [60].

They also tested the selectivity upon different pesticides of a functionalized M13 bacteriophage-SERS sensor using bipyridylium, which is a common herbicide. The WHW-type M13 bacteriophage-functionalized SERS sensor showed a better Raman signal when PQ was detected than bipyridylium at the same concentration. In Figure 9, using a hand-held Raman spectrometer, they could detect PQ concentrations as low as 0.1 $\mu g/cm^3$, which is far below the guidelines of most countries. Raman intensity changes before and after the washing step showed best results for WHW type M13 bacteriophage sensors. It also showed better efficiency when compared with wild type and nanowire without M13 bacteriophage. By measuring different pesticides (diquat (DQ) and difenzoquat (DIF)), PQ was the most suitable pesticide. [60,64]. This practical research showed detection results using a hand-held Raman spectrometer and agricultural product without pretreatment.

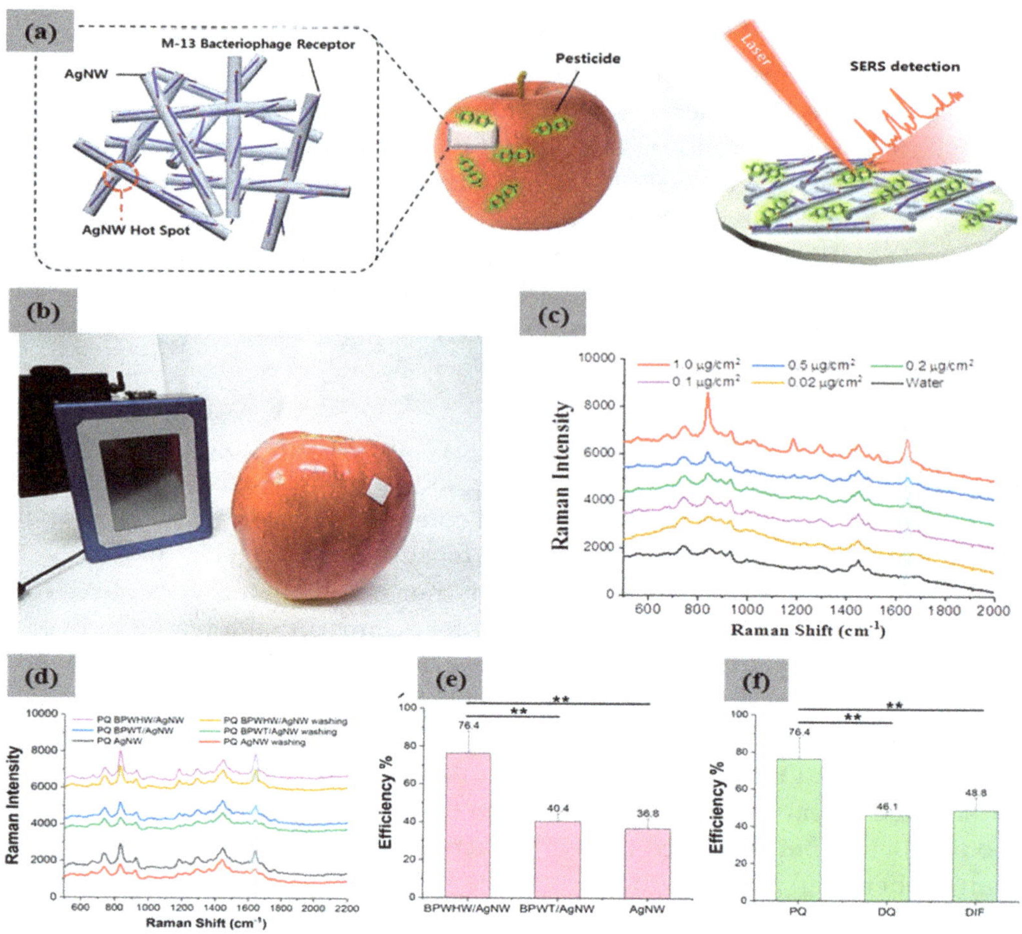

Figure 9. (**a**) Illustration of surface-enhanced Raman spectroscopy (SERS) by a functionalized M13 bacteriophage, (**b**) practical application of an M13 bacteriophage-based SERS sensor on an apple surface, (**c**) Raman spectra of paraquat (PQ) at different PQ concentrations on an apple surface, (**d**) Raman spectra of PQ using tryptophan–histidine–tryptophan (WHW), wild type of bacteriophage and without M13 bacteriophage before and after the washing step, (**e**) efficiency comparison of WHW, wild type and nanowire only, and (**f**) efficiency comparison of different pesticide (PQ, diquat (DQ), and difenzoquat (DIF)) using WHW M13 bacteriophage. Reproduced from [60], with permission from American Chemical Society, 2018.

4. M13 Bacteriophage-Based Color Sensor

The color sensor is one of the fields of M13 bacteriophage research, which has great potential in terms of sensitivity, selectivity, and portability [4,65,66]. Endocrine-disrupting chemicals were detected using an M13 bacteriophage-based color sensor. Genetically engineered M13 bacteriophages can be manufactured with a color sensor that has a structural color using a simple pulling technique, and applied to endocrine disrupting substance detection [65]. Although existing research focused on one specific substance, such as TNT [NC 13], this research focused on classifying various types of endocrine disruptors with similar structures. In this study, polychlorinated biphenyl (PCB) and phthalate compounds, traditional flame retardants, and plasticizers, were selected. These substances, which have been selected as regulated substances, are still included in pre-regulated polymer products, which act as environmental pollutants for use and disposal [65,67]. The M13 bacteriophage-based structural color sensor was modified with phage display technology. The protein sequence genetically engineered on the pVIII major coat protein is specific for covalent binding and is effective in the detection of aromatic compounds. In this study, four phthalate compounds and five PCB compounds were distinguished with high sensitivity, and the entire phthalate and PCB compound groups were distinguished with high accuracy (Figure 10) [65]. The detection limit of this color sensor was approximately 100 ppm, and the first two principal components of PCA analysis accounted for 54.89% of phthalate derivatives and 67.38% of PCB derivatives. The linear discriminant analysis (LDA) error between phthalate and PCB derivatives was 3.7% [65,68].

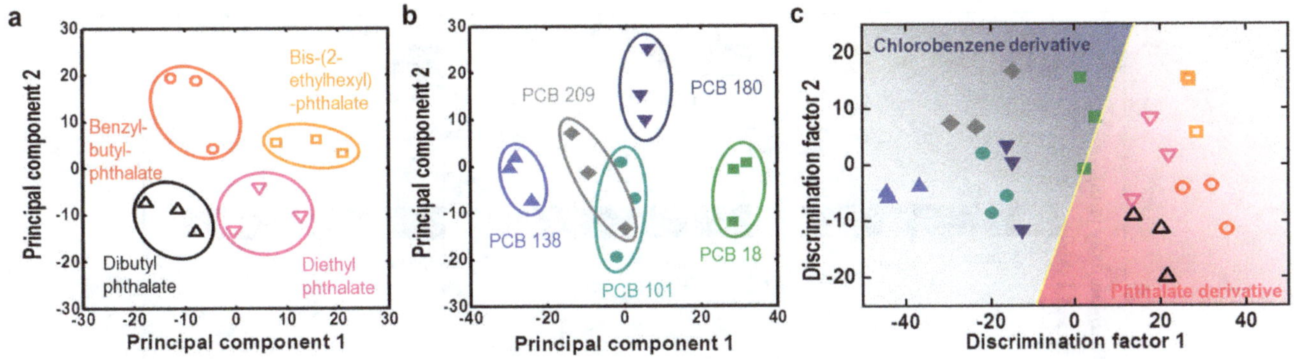

Figure 10. (**a**) PCA analysis of phthalate derivatives, (**b**) PCB derivatives, and (**c**) LDA analysis of phthalate and PCB derivatives using a M13 bacteriophage-based color sensor. Reproduced from [65], with permission from Elsevier, 2017.

Drug abuse causes serious health and environment problems [69–71]. Color sensor systems using M13 bacteriophages have shown promise for the detection of toxic substances [4,65]. Owing to the specific selectivity and sensitivity to target materials, M13 bacteriophage-based color sensors have the potential to detect various drugs [69]. Oh et al. discussed the structural color sensor to distinguish a range of antibiotics. The M13 bacteriophage-based structural color sensor was manufactured by inducing self-assembly using a simple pulling technique. Briefly, substrate (Au coated Si wafer) was deepen into M13 bacteriophage aqueous solution and pulled out slowly. By controlling the concentration of solution and pulling speed, a structural color band could be achieved through evaporation of M13 bacteriophage solution [4,65,69]. When the color sensor was exposed to an organic solvent, the bacteriophage bundle nanostructure constituting the microstructure of the sensor expanded, resulting in a structural color change. A genetically engineered M13 bacteriophage surface protein structure called WHWQ in this study could clearly distinguish three types of commercial antibiotics and three types of core ingredients of these antibiotics, which were confirmed by color and PCA analysis. Cefadroxil, amoxicillin, and rifampicin are some of the most popular antibiotics that a large number of pharmaceutical companies produce, and an absolute majority of patients consume. In Figure 11, PCA analysis produced clear discrimination between the three types of antibiotics as well as commercial and reagent grade. The first two discriminant factors accounted for 86.89% in this study [69].

An M13 bacteriophage-based color sensor also could be applied to microbial detection. Oh et al. reported a color sensor capable of detecting cancer cell types based on the M13 bacteriophage [72]. The M13 bacteriophage-based color sensor works in the form of a photonic nose, which functions similarly to other researched electronic noses, and shows structural color-based color changes by contact with the specific target materials. The target material in this study was a specific volatile compound that is unusually abundant in various cancer cells. The respiratory by-products of cancer cells, which are distinguished uniquely from normal cells, include hydrazine, xylene, ethylbenzene, ethanol, and toluene [72,73]. Therefore, they measured a range of organic reagents, populations of *E. coli*, and CO_2 concentration using the M13 bacteriophage-based color sensor. Here, the M13 bacteriophage-based color sensor showed significant sensitivity and selectivity for organic solvents, cell population, and CO_2 concentration. In this study, selectivity experiment has been done for hydrazine, xylene, ethylbenzene, ethanol, and toluene, volatile organic chemicals. These chemicals can be found in abnormally high concentration of in the lung cancer patient's respiration [72–75]. These organic compounds can act like an analytic marker for different cancer cells when released through respiration of cancer cells. Therefore, selective detection of these organic solvents was important to the selective detection of various cancer cells. Finally, each cancer cell's respiration by-product was analyzed using adjacent color sensors while culturing various types of cancer cells. As a result, they showed excellent discrimination through PCA and color analysis, with the first two discriminant factors accounting for 99.8% (Figure 12) [72].

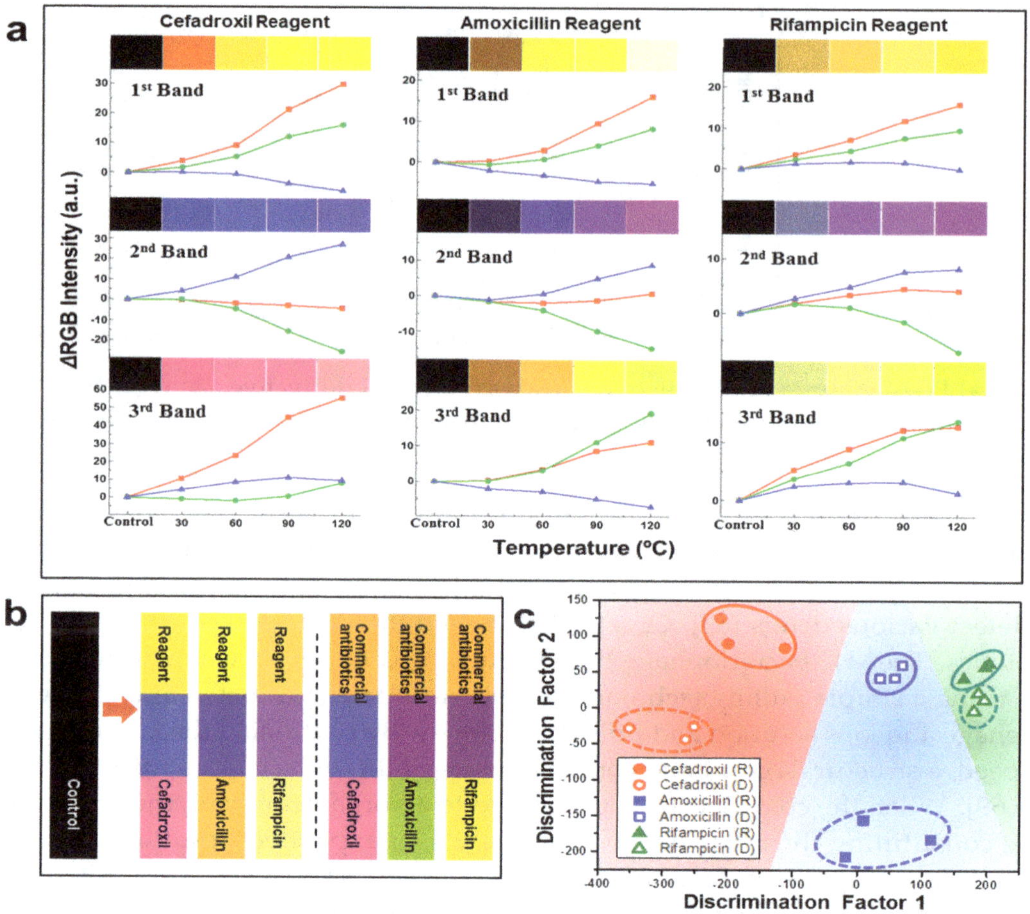

Figure 11. Measurement of various antibiotics using M13 bacteriophage-based color sensors. (**a**) Sensitivity measurement at different temperatures using fixed amount of samples, (**b**) color change translation of samples, and (**c**) PCA analysis of antibiotic sensing. Reproduced from [69], with permission from Wiley-VCH, 2016.

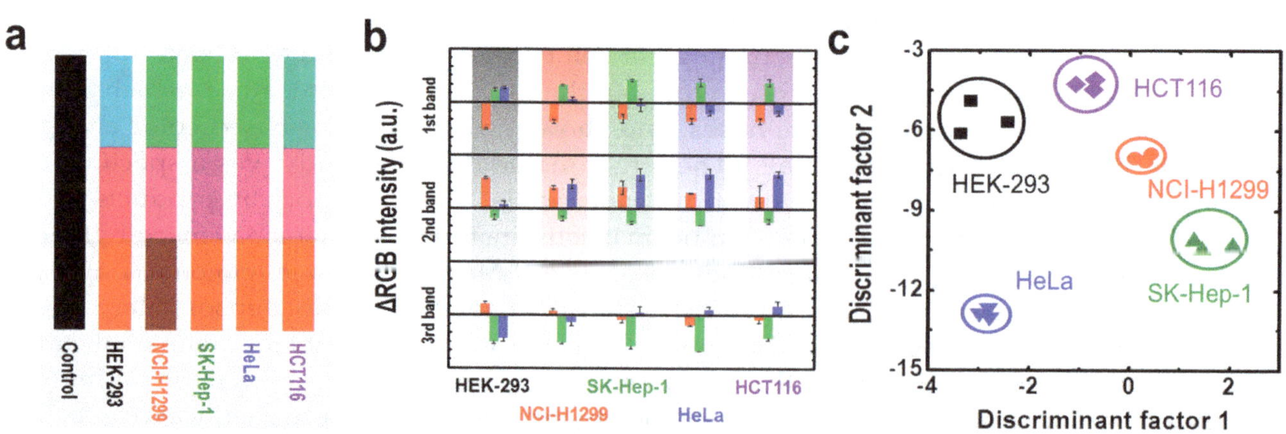

Figure 12. Cancer cell type identification using M13 bacteriophage-based color sensors. (**a**) Red, Green, and blue (RGB) color patterns of each cancer cell, (**b**) RGB intensity change, and (**c**) PCA analysis of samples. Reproduced from [72], with permission from The Royal Society of Chemistry, 2017.

5. Expansion of Analytic Method from Direct Sensing Using M13 Bacteriophage

A range of variations of WHW type M13 bacteriophages have common specific affinity to TNT molecules [4]. A π-π interaction was assumed among TNT molecules, histidine, and tryptophan based on experiment [4,6]. On the other hand, there is a lack of consideration as to whether WHW is really the best structure for TNT molecules or how this specific interaction occurs systemically and theoretically. Recently, Lee et al. reported the results of computational calculations of this interaction among TNT molecules, histidine, and tryptophan, using quantum mechanics (QM) calculations [76]. The theoretical candidates for comparison were WHW-, WAW-, WHA-, and AHW-simulated peptide sequences (A: Alanine). The specific binding affinity between the WHW and TNT molecules was confirmed experimentally using an SPR sensor system due to the significantly lower detection limit (500 fM) than any other sensor system based on M13 bacteriophages [4,77–80]. The dissociation constants for TNT molecules using the WHW-, WAW-, WHA-, and AHW-type M13 bacteriophages were 7.0×10^{-12}, 2.7×10^{-11}, 1.1×10^{-10}, and 3.3×10^{-8}, respectively. By QM calculations using Jaguar v8.4 software with the M06-2X/6-31G ** level of density functional theory, the binding energy of WHW-TNT was calculated to be 22.7 kcal mol^{-1}. Figure 13 shows the active binding mechanism for WHW and alanine substituted peptide sequences. For WAW, WHA, and AHW, the binding energy was 19.9, 16.0, and 18.8 kcal mol^{-1}, respectively. Finally, they confirmed the strong linear correlation between the experimental Gibbs free energy and calculated binding energy [76].

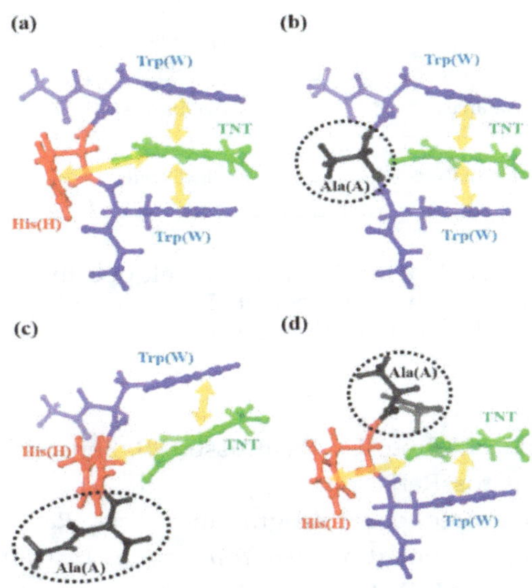

Figure 13. The stable constructions of TNT-specific peptides and their reactivity (red: histidine, blue: tryptophan, black: alanine, green: TNT, yellow, and arrow: π-π interaction). (**a**) WHW–TNT, (**b**) WAW–TNT, (**c**) WHA–TNT, and (**d**) AHW–TNT. Reproduced from [76], with permission from The Royal Society of Chemistry, 2019.

6. Conclusions

M13 bacteriophages have attracted considerable attention because of their simplicity of structure and functionality. The additional merit of M13 bacteriophages is the customized functionality, which can be provided easily through simple genetic engineering. In Table 1, large numbers of research groups have examined various fields using M13 bacteriophages. In particular, a biosensor using M13 bacteriophages has attracted serious interest because of the portability and intuitiveness. M13 bacteriophage has been utilized as an electrode support material for cysteine and *Salmonella* spp. detection [39,44,48]. Functionalized M13 bacteriophages have the potential to detect various cancer cells effectively [44,72]. Acoustic and SERS signals could be amplified by adding M13 bacteriophages to existing sensor systems [52,60]. In addition, M13 bacteriophage-based color sensors have been evaluated as a key tool to apply M13 bacteriophages to sensor systems [65,69,72]. Application of the M13 bacteriophage is still in its infancy. There remain many more protein sequences to be found, and low dimensional fabrication control should be studied. Mass production and reproducibility of nanostructures also need to be studied and overcome. Active theoretical studies on the function of genetically engineered M13 bacteriophages for specific target materials will open more systematic approaches as well as novel applications of M13 bacteriophages [12,76].

Table 1. Biosensor applications of M13 bacteriophages.

M13 Bacteriophage Immobilizations	Detection Technique	Analytes	Real Sample	Ref.
M13/Au coated Si wafer	Color analysis	Nitrotoluene	TNT, DNT, MNT	[4]
M13/donor and acceptor dye	FRET	Intracellular pH	RAW264.7 macrophage	[30]
M13/Au coated glass	SPR	Cell proliferation signal	NIH3T3 mouse fibroblast	[31]
M13/Au@Ag NPs/Raman dye/DNA	SERS	Antibody concentration	Sterptavidin/anti-goat IgG	[32]
M13/zinc phthalocyanine/methyl viologen	Fluorescence image	Breast cancer cells	SKBR-3 cell line	[35]
M13/CNF/GCE	Electrochemical	Cysteine	L-Cysteine solution in PBS	[39]
M13/scFv	Optical/surgical	CEA tumor cells	MC38- and CT26-CEA	[44]
M13/Pty/Au electrode	Capacitive	Salmonella	Chicken	[48]
M13/antibody	Electroacoustic	Cell number	Sp245 cell line	[52]
M13/Ag nanowire	SERS	Pesticides	PQ, DQ, and DIF	[60]
M13/Au coated Si wafer	Color analysis	Endocrine disrupting chemicals	Phthalate and PCB derivatives	[65]
M13/Au coated Si wafer	Color analysis	Antibiotics	Commercial and reagent antibiotics	[69]
M13/Au coated Si wafer	Color analysis	Cancer cells	HEK-293, NCI-H1299, SK-Hep-1, HeLa, HCT116 cancer cell lines	[72]

Author Contributions: J.-S.M., E.J.C., N.-N.J., and J.-W.O. developed the idea and wrote of the review article. J.-R.S. and D.-W.H. revised and improved the manuscript. J.-S.M. and J.-W.O. supervised the manuscript. All authors have given approval to the final version of the manuscript.

References

1. Andersen, P.D.; Jørgensen, B.H.; Lading, L.; Rasmussen, B. Sensor foresight—Technology and market. *Technovation* **2004**, *24*, 311–320. [CrossRef]
2. Hollingum, J. Foresight launches three sensor programmes. *Sens. Rev.* **1999**, *19*, 116–120. [CrossRef]
3. Sensors: The next wave of infotech innovation. *Ten-Year Forecast* Institute for the future. **1997**, 115–122.
4. Oh, J.-W.; Chung, W.-J.; Heo, K.; Jin, H.-E.; Lee, B.Y.; Wang, E.; Zueger, C.; Wong, W.; Meyer, J.; Kim, C.; et al. Biomimetic virus-based colorimetric sensors. *Nat. Commun.* **2014**, *5*, 3043. [CrossRef]
5. Martin-Herranz, A.; Ahmad, A.; Evans, H.M.; Ewert, K.; Schulze, U.; Safinya, C.R. Surface functionalized cationic lipid-DNA complexes for gene delivery: PEGylated lamellar complexes exhibit distinct DNA-DNA Interaction Regimes. *Biophys. J.* **2004**, *86*, 1160–1168. [CrossRef]
6. Moon, J.S.; Kim, W.G.; Kim, C.; Park, G.-T.; Heo, J.; Yoo, S.Y.; Oh, J.-W. M13 bacteriophage-based self-assembly Structures and their functional capabili-ties. *Mini Rev. Org. Chem.* **2015**, *12*, 271–281. [CrossRef]
7. Chung, W.J.; Lee, D.Y.; Yoo, S.Y. Chemical modulation of M13 bacteriophage and its functional opportunities for nanomedicine. *Int. J. Nanomed.* **2014**, *9*, 5825–5836.
8. Lee, Y.J.; Yi, H.; Kim, W.-J.; Kang, K.; Yun, D.S.; Strano, M.S.; Ceder, G.; Belcher, A.M. Fabricating genetically engineered high-power lithium-ion batteries using multiple virus genes. *Science* **2009**, *324*, 1051–1055.
9. Lee, Y.; Kim, J.; Yun, D.S.; Nam, Y.S.; Yang, S.-H.; Belcher, A.M. Virus-templated Au and Au-Pt core-shell nanowires and their electrocatalytic activities for fuel cell applications. *Energy Sci.* **2012**, *5*, 8328–8334. [CrossRef]
10. Mao, C.; Liu, A.; Cao, B. Virus-based chemical and biological sensing. *Angew* **2009**, *48*, 6790–6810. [CrossRef]
11. Sundar, V.C.; Yablon, A.D.; Grazul, J.L.; Ilan, M.; Aizenberg, J. Fibre-optical features of a glass sponge. *Nature* **2003**, *424*, 899–900. [CrossRef]

12. Espinosa, H.D.; Luster, A.L.; Latourte, F.J.; Loh, O.Y.; Gregoire, D.; Zavattieri, P.D. Tablet-level origin of toughening in abalone shells and translation to synthetic composite materials. *Nat. Commun.* **2011**, *2*, 173. [CrossRef] [PubMed]
13. Aizenberg, J.; Weaver, J.C.; Thanawala, M.S.; Sundar, V.C.; Morse, D.E.; Fratzl, P. Skeleton of euplectella sp.: Structural hierarchy from the nanoscale to the macroscale. *Science* **2005**, *309*, 275–278. [CrossRef] [PubMed]
14. Chung, W.-J.; Oh, J.-W.; Kwak, K.; Lee, B.Y.; Meyer, J.; Wang, E.; Hexemer, A.; Lee, S.-W. Biomimetic self-templating supramolecular structure. *Naure* **2011**, *478*, 364–368. [CrossRef] [PubMed]
15. Dogic, Z.; Fraden, S. Ordered phases of filamentous viruses. *Curr. Opin. Colloid Interface Sci.* **2006**, *11*, 47–55. [CrossRef]
16. Yang, S.H.; Chung, W.J.; Mcfarland, S.; Lee, S.W. Assembly of Bacteriophage into Functional Materials. *Chem. Rec.* **2013**, *13*, 43–59. [CrossRef]
17. Gimenez, S.; Lana-Villarreal, T.; Gómez, R.; Agouram, S.; Muñoz-Sanjosé, V.; Mora-Seró, I. Determination of limiting factors of photovoltaic efficiency in quantum dot sensitized solar cells: Correlation between cell performance and structural properties. *J. Appl. Phys.* **2010**, *108*, 064310. [CrossRef]
18. Karpan, V.M.; Khomyakov, P.A.; Starikov, A.A.; Giovannetti, G.; Zwierzycki, M.; Talanana, M.; Brocks, G.; van den Brink, J.; Kelly, P.J. Theoretical prediction of perfect spin filtering at interfaces between close-packed surfaces of Ni or Co and graphite or graphene. *Phys. Rev. B* **2008**, *78*, 195419. [CrossRef]
19. Smith, G.P.; Petrenko, V.A. Phage display. *Chem. Rev.* **1997**, *97*, 391–410. [CrossRef]
20. Smith, G.P. Filamentous fusion phage: Novel expression vectors that display cloned antigens on the virion surface. *Science* **1985**, *228*, 1315–1317. [CrossRef]
21. Cao, B.; Mao, C. *Phage Nanobiotechnology*; Petrenko, V., Smith, G.P., Eds.; RSC publishing: Cambridge, UK, 2011.
22. Brissette, R.; Goldstein, N.I. *Methods in Molecular Biology*; Fisher, P., Ed.; Humana: Totowa, NJ, USA, 2007.
23. Flynn, C.E.; Lee, S.W.; Peelle, B.R.; Belcher, A.M. Viruses as vehicles for growth, organization and assembly of materials. *Acta Mater.* **2003**, *51*, 5867–5880. [CrossRef]
24. Cui, Y.; Kim, S.N.; Jones, S.E.; Wissler, L.L.; Naik, R.R.; McAlpine, M.C. Chemical functionalization of graphene enabled by phage displayed peptides. *Nano Lett.* **2010**, *10*, 4559–4565. [CrossRef] [PubMed]
25. Kim, S.N.; Kuang, Z.; Slocik, J.M.; Jones, S.E.; Cui, Y.; Farmer, B.L.; McAlpine, M.C.; Naik, R.R. Preferential binding of peptides to graphene edges and planes. *J. Am. Chem. Soc.* **2011**, *133*, 14480–14483. [CrossRef] [PubMed]
26. Mao, C.; Solis, D.J.; Reiss, B.D.; Kottmann, S.T.; Sweeney, R.Y.; Hayhurst, A.; Georgiou, G.; Iverson, B.; Belcher, A.M. Virus-based toolkit for the directed synthesis of magnetic and semiconducting nanowires. *Science* **2004**, *303*, 213–217. [CrossRef] [PubMed]
27. Mio, C.; Flynn, C.E.; Hayhurst, A.; Sweeney, R.; Qi, J.; Georgiou, G.; Iverson, B.; Belcher, A.M. Viral assembly of oriented quantum dot nanowires. *Proc. Natl. Acad. Sci. USA* **2003**, *100*, 6946–6951. [CrossRef] [PubMed]
28. Huang, Y.; Chiang, C.-Y.; Lee, S.K.; Gao, Y.; Hu. E., L.; De Yoreo, J.; Belcher, A.M. Programmable assembly of nanoarchitectures using genetically engineered viruses. *Nano Lett.* **2005**, *5*, 1429–1434. [CrossRef] [PubMed]
29. Deo, S.; Godwin, H.A. A selective, ratiometric fluorescent sensor for $Pb2^{+}$. *J. Am. Chem. Soc.* **2000**, *122*, 174–175. [CrossRef]
30. Chen, L.; Wu, Y.; Lin, Y.; Wang, Q. Virus-templated FRET platform for the rational design of ratiometric fluorescent nanosensors. *Chem. Commun.* **2015**, *51*, 10190–10193. [CrossRef]
31. Yoo, S.Y.; Oh, J.-W.; Lee, S.W. Phage-chips for novel optically readable tissue engineering assays. *Langmuir* **2012**, *28*, 2166–2172. [CrossRef]
32. Lee, J.H.; Xu., P.F.; Domaille, D.W.; Choi, C.; Jin, S.; Cha, J.N. M13 Bacteriophage as materials for amplified surface enhanced Raman scattering protein sensing. *Adv. Funct. Mater.* **2014**, *24*, 2079–2084. [CrossRef]
33. Kawakami, K.; Nishihara, Y.; Hirano, K. Effect of hydrophilic polymers on physical stability of liposome dispersions. *J. Phys. Chem. B* **2001**, *105*, 2374–2385. [CrossRef]
34. Ruysschaert, T.; Germain, M.; Gomes, J.F.; Fournier, D.; Sukhorukov, G.B.; Meier, W.; Winterhalter, M. Liposome-based nanocapsules. *IEEE Trans. Nanobiosci.* **2004**, *3*, 49–55. [CrossRef]
35. Ngweniform, P.; Abbineni, G.; Cao, B.; Mao, C. Self-assembly of drug-loaded liposomes on genetically engineered target-recognizing M13 phage: A novel nanocarrier for targeted drug delivery. *Small* **2009**, *5*, 1963–1969. [CrossRef] [PubMed]

36. Jaworski, J.W.; Raorane, D.; Huh, J.H.; Majumdar, A.; Lee, S.-W. Evolutionary screening of biomimetic coatings for selective detection of explosives. *Langmuir* **2008**, *24*, 4938–4943. [CrossRef] [PubMed]
37. Jin, H.; Won, N.; Ahn, B.; Kwang, J.; Heo, K.; Oh, J.-W.; Sun, Y.; Cho, S.G.; Lee, S.-W.; Kim., S. Quantum dot-engineered M13 virus layer-by-layer composite films for highly selective and sensitive turn-on TNT sensors. *Chem. Commun.* **2013**, *49*, 6045–6047. [CrossRef]
38. Lee, B.Y.; Zhang, J.; Zueger, C.; Chung, W.-J.; Yoo, S.Y.; Wang, E.; Meyer, J.; Ramesh, R.; Lee, S.-W. Virus-based piezoelectric energy generation. *Nat. Nanotechnol.* **2012**, *7*, 351–356. [CrossRef]
39. Szot-Karpińska, K.; Leśniewski, A.; Jönsson-Niedziółka, M.; Marken, F.; Niedziółka-Jönsson, J. Electrodes modified with bacteriophages and carbon nanofibres for cysteine detection. *Sens. Actuat B Chem.* **2019**, *287*, 78–85. [CrossRef]
40. Szot, K.; Lesniewski, A.; Niedziolka, J.; Jönsson, M.; Rizzi, C.; Gaillon, L.; Marken, F.; Rogalski, J.; Opallo, M. Sol–gel processed ionic liquid—Hydrophilic carbon nanoparticles multilayer film electrode prepared by layer-by-layer method. *J. Electroanal. Chem.* **2008**, *623*, 170–176. [CrossRef]
41. Macdonald, S.; Szot, K.; Niedziolka, J.; Marken, F.; Opallo, M. Introducing hydrophilic carbon nanoparticles into hydrophilic sol-gel film electrodes. *J. Solid State Electr.* **2008**, *12*, 287–293. [CrossRef]
42. Bajenova, O.; Chaika, N.; Tolkunova, E.; Davydov-Sinitsyn, A.; Gapon, S.; Thomas, P.; O'Brien, S. Carcinoembryonic antigen promotes colorectal cancer progression by targeting adherens junction complexes. *Exp. Cell Res.* **2014**, *324*, 115–123. [CrossRef]
43. Bajenova, O.; Gorbunova, A.; Evsyukov, I.; Rayko, M.; Gapon, S.; Bozhokina, E.; Shishkin, A.; O'Brien, S.J. The genome-wide analysis of carcinoembryonic antigen signaling by colorectal cancer cells using RNA sequencing. *PLoS ONE* **2016**, *11*, e0161256. [CrossRef]
44. Murgas, P.; Bustamante, N.; Araya, N.; Cruz-Gómez, S.; Durán, E.; Gaete, D.; Oyarce, C.; López, E.; Herrada, A.A.; Ferreira, N.; et al. A filamentous bacteriophage targeted to carcinoembryonic antigen induces tumor regression in mouse models of colorectal cancer. *Cancer Immunol. Immun.* **2018**, *67*, 183–193. [CrossRef] [PubMed]
45. Ordoñez, C.; Screaton, R.A.; Ilantzis, C.; Stanners, C.P. Human carcinoembryonic antigen functions as a general inhibitor of anoikis. *Can. Res.* **2000**, *60*, 3419–3424.
46. Taheri, M.; Saragovi, H.U.; Stanners, C.P. The adhesion and differentiation-inhibitory activities of the immunoglobulin super-family member, carcinoembryonic antigen, can be independently blocked. *J. Biol. Chem.* **2003**, *278*, 14632–14639. [CrossRef] [PubMed]
47. Rondot, S.; Koch, J.; Breitling, F.; Dubel, S. A helper phage to improve single-chain antibody presentation in phage display. *Nat. Biotechnol.* **2001**, *19*, 75–78. [CrossRef]
48. Niyomdecha, S.; Limbut, W.; Numnuam, A.; Kanatharana, P.; Charlermroj, R.; Karoonuthaisiri, N.; Thavarungkul, P. Phage-based capacitive biosensor for Salmonella detection. *Talanta* **2018**, *188*, 658–664. [CrossRef]
49. Wolffs, P.F.G.; Glencross, K.; Thibaudeau, R.; Griffths, M.W. Direct quantitation and detection of salmonellae in biological samples without enrichment, using two-step filtration and real-time PCR. *Appl. Environ. Microbiol.* **2006**, *72*, 3896–3900. [CrossRef]
50. Bennett, A.R.; Greenwood, D.; Tennant, C.; Banks, J.G.; Betts, R.P. Rapid and definitive detection of Salmonella in foods by PCR. *Lett. Appl. Microbiol.* **1998**, *26*, 437–441. [CrossRef]
51. Herikstad, H.; Motarjemi, Y.; Tauxe, R.V. Salmonella surveillance: A global survey of public health serotyping. *Epidemiol. Infect.* **2002**, *129*, 1–8. [CrossRef]
52. Guliy, O.; Zaitsev, B.D.; Borodina, I.A.; Shikhabudinov, A.M.; Teplykh, A.A.; Staroverov, S.A.; Fomin, A.S. The biological acoustic sensor to record the interactions of the microbial cells with the phage antibodies in conducting suspensions. *Talanta* **2018**, *178*, 569–576. [CrossRef]
53. Kanoatov, M.; Krylov, S.N. Analysis of DNA in phosphate buffered saline using kinetic capillary electrophoresis. *Anal. Chem.* **2016**, *88*, 7421–7428. [CrossRef]
54. Smith, G.P.; Scott, J.K. Libraries of peptides and proteins displayed on filamentous phage. *Methods Enzymol.* **1993**, *217*, 228–257. [PubMed]
55. Yan, Y.; Zhang, M.; Moon, C.H.; Su, H.C.; Myung, N.V.; Haberer, E.D. Viral-templated gold/polypyrrole nanopeapods for an ammonia gas sensor. *Nanotechnology* **2016**, *27*, 325502. [CrossRef] [PubMed]
56. Oh, D.; Dang, X.; Yi, H.; Allen, M.A.; Xu, K.; Lee, Y.J.; Belcher, A.M. Graphene sheets stabilized on genetically engineered M13 viral templates as conducting frameworks for hybrid energy-storage materials. *Small* **2012**, *8*, 1006–1011. [CrossRef] [PubMed]

57. Gustafsson, G.; Lundström, I. The effect of ammonia on the physical properties of polypyrrole. *Synth. Met.* **1987**, *21*, 203–208. [CrossRef]
58. Gustafsson, G.; Lundström, B.; Liedberg, I.; Wu, C.R.; Inganäs, O.; Wennerström, O. The interaction between ammonia and poly(pyrrole). *Synth. Met.* **1989**, *31*, 163–179.
59. Choi, J.; Hormes, J.; Kahol, P.K. Study of ammonia-gas-induced irreversibility in polypyrrole films. *Appl. Phys. Lett.* **2003**, *83*, 2288–2290. [CrossRef]
60. Koh, E.H.; Mun, C.; Kim, C.; Park, S.G.; Choi, E.J.; Kim, S.H.; Dang, J.; Choo, J.; Oh, J.-W.; Kim, D.H.; et al. M13 bacteriophage/silver nanowire surface-enhanced Raman scattering sensor for sensitive and selective pesticide detection. *ACS Appl. Mater. Interfaces* **2018**, *10*, 10388–10397. [CrossRef]
61. Park, S.-G.; Mun, C.W.; Lee, M.K.; Jeon, T.Y.; Shim, H.-S.; Lee, Y.-J.; Kwon, J.-D.; Kim, C.S.; Kim, D.-H. 3D Hybrid Plasmonic Nanomaterials for Highly Efficient Optical Absorbers and Sensors. *Adv. Mater.* **2015**, *27*, 4290–4295. [CrossRef]
62. Nagy-Simon, T.; Tatar, A.-S.; Craciun, A.-M.; Vulpoi, A.; Jurj, M.-A.; Florea, A.; Tomuleasa, C.; Berindan-Neagoe, I.; Astilean, S.; Boca, S. Antibody Conjugated, Raman Tagged Hollow Gold–Silver Nanospheres for Specific Targeting and Multimodal Dark-Field/SERS/Two Photon-FLIM Imaging of CD19(+) B Lymphoblasts. *ACS Appl. Mater. Interfaces* **2017**, *9*, 21155–21168. [CrossRef]
63. Zhao, J.; Zhang, K.; Li, Y.; Ji, J.; Liu, B. High-Resolution and universal visualization of latent fingerprints based on aptamer-functionalized core–shell nanoparticles with embedded SERS reporters. *ACS Appl. Mater. Interfaces* **2016**, *8*, 14389–14395. [CrossRef]
64. Fang, H.; Zhang, X.; Zhang, S.J.; Liu, L.; Zhao, Y.M.; Xu, H.J. Ultrasensitive and quantitative detection of paraquat on fruits skins via surface-enhanced Raman spectroscopy. *Sens. Actuators B* **2015**, *213*, 452–456. [CrossRef]
65. Moon, J.-S.; Park, M.; Kim, W.-G.; Kim, C.; Hwang, J.; Seol, D.; Kim, C.-S.; Sohn, J.-R.; Chung, H.; Oh, J.-W. M-13 bacteriophage based structural color sensor for detecting antibiotics. *Sens. Actuators B* **2017**, *240*, 757–762. [CrossRef]
66. Slavik, R.; Homola, J. Ultrahigh resolution long range surface plasmon-based sensor. *Sens. Actuators B* **2007**, *123*, 10–12. [CrossRef]
67. Ferber, D. WHO advises kicking the livestock antibiotic habit. *Science* **2003**, *301*, 1027. [CrossRef] [PubMed]
68. Nam, K.T.; Kim, D.-W.; Yoo, P.J.; Chiang, C.-Y.; Meethong, N.; Hammond, P.T.; Chiang, Y.-M.; Belcher, A.M. Virus-enabled synthesis and assembly of nanowires for lithium ion battery electrodes. *Science* **2006**, *312*, 885–888. [CrossRef] [PubMed]
69. Moon, J.S.; Lee, Y.; Shin, D.M.; Kim, C.; Kim, W.G.; Park, M.; Han, J.; Song, H.; Kim, K.; Oh, J.-W. Identification of endocrine disrupting chemicals using a virus-based colorimetric sensor. *Chem. Asian J.* **2016**, *11*, 3097–3101. [CrossRef]
70. Sadik, O.A.; Witt, D.M. Peer reviewed: Monitoring endocrine-disrupting chemicals. *Environ. Sci. Technol.* **1999**, *33*, 368A–374A. [CrossRef]
71. Yoon, Y.; Westerhoff, P.; Snyder, S.A.; Esparza, M. HPLC-fluorescence detection and adsorption of bisphenol A, 17beta-estradiol, and 17alpha-ethynyl estradiol on powdered activated carbon. *Water Res.* **2003**, *37*, 3530–3537. [CrossRef]
72. Moon, J.-S.; Kim, W.-G.; Shin, D.-M.; Lee, S.Y.; Kim, C.; Lee, Y.; Han, J.; Kim, K.; Yoo, S.Y.; Oh, J.-W. Bioinspired M-13 bacteriophage-based photonic nose for differential cell recognition. *Chem. Sci.* **2017**, *8*, 921–927. [CrossRef]
73. Peng, G.; Tisch, U.; Adams, O.; Hakim, M.; Shehada, N.; Broza, Y.Y.; Billan, S.; Bortnyak, R.R.; Kuten, A.; Haick, H. Diagnosing lung cancer in exhaled breath using gold nanoparticles. *Nat. Nanotechnol.* **2009**, *4*, 669–673. [CrossRef]
74. Yun, J.M.; Kim, K.N.; Kim, J.Y.; Shin, D.O.; Lee, W.J.; Lee, S.H.; Lieberman, M.; Kim, S.O. DNA origami nanopatterning on chemically modified graphene. *Angew* **2011**, *51*, 912–915. [CrossRef] [PubMed]
75. Yun, J.M.; Ganesan, R.; Choi, J.-H.; Kim, J.-B. Local pH-Responsive Diazoketo-Functionalized Photoresist for Multicomponent Protein Patterning. *ACS Appl. Mater. Interfaces* **2013**, *5*, 10253–10259. [CrossRef] [PubMed]
76. Kim, W.G.; Zueger, C.; Kim, C.; Wong, W.; Devaraj, V.; Yoo, H.W.; Hwang, S.; Oh, J.-W.; Lee, S.W. Experimental and numerical evaluation of a genetically engineered M13 bacteriophage with high sensitivity and selectivity for 2,4,6-trinitrotoluene. *Org. Biomol. Chem.* **2019**, *17*, 5666–5670. [CrossRef] [PubMed]

77. Blaik, R.A.; Lan, E.; Huang, Y.; Dunn, B. gold-coated m13 bacteriophage as a template for glucose oxidase biofuel cells with direct electron transfer. *ACS Nano* **2015**, *10*, 324–332. [CrossRef] [PubMed]
78. Aguilar, A.D.; Forzani, E.S.; Leright, M.; Tsow, F.; Cagan, A.; Iglesias, R.A.; Nagahara, L.A.; Amlani, I. A hybrid nanosensor for tnt vapor detection. *Nano Lett.* **2010**, *10*, 380–384. [CrossRef]
79. Rose, A.; Zhu, Z.; Madigan, C.F.; Swager, T.M.; Bulović, V. Sensitivity gains in chemosensing by lasing action in organic polymers. *Nature* **2005**, *434*, 876–879. [CrossRef]
80. Riskin, M.; Tel-Vered, R.; Lioubashevski, O.; Willner, I. Ultrasensitive surface plasmon resonance detection of trinitrotoluene by a bis-aniline-cross-linked au nanoparticles composite. *J. Am. Chem. Soc.* **2009**, *131*, 7368–7378. [CrossRef]

2

Structure and Hierarchy of Influenza Virus Models Revealed by Reaction Network Analysis

Stephan Peter [1,2], Martin Hölzer [3,4], Kevin Lamkiewicz [3,4], Pietro Speroni di Fenizio [2], Hassan Al Hwaeer [5], Manja Marz [3,4], Stefan Schuster [6], Peter Dittrich [2,*] and Bashar Ibrahim [4,6,*]

[1] Ernst-Abbe University of Applied Sciences Jena, Carl-Zeiss-Promenade 2, 07745 Jena, Germany; stephan.peter@eah-jena.de

[2] Bio Systems Analysis Group, Department of Mathematics and Computer Science, University of Jena, Ernst-Abbe-Platz 2, 07743 Jena, Germany; pietros@gmail.com

[3] RNA Bioinformatics and High-Throughput Analysis, Friedrich Schiller University Jena, Leutragraben 1, 07743 Jena, Germany; martin.hoelzer@uni-jena.de (M.H.); Kevin.Lamkiewicz@uni-jena.de (K.L.); manja@uni-jena.de (M.M.)

[4] European Virus Bioinformatics Center, Leutragraben 1, 07743 Jena, Germany

[5] Mathematics and Computer Applications Department, Al-Nahrain University, Al-Jadriya, Baghdad 10072, Iraq; hassan1167@yahoo.com

[6] Chair of Bioinformatics, Matthias-Schleiden-Institute, University of Jena, Ernst-Abbe-Platz 2, 07743 Jena, Germany; stefan.schu@uni-jena.de

* Correspondence: peter.dittrich@uni-jena.de (P.D.); bashar.ibrahim@uni-jena.de (B.I.)

Abstract: Influenza A virus is recognized today as one of the most challenging viruses that threatens both human and animal health worldwide. Understanding the control mechanisms of influenza infection and dynamics is crucial and could result in effective future treatment strategies. Many kinetic models based on differential equations have been developed in recent decades to capture viral dynamics within a host. These models differ in their complexity in terms of number of species elements and number of reactions. Here, we present a new approach to understanding the overall structure of twelve influenza A virus infection models and their relationship to each other. To this end, we apply chemical organization theory to obtain a hierarchical decomposition of the models into chemical organizations. The decomposition is based on the model structure (reaction rules) but is independent of kinetic details such as rate constants. We found different types of model structures ranging from two to eight organizations. Furthermore, the model's organizations imply a partial order among models entailing a hierarchy of model, revealing a high model diversity with respect to their long-term behavior. Our methods and results can be helpful in model development and model integration, also beyond the influenza area.

Keywords: chemical organization theory; influenza A; virus dynamics modeling; complex networks analysis

---

1. Introduction

Influenza is an infectious respiratory disease, annually infecting 5–15% of the human population and causing epidemics that result in 3–5 million severe cases with 300,000–500,000 deaths each year [1]. The annual recurrence of epidemics is caused by the continuous alteration of seasonal influenza viruses, which enables them to efficiently escape the immune system even due to previous infections or vaccinations [2]. The major burden of disease in humans is caused by seasonal influenza A (IAV) and influenza B viruses, causing symptoms varying from mild respiratory disease characterized by fever, sore throat, headache and muscle pain to severe and in some cases lethal pneumonia and secondary bacterial infections [3].

The long-term spread of influenza viruses in the human population and the acute nature of influenza virus epidemics is driven by the global movement of these viruses. Differences in seasonal epidemics caused by influenza viruses are mainly driven by differences in the rates of virus evolution. The single-stranded RNA segments of influenza viruses, which are located inside the virus particle (or virion), evolve rapidly and thus can escape the host's immune response very efficiently.

Several ordinary differential equations (ODEs) models have been developed to provide insight into within-host dynamics of influenza A virus infections (for reviews, see [4–8]). These models work in a time scale of days and describe the concentration dynamics of target cells, immune system components, viral load, and sometimes co-infecting pathogens. The models differ in terms of complexity and state space dimensions, which are between 3 and 15 for the models examined here. While the low-dimensional models can be analyzed completely and in a straightforward way (e.g., by calculating their fixed points and stability analysis), the characterization of the entire behavioral spectrum of complex models is more difficult (see for example [9]).

Here, we present an approach to understand the overall structure of these models that allows them to be related to each other in a simple way. To this end, we apply chemical organization theory [10,11] to obtain a hierarchical decomposition of each model into chemical organizations. A chemical organization is a sub-set of species (i.e., dimensions or model components, like, uninfected cells or viruses) that cannot generate any other species (property of closure) and that can self-maintain its own species, i.e., any species consumed by a process within the organization can be regenerated by a process within the organization. The organizations of an ODE model are rigorously related to its long-term dynamics in the following way: Given a stationary state of the ODE model, the set of species with strictly positive concentrations must be an organization [10]. The same is true for all practically relevant periodic and chaotic attractors [12]. Note that the advantage of this approach is that decomposition into organizations is based solely on the model structure (i.e., reaction rules) and thus is independent of kinetic details, like rate constants. The relation between measured data, ODE model, and organizations is depicted in Figure 1.

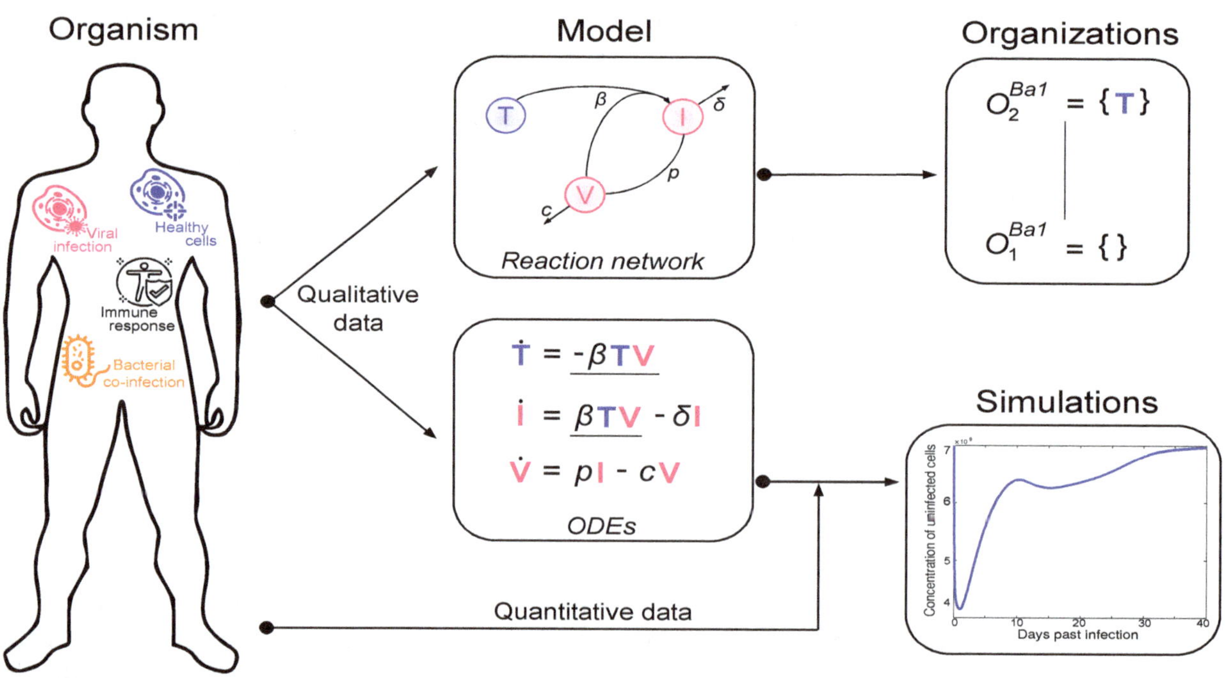

Figure 1. Relation between measured data, ordinary differential equations (ODE) model, and hierarchy of organizations.

By applying the method to twelve models of influenza A virus infection, we found different types of model structures ranging from two to eight organizations. Furthermore, the models' organizations imply a partial order among models. The resulting hierarchy of models can help to select a suitable model for certain data or serve as a framework for further model development.

We provide reaction network files for all models and a software tool for computing their organizations (https://github.com/stephanpeter/orgsflu).

2. Materials and Methods: Procedure for the Organizational Analysis

To illustrate our method, we follow a basic ODE model of influenza dynamics, namely the target cell limited model by Baccam et al. [13], called Baccam Model in the following. Note that we refer throughout this paper to a model by the first author's name of the respective publication.

The Baccam Model is based on in vivo experimental data and contains three variables: the number of susceptible and uninfected target (epithelial) cells T, the number of infected cells I, and the number of infectious-viral titer V. The dynamical behavior of the variables is given by the ODEs shown in Figure 2a. That is, target cells become infected and thus converted to infected cells at a rate βTV, infected cells die spontaneously at rate δI, virus proliferates at a rate pI and dies at a rate cV. The parameters β, δ, p and c are, as usual, positive real numbers (cf. [13] for actual values).

$$\dot{\mathbf{T}} = \underline{-\beta \mathbf{TV}}$$
$$\dot{\mathbf{I}} = \underline{\beta \mathbf{TV}} - \delta \mathbf{I}$$
$$\dot{\mathbf{V}} = p\mathbf{I} - c\mathbf{V}$$

(**a**) ODEs model

$R_1: \mathrm{T} + \mathrm{V} \rightarrow \mathrm{I} + \mathrm{V}$
$R_2: \mathrm{I} \rightarrow \varnothing$
$R_3: \mathrm{I} \rightarrow \mathrm{I} + \mathrm{V}$
$R_4: \mathrm{V} \rightarrow \varnothing$

(**b**) Reaction network

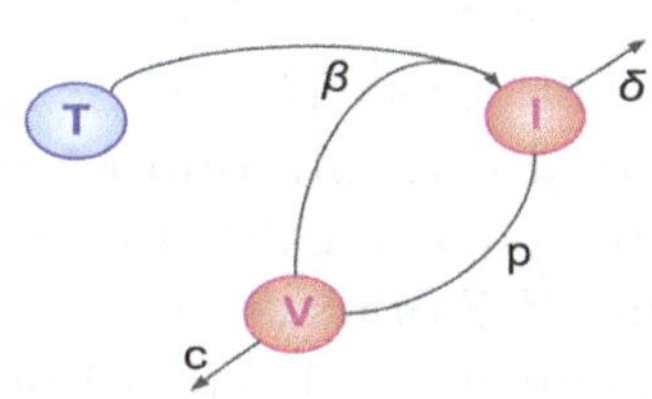

(**c**) Network picture

$O_2^{Ba1} = \{\mathbf{T}\}$
$O_1^{Ba1} = \{\}$

(**d**) Organizations

Figure 2. The **Baccam Model** [13] with three variables: uninfected (susceptible) target cells (**T**), infected cells (**I**) and infectious-viral titer (**V**).

Throughout this paper, the following coloring scheme for particular classes of species is used to improve readability:

- Uninfected (target) cells or those resistant/refractory to infection are marked in blue, e.g., T.
- Infected cells, partially or latently infected cells, and viruses are marked in magenta, e.g., I and V.
- Species belonging to the active immune response are marked in green. It is worth noting that the first two models analyzed in this paper [13,14] do not explicitly have immune system components.
- Bacterial co-infection species are marked in orange. These species are only occurring in Smith's model [15].
- Text referring to any other species is marked in black, e.g., transient target cell states, passive immune system, or dead cells.

For simplicity, the models' variable names also denote the related *species*. For example, V denominates not only the number of viruses in the ODE model (Figure 2a), but also the virus itself (e.g., Figure 2b).

2.1. Deriving the Reaction Network from the ODE System

In a first step, we need to obtain the reaction network underlying the ODE model. A reaction represents, for example, a cell infection by a virus, the generation of new viruses from an infected cell or the spontaneous death of a cell. The reaction rules can be derived from the ODEs in a straightforward way [16]. This step can also be performed by an online tool presented by Soliman and colleagues [16]. Note that in modeling one first creates the network and then derives the ODEs. For our analysis, we have to take the other direction.

For this purpose, we have to investigate the kinetic terms (kinetic laws) of the ODE (Figure 2a):

- The term βTV represents the reaction $R_1 : T + V \rightarrow I + V$, which in turn denotes the transformation of an uninfected target cell T to an infected cell I catalysed by the virus V.
- The terms $-\delta I$ and $-cV$ represent reactions $R_2 : I \rightarrow \emptyset$ and $R_4 : V \rightarrow \emptyset$ which are the outflow of infected cells I resp. virus V.
- The term pI represents the reaction $R_3 : I \rightarrow I + V$ which is the production of viruses V catalysed by infected cells I.

The set of species $\mathcal{S} = \{T, I, V\}$ together with the set of reactions $\mathcal{R} = \{R_1, R_2, R_3, R_4\}$ constitute the so-called *reaction network* $(\mathcal{S}, \mathcal{R})$ associated with the model. The set of reactions together with their kinetic parameters are depicted in Figure 2c. Note that for clarity we use different types of underlining to highlight certain recurring kinetic terms in the ODE:

- Single underline for the transformation of uninfected cells into infected ones by the action of viruses.
- Double underline for kinetic terms involving interferon.

We call the species on the left-hand side (LHS) of a reaction R *support* of R and write $supp(R)$, e.g., $supp(R_1) = \{T, V\}$ (see Figure 2d). Analogously, we call the set of species occurring on the right-hand side (RHS) of a reaction R *products* of R and denominate this set by $prod(R)$, e.g., $prod(R_1) = \{I, V\}$.

Furthermore, let us recall the *stoichiometric matrix* $\mathbf{N} = (n_{ij})$ of a reaction network [17]. The element n_{ij} in the i-th line and j-th column of $\mathbf{N}$ denotes the net-production of the i-th species in reaction R_j. The net-production n_{ij} is the difference between the number of occurrences (stoichiometric coefficient) of species i in the RHS of reaction R_j minus the number of occurrences of species i in the LHS of reaction R_j. For example, $n_{21} = 1 - 0 = 1$ because the second species (I) appears in the first R_1 once as a reactant in the support of R_1 (LHS) but does not appear in R_1 as a product (RHS). For our example in Figure 2, the stoichiometric matrix becomes:

$$\mathbf{N} = \begin{pmatrix} -1 & 0 & 0 & 0 \\ 1 & -1 & 0 & 0 \\ 0 & 0 & 1 & -1 \end{pmatrix}. \tag{1}$$

2.2. Computing the Organizations from the Reaction Network

From the reaction network, we can compute the (chemical) *organizations* of the model. Each organization is a subset of species that is *closed* and *self-maintaining* [10,18]. In the following, let $S \subseteq \mathcal{S}$ be a subset of species and n be the total number of reactions of the reaction network ($n = 4$ in our example).

We call S ***closed*** if and only if all reactions $R \in \mathcal{R}$ with $supp(R) \subseteq S$ fulfill $prod(R) \subseteq S$ too [10,18]. This means that the products of a reaction R with support in S are also in S. In other words, no species outside of S can be produced by the reactions "running on" S. As an example, we assume $S = \{T, I\}$. The reactions with support in S are R_2 and R_3. However, R_3 produces species V, which is not in S. Thus, S is not closed.

We call a vector $\mathbf{v} \in \mathbb{R}^n$ *flux vector for* S if and only if it fulfills

$$\mathbf{v}_j \begin{cases} > 0, & \text{if} \quad supp(R_j) \subseteq S, \\ = 0, & \text{if} \quad supp(R_j) \not\subseteq S. \end{cases} \tag{2}$$

Thus, all flux vectors for S have in common that those components are strictly positive which correspond to reactions that can run on S, while all other entries are zero. As an example, consider $S_{Example} = \{T, I\}$ again. We know that the reactions R_2 and R_3 can "run on" it, i.e., they have support in $S_{Example}$. Thus, $\mathbf{v}_1 = (0, 1, 3, 0)^T$ or $\mathbf{v}_2 = (0, 5, 2, 0)^T$ are example flux vectors for $S_{Example}$.

We call S ***self-maintaining*** if and only if there exists (at least one) flux vector $\mathbf{v} \in \mathbb{R}^n$ for S that fulfills

$$\mathbf{N} \cdot \mathbf{v} \geq \mathbf{0}, \tag{3}$$

i.e., $(\mathbf{N} \cdot \mathbf{v})_i \geq 0$ for all $i = 1, \ldots, n$, where n is again the total number of reactions [10,19–21]. Roughly speaking, if S is self-maintaining, it has the potential to sustain the amount of its species above a certain level. Our example set $S_{Example}$ is not self-maintaining because $(\mathbf{N} \cdot \mathbf{v})_2 < 0$ for all flux vectors for $S_{Example}$. That is, species number 2 (the infected cells I) can not be maintained by this set.

As mentioned in the beginning of this section, we call S an *organization* if and only if it is both closed and self-maintaining. Clearly $S_{Example}$ is not an organization, as it has neither of these properties. In Figure 2d, the so-called *Hasse diagram* of organizations of this model can be seen. In it, two organizations are linked by a line if the lower one is a subset of the upper one and there is no organization in between. The Hasse diagram for the Baccam Model contains only the two organizations $O_1^B = \emptyset$ and $O_2^B = \{T\}$ (see Figure 2d). The superscript "B" stands for Baccam Model and with the subscripts we refer to the organizations within a model. The organization O_2^B represents an organism without any influenza A virus infection. Note that there is no organization with all species, i.e., representing the infected body.

2.3. The Role Organizations Play in the Dynamics

Given a fixed point x of an ODE describing a reaction system, the set of species with strictly positive concentrations in x is an organization. This is shown in [10]. This in turn means that, if a subset of species is not an organization, then the system does not have a fixed point with exactly the chosen subset of species. This is not only true for fixed points but also for other attractors [12]. Attractors are those states that a system approaches in the long-run and once reached never leaves anymore. Besides converging towards a fixed point, the long-term behavior can be also periodic oscillations or chaotic trajectories. In particular, it was shown that the long-term behavior of the system always tends at least to one organization [12]. Thus, organizations rule the long-term behavior of such dynamical reaction system. Note that the case of a system tending towards a fixed point is included in this statement as a special case. For example, the *simulation* results of the Baccam model (Figure 2a and Ref. [13]) suggest that after about two to three days the species begin to decay to finally arrive in an organization namely the empty set.

3. Results and Discussion

In the literature, there exist several mathematical models of IAV dynamics that are derived from experimental data, reviewed in Refs. [4–7]). These models differ in their complexity, e.g., the number of reactions and the number of species, depending on the available experimental data used for parameter fitting and questions to be answered. For example, models can include eclipse phases, an innate immune response, or an adaptive immune response.

After having exemplified our method in Section 2 by an analysis of the basic target cell limited model by Baccam et al. [13], we present now the full analysis for eleven additional more detailed influenza models of IAV dynamics, with up to 15 variables (species) and 45 reactions (cf. Table 1 for an overview at the end). Note that for our analysis we abstract from kinetic details, that is, the organizations are independent of particular settings of parameter values.

3.1. Target Cell Limited Model by Miao et al. (Miao Model, M, 2010)

The models by Miao et al. [14] are designed to fit experimental in vivo data from mice [6,7]. The first one ([14], Equation (1)) depends on measured time-series. The second one ([14], Equation (2)) is a simplified version of the first one, neglecting the terms depending on those time-series and still leading to a good fit within the first five days after infection [14]. Thus, we analyze this second model (Miao Model).

Compared to the basic Baccam Model, the Miao Model (Figure 3a) has the same three variables (named differently) and one additional reaction, $E_P \rightarrow 2E_P$. This reaction represents the self-replication of target cells E_P taking place at a rate $\rho_E E_P$. The full set of reactions can be found in the Appendix A (Figure A2).

In the Hasse diagram of organizations (Figure 3b), a new "full" organization O_3^M appears, which contains all three species. Thus, organization O_3^M reflects the slight difference between the two models: in the Baccam Model, uninfected target cells T are only the susceptible ones and can not increase in number, but in the Miao Model uninfected cells E_P are reproduced repeatedly by the organism. Thus, in the Baccam Model, infection is limited inherently by the limited number of uninfected target cells, while in the Miao Model the limitation of an infection in time and number of infected cells and viruses depends on other mechanisms:

$$\dot{\mathbf{E}}_\mathbf{P} = \rho_E \mathbf{E_P} - \underline{\beta_\alpha \mathbf{E_P V}}$$

$$\dot{\mathbf{E}}^*_\mathbf{P} = \underline{\beta_\alpha \mathbf{E_P V}} - \delta_{E^*} \mathbf{E^*_P}$$

$$\dot{\mathbf{V}} = \pi_\alpha \mathbf{E^*_P} - c_V \mathbf{V}$$

(**a**) ODE model

$$O_3^M = \{all\ species\} = \{\mathbf{E_P}, \mathbf{E^*_P}, \mathbf{V}\}$$

$$O_2^M = \{\mathbf{E_P}\}$$

$$O_1^M = \{\}$$

(**b**) Hasse diagram of organizations

Figure 3. The **Miao Model** [14] with three variables: uninfected target cells ($\mathbf{E_P}$), productively infected cells ($\mathbf{E^*_P}$) and free infectious influenza viruses ($\mathbf{V}$).

3.2. *Target Cell Limited Model with Delayed Virus Production (Baccam II Model, Ba2, 2006)*

The Baccam II Model [13,22] contains one more species than the Baccam Model presented in the methods section above. That is, there are now *two* types of infected cells: those which do not yet produce viruses I_1 and those which actively produce viruses I_2. In addition, there is only one new reaction, which transforms infected cells of type I_1 into type I_2 at rate kI_1 (Figure 4a). However, the Hasse diagram of organizations remains the same when compared with the basic Baccam Model [13].

$$\dot{\mathbf{T}} = \underline{-\beta \mathbf{TV}}$$

$$\dot{\mathbf{I}}_1 = \underline{\beta \mathbf{TV}} - k\mathbf{I_1}$$

$$\dot{\mathbf{I}}_2 = k\mathbf{I_1} - \delta \mathbf{I_2}$$

$$\dot{\mathbf{V}} = p\mathbf{I_2} - c\mathbf{V}$$

(**a**) ODE model

$$O_2^{Ba2} = \{\mathbf{T}\}$$

$$O_1^{Ba2} = \{\}$$

(**b**) Hasse diagram of organizations

Figure 4. The **Baccam II Model** [13] with delayed virus production and four variables: uninfected (susceptible) target cells ($\mathbf{T}$), infected cells not yet producing virus ($\mathbf{I_1}$), infected cells actively producing virus ($\mathbf{I_2}$) and infectious-viral titer ($\mathbf{V}$).

3.3. *Innate and (Simple) Adaptive Immune Response (Pawelek Model, P, 2012)*

The Pawelek Model [23] contains 11 parameters and was designed to fit in vivo experimental data of horses [6,7]. The model has five variables and nine reactions. Like the basic Baccam Model, it contains uninfected target cells (T), infected cells (I), and viruses (V). Furthermore, there are two new species: interferon (F) and uninfected cells that are refractory to infections (R) because of the antiviral effect induced by interferons.

Investigating the reaction network (Figure A4 in the Appendix A) derived from the differential equations (Figure 5a), we can see that, like in the basic Baccam Model, self-replication of uninfected cells T is missing. However, due to the two new species R and F, we have five new reactions, which are neither included in the Baccam Model nor in the Miao Model. One of these five reactions is the spontaneous decay of interferon F at a rate dF. The other four new reactions describe interactions between different species:

- The rate term ϕFT represents the transformation of uninfected target cells to refractory cells catalysed by interferon.
- The reverse shift back from refractory to simple uninfected cells is represented by the term ρR.
- Furthermore, infected cells are deleted by the action of interferon at a rate κIF.
- Interferon is produced in the presence of infected cells at a rate qI.

Even though we have more species and more reactions, we get the same small pattern of organizations as in the basic Baccam Model (Figure 5b). Both models have in common that there is no self-replication of target cells. This might be one reason for the missing of other and bigger organizations which could contain species related to infection and/or immune response. This in turn means that, like the Baccam Model, this model implicitly treats infections and immune responses as phenomena that can only appear in a limited (transient) time span. The Hasse diagram of organizations (Figure 5b) tells us that the system necessarily tends towards a state of healthiness, which is represented by the organizations $O_1^P = \{\}$ and $O_2^P = \{T\}$, showing absence of infection and immune response.

$$\dot{\mathbf{T}} = -\underline{\beta \mathbf{VT}} - \underline{\underline{\phi \mathbf{FT}}} + \rho \mathbf{R}$$

$$\dot{\mathbf{I}} = \underline{\beta \mathbf{VT}} - \delta \mathbf{I} - \underline{\underline{\kappa \mathbf{IF}}}$$

$$\dot{\mathbf{R}} = \underline{\underline{\phi \mathbf{FT}}} - \rho \mathbf{R}$$

$$\dot{\mathbf{V}} = p\mathbf{I} - c\mathbf{V}$$

$$\dot{\mathbf{F}} = q\mathbf{I} - d\mathbf{F}$$

(**a**) ODE model

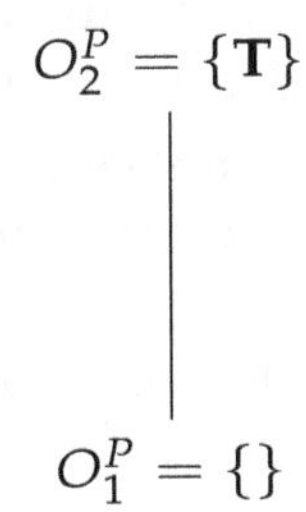

(**b**) Hasse diagram of organizations

Figure 5. The **Pawelek Model** [23] with five variables: (uninfected) target cells (**T**), productively infected cells (**I**), uninfected cells refractory to infections (**R**), free viruses (**V**) and interferon (**F**).

3.4. A Model Including Bacterial Co-Infection (Smith Model, Sm, 2016)

The Smith Model [15] contains 15 parameters and compared to experimental in vivo data from mice. It has five variables and 12 reactions. Like in the previous models, we have susceptible target cells (T) and viruses (V). Note that T is only consumed in this model but not produced. Contrarily to previous models, we have two kinds of infected cells (I_1 and I_2) here as well as bacteria (P). Bacteria P represent *bacterial co-infection* during or after virus infection. Infection is modeled as a transformation of susceptible target cells T into infected cells I_1 catalyzed by viruses V (see underlined terms in Figure 6a). Infected cells I_1 in turn spontaneously transform into I_2 at rate kI_1. Only infected cells I_2 produce viruses V at a rate pI_2. Furthermore, infected cells I_2 produce viruses V together with bacteria P. Bacteria P are self-replicating (rate term rP). Viruses V are the only species influencing bacteria P.

Figure 6b shows the Hasse diagram of organizations. The smallest one is the empty set. The biggest one is O_4^{Sm}, which contains susceptible target cells T and bacteria P. It represents an organism without viral but with bacterial infection. Between those two extreme organizations, we find $O_2^{Sm} = \{\mathbf{T}\}$ and $O_3^{Sm} = \{\mathbf{P}\}$. Thus, O_2^{Sm} represents the healthy organism without any infection. $O_3^{Sm} = \{\mathbf{P}\}$ could mark the state after a viral-bacterial co-infection: after viral infection and because of the death of all target cells as well as all viruses only bacteria remain.

$$
\begin{aligned}
\dot{\mathbf{T}} &= -\beta\mathbf{TV} \\
\dot{\mathbf{I_1}} &= \beta\mathbf{TV} - k\mathbf{I_1} - \mu\mathbf{PI_1} \\
\dot{\mathbf{I_2}} &= k\mathbf{I_1} - \delta\mathbf{I_2} - \mu\mathbf{PI_2} \\
\dot{\mathbf{V}} &= p\mathbf{I_2}(1 + \mathbf{P}^Z) - c\mathbf{V} \\
\dot{\mathbf{P}} &= r\mathbf{P}(1 - \frac{\mathbf{P}}{K_P(1 + \psi\mathbf{V})}) \\
&\quad - \gamma_{M_A} n^2 \frac{M_A}{\mathbf{P}^2 + n^2 M_A} M_A^* \mathbf{P}(1 - \frac{\phi\mathbf{V}}{K_{PV} + \mathbf{V}})
\end{aligned}
$$

(**a**) ODE model

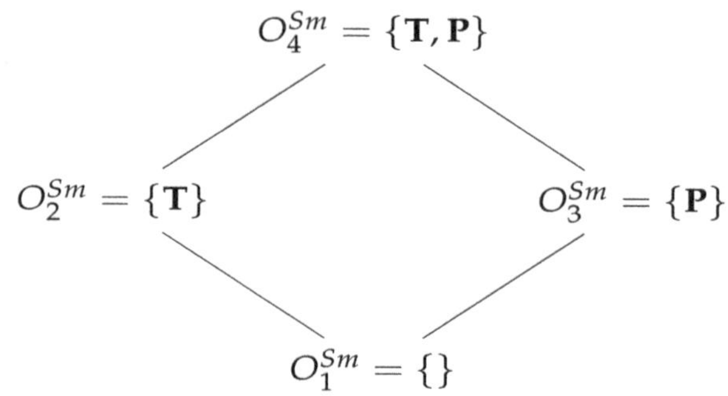

(**b**) Hasse diagram of organizations

Figure 6. The **Smith Model** [15] with five variables: susceptible target cells (**T**), two classes of infected cells ($\mathbf{I_1}$ and $\mathbf{I_2}$), free viruses (**V**), and bacteria (**P**).

3.5. *Innate and Adaptive Immune Response (Handel Model, Ha, 2009)*

The Handel Model [24] contains eight parameters and was designed to fit experimental in vivo data from mice [6,7,25]. It has seven variables (see Figure 7) and only 12 reactions (see Figure 7a):

- *Infection* is catalyzed by viruses V and transforms uninfected cells U to latently infected cells E and viruses V are consumed thereby. Latently infected cells E transform into infected cells I autonomously, which in turn transform into dead cells D autonomously too. Finally, the transformation of dead cells D into non-infected cells U closes the *circle*.
- The *remaining three species* V, F and X form an almost totally separate *subsystem* since the only interaction with the four species from the "circle" mentioned above is the catalysis of the infection by viruses V.
- The *interactions within the subsystem* $\{V, F, X\}$ consisting of viruses V and immune responses F and X are as follows:

  - Viruses V catalyze the proliferation of F and X. In the Hernandez model, proliferation of interferon F is catalyzed by infected cells instead of viruses.
  - There is no direct interaction between innate immune response F and adaptive immune response X.
  - The adaptive immune response X deletes viruses directly. Innate immune response F inhibits the self-replication of the viruses which is represented by the denominator of the fraction $\frac{p\mathbf{I}}{1+\kappa\mathbf{F}}$. We ignore the inhibition because whether the rate is zero or not is independent of F.

The Hasse diagram Figure 7b shows five organizations. For the first time, it contains the empty set as well as the set of all species at the same time. Between these extremes, we find $O_7^{Ha} = \{\mathbf{U}\}$ representing the healthy organism. The Baccam, Miao, and Pawelek models exhibit the same organization. Structurally, the Hasse diagram of the Handel Model is very similar to that of the Smith Model (Figure 6b). The first reveals an autonomy of the adaptive immune response X, whereas the latter does this same for bacteria P.

$$\dot{\mathbf{U}} = \lambda D - \underline{b\mathbf{UV}}$$
$$\dot{\mathbf{E}} = \underline{b\mathbf{UV}} - g\mathbf{E}$$
$$\dot{\mathbf{I}} = g\mathbf{E} - d\mathbf{I}$$
$$\dot{D} = d\mathbf{I} - \lambda D$$
$$\dot{\mathbf{V}} = \frac{p\mathbf{I}}{\underline{\underline{1+\kappa\mathbf{F}}}} - c\mathbf{V} - \gamma b\mathbf{UV} - k\mathbf{VX}$$
$$\dot{\mathbf{F}} = w\mathbf{V} - \delta\mathbf{F}$$
$$\dot{\mathbf{X}} = f\mathbf{V} + r\mathbf{X}$$

(**a**) ODE model

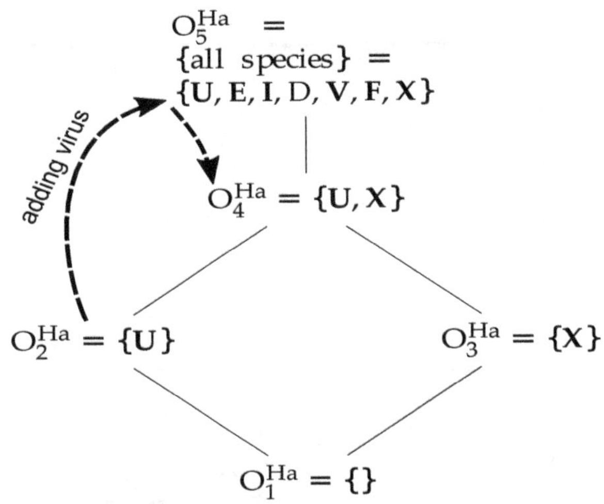

(**b**) Hasse diagram of organizations

Figure 7. The **Handel model** [24] with seven variables: uninfected cells (**U**), latently infected cells (**E**), productively infected cells (**I**), dead cells (D), free viruses (**V**), innate immune response (**F**) and adaptive immune response (**X**). The dotted arrows denote the projection of the dynamics shown in Figure 8.

Temporal Dynamics

For the Handel Model, we perform dynamical simulation in order to show how the organizational hierarchy helps also to understand transient short-term behavior. We start at $t = -20d$ with an uninfected state, i.e., 7×10^9 uninfected cells. After 20 days, at $t = 0$, we add 10^4 virus particles. The resulting seven-dimensional trajectory in state space is shown in Figure 8. Projecting this trajectory to organizations results in a more abstract view of the dynamics, shown as a dashed curved arrow in Figure 7b. The system starts in organization O_2^{Ha} (uninfected organization), moves after adding virus particles at $t = 0$ into organization O_5^{Ha} (infected organization with immune response), and drops after 37 days into organization O_4^{Ha} (immune response active, virus absent).

The projection of a state x to an organization O follows the procedure suggested by Dittrich and Speroni d.F. [10]: First, we generate a set S of those species whose concentration is greater than a particular threshold (here: $10^0 = 1$). Then, we generate the closure of this set by adding all species that can be produced from the set. Finally, we take the largest organization O that is a subset of that closure. For example: At $t = 0$ by adding viruses to the system, we have $S = \{U, V\}$, whose closure is the set of all species, which is also an organization, here; thus, the state at $t = 0$ is projected to organization O_5^{Ha}. At $t = 60d$, we have $S = \{U, X, D\}$, whose closure is again $\{U, X, D\}$ and the largest organization contained is $O_4^{Ha} = \{U, X\}$. Thus, as can be seen in Figure 8, the system state is projected to organization O_4^{Ha}, in which it stays for $t > 37d$.

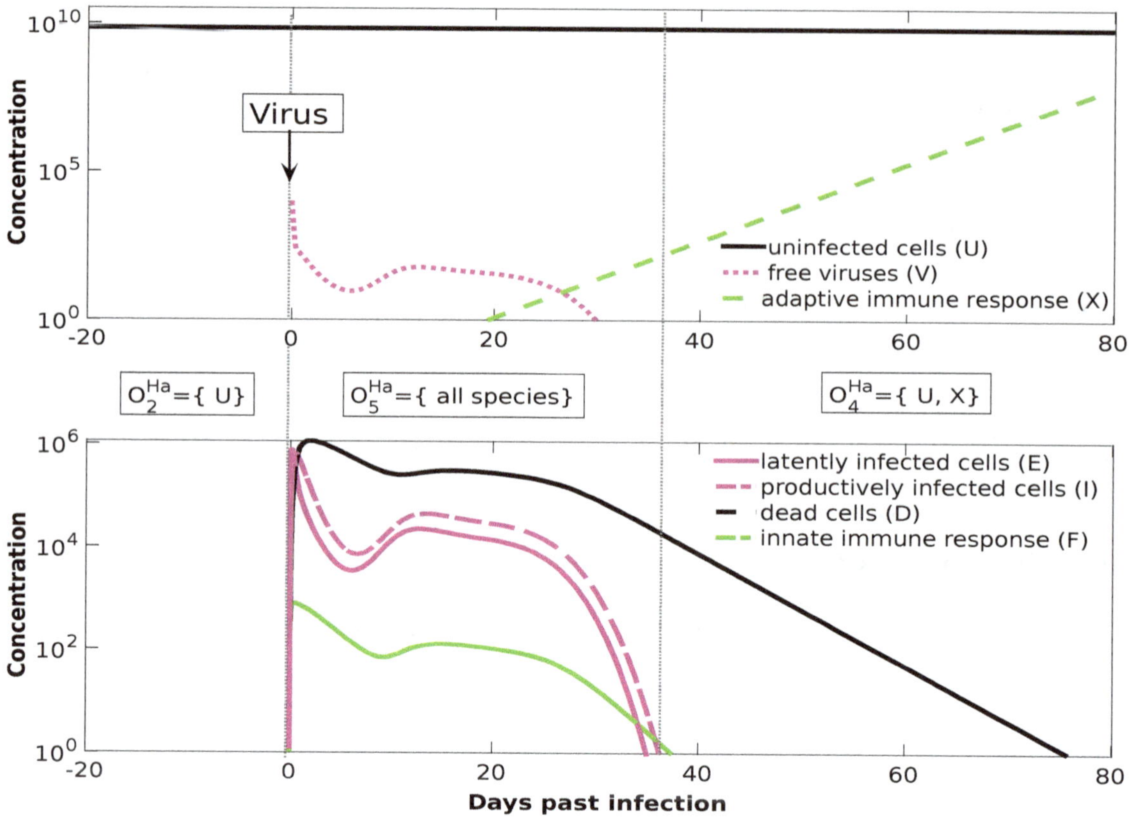

Figure 8. Temporal dynamics of the Handel model. By projecting the seven-dimensional trajectory to organizations (dotted arrows in Figure 7b) we find three phases: (Phase 1) Until day number 0, there are solely 7×10^9 uninfected cells **U** in the system represented by the organization $O_2^{Ha} = \{\mathbf{U}\}$. (Phase 2) At day 0, **infection** is simulated by adding $V(0) = 10^4$ virus particles to the system. The resulting state $\{\mathbf{U}, \mathbf{V}\}$ is projected to organization O_5^{Ha} (all species). (Phase 3) Lastly, at day t = 37d past infection the system settles in the final organization, namely $O_4^{Ha} = \{\mathbf{U}, \mathbf{X}\}$, which is generated by the set $\{\mathbf{U}, \mathbf{X}, D\}$ (see text). The values of the model parameters are (from [24]): $\lambda = 0.25$, $b = 2.1 \times 10^{-7}$, $g = 4$, $d = 2$, $p = 5 \times 10^{-2}$, $\kappa = 1.8 \times 10^{-2}$, $c = 10$, $\gamma = 7.5 \times 10^{-4}$, $k = 1.8$, $w = 1$, $\delta = 0.4$, $f = 2.7 \times 10^{-6}$, and $r = 0.3$. Note that the number of uninfected cells **U** is not constant after infection as it may seem from the figure. In fact, after infection, the number of uninfected cells first decreases and than rises again [24].

3.6. Innate Immune Response and Resistance to Infection (Hernandez Model, He, 2012)

The 13 parameters of the Hernandez Model [26] were fitted to data from many different sources. The model contains seven variables and 16 reactions (see Figure 9). The species refer to viruses (V), interferon (F) and natural killer cells (K) as well as four types of respiratory tract epithelial cells: healthy/uninfected (U_H), partially infected (U_E), infected (U_I) and resistant to infection (U_R). Compared to the Pawelek Model, there are two qualitatively new species: partially infected cells U_E and killers K.

Next, we state some remarks about the reactions:

- There is an *infection* reaction catalyzed by viruses like in all previous models but with one difference: during infection, healthy cells U_H first transform to *partially infected* cells U_E and only after that they transform spontaneously to infected cells U_I at a rate $k_e U_E$.
- *Interferon* catalyzes the transformation of healthy cells to resistant cells U_R, like in the Pawelek Model. However, in the Pawelek Model, interferon removes infected cells. Here, interferon's production is catalyzed by infected cells U_I at a rate $a_I U_I$. There is no further influence of interferon on any other species.

- Infected cells are removed by natural *killers* K, which also delete partially infected cells in this model. The production of *killers* K is catalyzed by infected cells U_I at a rate $\Phi_K U_I$.
- Note that here we have an constant *inflow of healthy cells* U_H at a rate S_H (first differential equation). Thus, healthy cells cannot converge to zero.

The Hasse diagram of organizations consists only of two organizations (Figure 9b). For the first time, the empty set is not an organization because the empty set is not closed due to a constant inflow of healthy cells U_H and killers K, represented by the reaction $\emptyset \to U_H$ and $\emptyset \to K$, respectively. The smallest organization $O_1^{He} = \{U_H, K\}$ can be regarded as a state of healthiness. Contrarily, the second organization O_2^{He} contains all species (as in the Miao Model) and therefore can be interpreted as the infected organism exhibiting immune response to infection.

$$
\begin{aligned}
\dot{\mathbf{U}}_{\mathbf{H}} &= S_H - \underline{k_I \mathbf{U_H V}} - \underline{\underline{k_R \mathbf{U_H F}}} - \delta_H \mathbf{U_H} \\
\dot{\mathbf{U}}_{\mathbf{E}} &= \underline{k_I \mathbf{U_H V}} - k_E \mathbf{U_E} - q_K \mathbf{U_E K} \\
\dot{\mathbf{U}}_{\mathbf{I}} &= k_E \mathbf{U_E} - \delta_I \mathbf{U_I} - q_K \mathbf{U_I K} \\
\dot{\mathbf{U}}_{\mathbf{R}} &= \underline{\underline{k_R \mathbf{U_H F}}} - \delta_R \mathbf{U_R} \\
\dot{\mathbf{V}} &= \rho_V \mathbf{U_I} - \delta_V \mathbf{V} \\
\dot{\mathbf{F}} &= a_I \mathbf{U_I} - \delta_F \mathbf{F} \\
\dot{\mathbf{K}} &= S_K + \Phi_K \mathbf{U_I} - \delta_K \mathbf{K}
\end{aligned}
$$

(**a**) ODE model

$$O_2^{He} = \{all\ species\} = \{\mathbf{U_H, U_E, U_I, U_R, V, F, K}\}$$

$$|$$

$$O_1^{He} = \{\mathbf{U_H, K}\}$$

(**b**) Hasse diagram of organizations

Figure 9. The **Hernandez Model** [26] with seven variables: healthy cells ($\mathbf{U_H}$), partially infected cell ($\mathbf{U_E}$), infected cells ($\mathbf{U_I}$), cells resistant to infection ($\mathbf{U_R}$), virus particles ($\mathbf{V}$), interferon ($\mathbf{F}$) and natural killers ($\mathbf{K}$).

3.7. A Model with More Detailed Immune Response (Cao Model, C, 2015)

The Cao Model [27] consists of 20 parameters and has been derived by referring to experimental data from ferrets. The model has seven variables and 26 reactions. Like in most of the previous models, we have (uninfected) target cells (T), infected cells (I), viruses (V), resistant cells (R), and interferon (F). Furthermore, there are B cells B and antibodies A.

According to the ODE shown in Figure 10a, *B cells* are only influenced by viruses: viruses support the production of B cells (rate term $m_1 V$), but the more B cells that are present, the more of them are destroyed by viruses again (term: $m_1 VB$). B cells influence only one other species, namely antibodies B, which they produce. *Antibodies A* influence only one other species, namely viruses, which are destroyed by this reaction at rate μVA. Antibodies in turn are influenced by B cells positively and by viruses negatively.

There are three organizations in this model (Figure 10b): the empty set, the healthy organism without any infection and without any immune response ($O_2^C = \{\mathbf{T}\}$) and the organization O_3^C containing all species.

$$
\begin{aligned}
\dot{\mathbf{T}} &= g(\mathbf{T}+\mathbf{R})(1-\frac{\mathbf{T}+\mathbf{R}+\mathbf{I}}{C_t}) - \beta'\mathbf{VT} + \rho\mathbf{R} - \underline{\underline{\phi\mathbf{FT}}}\\
\dot{\mathbf{I}} &= \beta'\mathbf{VT} - \delta\mathbf{I} - \underline{\underline{\kappa\mathbf{IF}}}\\
\dot{\mathbf{V}} &= \frac{p\mathbf{I}}{\underline{\underline{1+s\mathbf{F}}}} - c\mathbf{V} - \mu\mathbf{VA} - \beta\mathbf{VT}\\
\dot{\mathbf{R}} &= \underline{\underline{\phi\mathbf{FT}}} - \rho\mathbf{R} - \xi\mathbf{R}\\
\dot{\mathbf{F}} &= q\mathbf{I} - d\mathbf{F}\\
\dot{\mathbf{B}} &= m_1\mathbf{V}(1-\mathbf{B}) - m_2\mathbf{B}\\
\dot{\mathbf{A}} &= m_3\mathbf{B} - r\mathbf{A} - \mu'\mathbf{VA}
\end{aligned}
$$

(**a**) ODE model

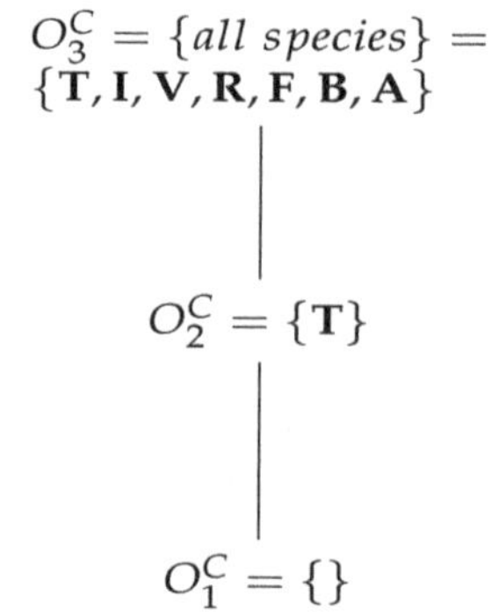

(**b**) Hasse diagram of organizations

Figure 10. The **Cao Model** [27] with seven variables: target cells (**T**), infected cells (**I**), viruses (**V**), resistant cells (**R**), interferon (**F**), B cells (**B**), and antibodies (**A**).

3.8. Innate Immune Response and Eclipse Phase (Saenz Model, Sa, 2010)

The Saenz Model [28] requires 12 parameters and was designed to fit experimental in vivo data from horses [6,7]. The model contains eight variables and 12 reactions (Figure 11a). There are no adaptive immune response, no dead cells, and no natural killer cells. However, the model contains viruses V and interferon F. There is an eclipse phase (E_1 and E_2) here as well as prerefractory and refractory cells. In particular, epithelial cells are represented by six species: susceptible (T), eclipse phases (E_1 *and* E_2), infectious (I), prerefractory (W), and refractory (R). Thus, the new features are the inclusion of two eclipse phases and three steps for the transformation of uninfected cells to refractory cells.

$$
\begin{aligned}
\dot{\mathbf{T}} &= -\underline{\beta\mathbf{VT}} - \underline{\underline{\phi\mathbf{FT}}}\\
\dot{\mathbf{E}}_1 &= \underline{\beta\mathbf{VT}} - k_1\mathbf{E}_1\\
\dot{W} &= \underline{\underline{\phi\mathbf{FT}}} - m\beta\mathbf{V}W - a\mathbf{W}\\
\dot{E}_2 &= m\beta VW - k_2E_2\\
\dot{\mathbf{R}} &= aW\\
\dot{\mathbf{I}} &= k_1\mathbf{E}_1 + k_2E_2 - \delta\mathbf{I}\\
\dot{\mathbf{V}} &= p\mathbf{I} - c\mathbf{V}\\
\dot{\mathbf{F}} &= nqE_2 + q\mathbf{I} - d\mathbf{F}
\end{aligned}
$$

(**a**) ODE model

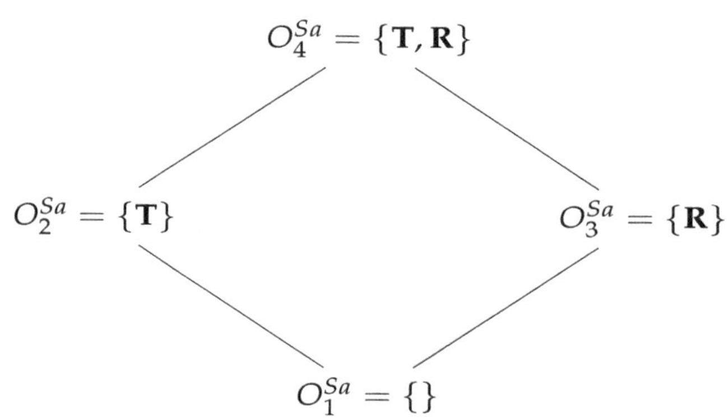

(**b**) Hasse diagram of organizations

Figure 11. The **Saenz Model** [28] with eight variables: Epithelial cells in one of the states: susceptible (**T**), eclipse phase ($\mathbf{E_1}$ and E_2), prerefractory (W), refractory (**R**) and infectious (**I**). The further variables are: virus cells (**V**) and interferon (**F**).

The Hasse diagram of organizations is composed by four organizations: $O_1^{Sa} = \{\}$: the empty set; $O_2^{Sa} = \{\mathbf{T}\}$: representing a healthy organism; $O_3^{Sa} = \{\mathbf{R}\}$: there is no consuming reaction for refractory cells R; $O_4^{Sa} = \{\mathbf{T}, \mathbf{R}\}$: also representing a healthy organism that contains refractory cells maybe as the remains of a previous infection.

The Hasse diagram is very similar to that from of the Handel Model. There are only two differences:

- The "full" organization is missing here. For sure, one of the reasons is that there is no reaction producing susceptible cells T. Thus, when viruses V or interferon F are present susceptible cells T can not survive and the "full" organization neither.

- Adaptive immune response is replaced by refractory cells in the organizations here.

3.9. Focusing on Innate and Adaptive Immunity (Hancioglu Model, Hcg, 2007)

The Hancioglu Model [29] contains 28 parameters, 10 variables, and 44 reactions. It has not been mathematically fitted to data but has been designed to meet specific general criteria [6,7]. The ODEs (Figure 12a) describe the dynamics of the following 10 species: viruses (V), healthy cells (H), infected cells (I), interferon (F) and resistant cells (R). The remaining species are new: antigen presenting cells (M), effector cells (E), plasma cells (P) antibodies (A) and antigenic distance (S). There are no species for an eclipse phase in this model.

Looking at the reaction network (Figure A8), we can see again a reaction for *infection*, i.e., the transformation of healthy cells H into infected cells I catalyzed by viruses V at a rate $\gamma_{HV}VH$ (single underlined in Figure 12a). Furthermore, *interferon* F is produced catalytically by antigen presenting M and infected cells I, decays spontaneously at a rate $a_F F$, and is additionally removed when converting healthy cells H into resistant cells R by the reaction $H + F \rightarrow R$ at rate $b_{HF}FH$ (double underlined in Figure 12a).

The Hancioglu Model has three organizations (Figure 12b):

- The smallest one is O_1^{Hcg}, which contains all the species responsible for the *immune response*.
- O_1^{Hcg} is a subset of O_2^{Hcg}, which additionally contains species H and R, representing the healthy organism without infection but with the immune response turned on.
- O_3^{Hcg} is the "full" organization containing all the species of the models and thus representing the organism with infection and immune response.

Thus, all the organizations represent meaningful states of the organism. However, there is no organization that only consists of healthy cells without any infection and immune response. Note that almost all the previous models except for Hernandez have such an organization.

$$\dot{\mathbf{H}} = b_{HD}(1 - \mathbf{H} - \mathbf{R} - \mathbf{I})(\mathbf{H} + \mathbf{R}) + a_R\mathbf{R} - \underline{\gamma_{HV}\mathbf{VH}} - \underline{\underline{b_{HF}\mathbf{FH}}}$$

$$\dot{\mathbf{I}} = \underline{\gamma_{HV}\mathbf{VH}} - b_{IE}\mathbf{EI} - a_I\mathbf{I}$$

$$\dot{\mathbf{V}} = \gamma_V\mathbf{I} - \gamma_{VA}S\mathbf{AV} - \gamma_{VH}\mathbf{HV} - \alpha_V\mathbf{V} - \frac{a_{V1}\mathbf{V}}{1 + a_{V2}\mathbf{V}}$$

$$\dot{\mathbf{R}} = \underline{\underline{b_{HF}\mathbf{FH}}} - a_R\mathbf{R}$$

$$\dot{M} = (b_{MD}(1 - \mathbf{H} - \mathbf{R} - \mathbf{I}) + b_{MV}\mathbf{V})(1 - M) - a_M M$$

$$\dot{\mathbf{F}} = b_F M + c_F\mathbf{I} - b_{FH}\mathbf{HF} - a_F\mathbf{F}$$

$$\dot{\mathbf{E}} = b_{EM}M\mathbf{E} - b_{EI}\mathbf{IE} + a_E(1 - \mathbf{E})$$

$$\dot{\mathbf{P}} = b_{PM}M\mathbf{P} + a_P(1 - \mathbf{P})$$

$$\dot{\mathbf{A}} = b_A\mathbf{P} - \gamma_{AV}S\mathbf{AV} - a_A\mathbf{A}$$

$$\dot{S} = r\mathbf{P}(1 - S)$$

(**a**) ODE model

$$O_3^{Hcg} = \{all\ species\} = \{\mathbf{H}, \mathbf{I}, \mathbf{V}, \mathbf{R}, M, \mathbf{F}, \mathbf{E}, \mathbf{P}, \mathbf{A}, S\}$$

$$O_2^{Hcg} = \{\mathbf{H}, M, \mathbf{F}, \mathbf{R}, \mathbf{E}, \mathbf{P}, \mathbf{A}, S\}$$

$$O_1^{Hcg} = \{M, \mathbf{F}, \mathbf{E}, \mathbf{P}, \mathbf{A}, S\}$$

(**b**) Hasse diagram of organizations

Figure 12. The **Hancioglu Model** [29] with 10 variables: viral load (**V**), healthy cells (**H**), infected cells (**I**), antigen presenting cells (M), interferon (**F**), resistant cells (**R**), effector cells (**E**), plasma cells (**P**), antibodies (**A**) and antigenic distance (S).

3.10. Model with Delay Differential Equations (Bocharov Model, Bo, 1994)

The Bocharov Model [30] contains 49 parameters and was designed to fit experimental in vivo data from humans [6,7]. It includes 10 variables and 51 reactions (Figure A9). Only here and in the Lee

Model (below) we have differential equations with *delay*, i.e., some rates depend on variable values from the past (Figure 13a). Because the delay does not matter in a steady-state, we can also neglect the delay when analyzing the chemical organizations of a delay differential equation model.

$$\dot{\mathbf{V}}_{\mathbf{f}} = \nu\mathbf{C} + nb_{CE}\mathbf{CE} - \underline{\underline{\gamma_{VF}\mathbf{V}_{\mathbf{f}}\mathbf{F}}} - \gamma_{VM}\mathbf{V}_{\mathbf{f}} - \gamma_{VC}\mathbf{V}_{\mathbf{f}}\mathbf{U}$$

$$\dot{\mathbf{C}} = \sigma\mathbf{V}_{\mathbf{f}}\mathbf{U} - b_{CE}\mathbf{CE} - b_m\mathbf{C}$$

$$\dot{m} = b_{CE}\mathbf{CE} + b_m\mathbf{C} - \alpha_m m$$

$$\dot{\mathbf{M}}_{\mathbf{V}} = \gamma_{MV}M^*\mathbf{V}_{\mathbf{f}} - \alpha_M\mathbf{M}_{\mathbf{V}}$$

$$\dot{\mathbf{H}}_{\mathbf{E}} = b_H^E[(1 - \frac{m}{C^*})\rho_H^E\mathbf{M}_{\mathbf{V}}(t - \tau_H^E)\mathbf{H}_{\mathbf{E}}(t - \tau_H^E) - \mathbf{M}_{\mathbf{V}}\mathbf{H}_{\mathbf{E}}] - b_P^{H_E}\mathbf{M}_{\mathbf{V}}\mathbf{H}_{\mathbf{E}}\mathbf{E} + \alpha_H^E(H_E^* - \mathbf{H}_{\mathbf{E}})$$

$$\dot{\mathbf{H}}_{\mathbf{B}} = b_H^B[(1 - \frac{m}{C^*})\rho_H^B\mathbf{M}_{\mathbf{V}}(t - \tau_H^B)\mathbf{H}_{\mathbf{B}}(t - \tau_H^B) - \mathbf{M}_{\mathbf{V}}\mathbf{H}_{\mathbf{B}}] - b_P^{H_B}\mathbf{M}\mathbf{H}_{\mathbf{B}}\mathbf{B} + \alpha_H^B(H_B^* - \mathbf{H}_{\mathbf{B}})$$

$$\dot{\mathbf{E}} = b_P^E[(1 - \frac{m}{C^*})\rho_E\mathbf{M}_{\mathbf{V}}((t - \tau_E))\mathbf{H}_{\mathbf{E}}(t - \tau_E)\mathbf{E}(t - \tau_E) - \mathbf{M}_{\mathbf{V}}\mathbf{H}_{\mathbf{E}}\mathbf{E}] - b_{EC}\mathbf{C}_{\mathbf{V}}\mathbf{E} + \alpha_E(E^* - \mathbf{E})$$

$$\dot{\mathbf{B}} = b_P^B[(1 - \frac{m}{C^*})\rho_B\mathbf{M}_{\mathbf{V}}(t - \tau_B)\mathbf{H}_{\mathbf{B}}(t - \tau_B)\mathbf{B}(t - \tau_B) - \mathbf{M}_{\mathbf{V}}\mathbf{H}_{\mathbf{B}}\mathbf{B}] + \alpha_E(B^* - \mathbf{B})$$

$$\dot{\mathbf{P}} = b_P^P(1 - \frac{m}{C^*})\rho_P\mathbf{M}_{\mathbf{V}}(t - \tau^P)\mathbf{H}_{\mathbf{B}}(t - \tau_P)B(t - \tau_P) + \alpha_P(P^* - P)$$

$$\dot{\mathbf{F}} = \rho_F\mathbf{P} - \gamma_{FV}\mathbf{F}\mathbf{V}_{\mathbf{f}} - \alpha_F\mathbf{F}$$

$$\dot{\mathbf{U}} = \frac{d}{dt}(C^* - \mathbf{C} - m) = -\sigma\mathbf{V}_{\mathbf{f}}\mathbf{U} + \alpha_m m$$

(**a**) ODE model

$$O_2^{Bo} = \{\mathbf{U}, \mathbf{H}_{\mathbf{E}}, \mathbf{H}_{\mathbf{B}}, \mathbf{E}, \mathbf{B}, \mathbf{P}, \mathbf{F}\}$$

$$|$$

$$O_1^{Bo} = \{\mathbf{H}_{\mathbf{E}}, \mathbf{H}_{\mathbf{B}}, \mathbf{E}, \mathbf{B}, \mathbf{P}, \mathbf{F}\}$$

(**b**) Hasse diagram of organizations

Figure 13. The **Bocharov Model** [30] with 10 variables: infective IAV particles ($\mathbf{V}_{\mathbf{f}}$), IAV-infected cells (**C**), destroyed epithelial cells (m), stimulated macrophages ($\mathbf{M}_{\mathbf{V}}$), activated helper T cells providing proliferation of cytotoxic T cells ($\mathbf{H}_{\mathbf{E}}$), activated helper T cells providing proliferation and differentiation of B cells B ($\mathbf{H}_{\mathbf{B}}$), activated CTL (E), B cells (**B**), plasma cells (**P**), antibodies to IAV (**F**), and uninfected epithelial cells (**U**). Note that, for clarity, we have added **U** as a state variable, which is only implicitly represented as $\mathbf{U} = C^* - \mathbf{C} - m$ in the original model by Bocharov et al.

Note that this is by far the oldest model analyzed here. The names of the variables are a bit particular when compared to those of the previously analyzed models. As in all the other models, we have viruses V_f and infected cells C. Furthermore, there are destroyed epithelial cells m as in the Handel Model. All other species belong to the immune response. Note that only in this model there is no state variable for uninfected, healthy cells. Bocharov et al. represent these healthy cells implicitly by subtracting the amounts of infected-cells C and destroyed epithelial cells m from the initial total amount of target epithelial cells C^*. Since all the other models analyzed here have a related variable, we inserted the variable $U = C^* - C - m$ for uninfected cells together with its ODE to make the model comparable to the others.

Due to the fact that the majority of the species belongs to the immune response, this is the case for most of the reactions too. These species of the immune response form exactly the organization O_1^{Bo}, only macrophages M_V are missing.

Similarly to the Hancioglus model, the smallest organization O_1^{Bo} is an organization with immune response but without infection (C, V_f). There is only one further organization that contains only one more species than O_1^{Bo}, namely uninfected cells U. This organization we already found in three of the previous models. However, for the first time, there is no bigger organization in this model. Thus, virus infection is necessarily transient.

3.11. *Complex Dual-Compartment Model (Lee Model, L, 2009)*

The Lee Model [9] is the most complex model considered here. It contains 48 parameters and was designed with respect to experimental in vivo data from mice [6,7]. It has 15 variables and 37 reactions (Figure A10). Like in the Bocharov Model, Lee et al. apply delay differential equations.

Note that this model is the only one analyzed here that distinguishes between *lung compartment* and *lymphatic compartment*. There is one species representing uninfected (healthy) cells $\mathbf{E_P}$ and three species for modelling *infection*: $\mathbf{E_P^*}$, $\mathbf{D}^*$ and viruses $\mathbf{V}$. The remaining species belong to the *immune response*, colored black when naive to infection, while colored green when activated for infection.

Note that we write a species in the *organizations* in Figure 14b in bold text, if it is "new", that is, not contained in neither of its subset organizations. The Hasse diagram contains eight organizations. The smallest one is $O_1^L = \{\mathbf{E_P}, \mathbf{D}, H_N, T_N, B_N\}$ and contains exactly the uninfected cells as well as the naive part of the immune response. The biggest organization contains all species. Between these two "extreme" organizations are six further organizations containing different parts of the activated part of the immune response.

$$\dot{\mathbf{E_P}} = \delta_E(E_0 - \mathbf{E_P}) - \underline{\beta_E \mathbf{E_P V}}$$

$$\dot{\mathbf{E_P^*}} = \underline{\beta_E \mathbf{E_P V}} - k_E \mathbf{E_P^*} \gamma \mathbf{T_E}(t - \tau_T) - \delta_{E^*} \mathbf{E_P^*}$$

$$\dot{\mathbf{V}} = \pi_V \mathbf{E_P^*} - c_V \mathbf{V} - k_V \mathbf{V A}(t)$$

$$\dot{\mathbf{D}} = \delta_D(D_0 - \mathbf{D}) - \beta_D \mathbf{DV}$$

$$\dot{\mathbf{D}}^* = \beta_D \mathbf{DV} - \delta_{D^*} \mathbf{D}^*$$

$$\dot{\mathbf{D_M}} = k_D \mathbf{D}^*(t - \tau_D) - \delta_{D_M} \mathbf{D_M}$$

$$\dot{H_N} = \delta_{H_N}(H_{N0} - H_N) - \pi_{H1} \frac{\mathbf{D_M}}{\mathbf{D_M} + \pi_{H2}} H_N$$

$$\dot{\mathbf{H_E}} = \pi_{H1} \frac{\mathbf{D_M}}{\mathbf{D_M} + \pi_{H2}} H_N + \rho_{H1} \frac{\mathbf{D_M}}{\mathbf{D_M} + \rho_{H2}} \mathbf{H_E} - \delta_{H1} \frac{\mathbf{D_M}}{\mathbf{D_M} + \delta_{H2}} \mathbf{H_E}$$

$$\dot{T_N} = \delta_{T_N}(T_{N0} - T_N) - \pi_{T1} \frac{\mathbf{D_M}}{\mathbf{D_M} + \pi_{T2}} T_N$$

$$\dot{\mathbf{T_E}} = \pi_{T1} \frac{\mathbf{D_M}}{\mathbf{D_M} + \pi_{T2}} H_N + \rho_{T1} \frac{\mathbf{D_M}}{\mathbf{D_M} + \rho_{T2}} \mathbf{T_E} - \delta_{T1} \frac{\mathbf{D_M}}{\mathbf{D_M} + \delta_{T2}} \mathbf{T_E}$$

$$\dot{B_N} = \delta_B(B_{N0} - B_N) - \pi_{B1} \frac{\mathbf{D_M}}{\mathbf{D_M} + \pi_{B2}} B_N$$

$$\dot{\mathbf{B_A}} = \pi_{B1} \frac{\mathbf{D_M}}{\mathbf{D_M} + \pi_{B2}} B_N + \rho_{B1} \frac{\mathbf{D_M} + h\mathbf{H_E}}{\mathbf{D_M} + h\mathbf{H_E} + \rho_{B2}} \mathbf{B_A} - \delta_{B_A} \mathbf{B_A} - \pi_S \mathbf{B_A} - \pi_L \mathbf{H_E B_A}$$

$$\dot{\mathbf{P_S}} = \pi_S \mathbf{B_A} - \delta_S \mathbf{P_S}$$

$$\dot{\mathbf{P_L}} = \pi_L \mathbf{H_E B_A} - \delta_L \mathbf{P_L}$$

$$\dot{\mathbf{A}} = \pi_{AS} \mathbf{P_S} + \pi - AL\mathbf{P_L} - \delta_A \mathbf{A}$$

(**a**) ODE model

Figure 14. *Cont.*

$$O_8^L = \{all\ species\} = \{\mathbf{E_P}, \mathbf{E_P^*}, \mathbf{V}, \mathbf{D}, \mathbf{D^*}, \mathbf{D_M}, H_N, \mathbf{H_E}, T_N, \mathbf{T_E}, B_N, \mathbf{B_A}, \mathbf{P_S}, \mathbf{P_L}, \mathbf{A}\}$$

$$O_7^L = \{\mathbf{E_P}, \mathbf{D}, \mathbf{D^*}, \mathbf{D_M}, H_N, \mathbf{H_E}, T_N, \mathbf{T_E}, B_N, \mathbf{B_A}, \mathbf{P_S}, \mathbf{P_L}, \mathbf{A}\}$$

$$O_6^L = \{\mathbf{E_P}, \mathbf{D}, H_N, \mathbf{H_E}, T_N, \mathbf{T_E}, B_N, \mathbf{B_A}, \mathbf{P_S}, \mathbf{P_L}, \mathbf{A}\}$$

$$O_5^L = \{\mathbf{E_P}, \mathbf{D}, H_N, \mathbf{H_E}, T_N, B_N, \mathbf{B_A}, \mathbf{P_S}, \mathbf{P_L}, \mathbf{A}\}$$

$$O_4^L = \{\mathbf{E_P}, \mathbf{D}, H_N, \mathbf{H_E}, T_N, \mathbf{T_E}, B_N\}$$

$$O_2^L = \{\mathbf{E_P}, \mathbf{D}, H_N, \mathbf{H_E}, T_N, B_N\} \qquad O_3^L = \{\mathbf{E_P}, \mathbf{D}, H_N, T_N, \mathbf{T_E}, B_N\}$$

$$O_1^L = \{\mathbf{E_P}, \mathbf{D}, H_N, T_N, B_N\}$$

(**b**) Hasse diagram of organizations

Figure 14. The **Lee model** [9] which contains 15 variables: uninfected epithelial cells ($\mathbf{E_P}$), infected epithelial cells ($\mathbf{E_P^*}$), virus titer (EID_{50}/ml) (**V**), immature dendritic cells (**D**), virus-loaded dendritic cells ($\mathbf{D^*}$), mature dendritic cells ($\mathbf{D_M}$), naive CD4+ T cells (H_N), effector CD4+ T cells ($\mathbf{H_E}$), naive CD8+ T cells (T_N), effector CD8+ T cells ($\mathbf{T_E}$), naive B cells (B_N), activated B cells ($\mathbf{B_A}$), short-lived plasma (antibody-secreting) B cells ($\mathbf{P_S}$), long-lived plasma (antibody-secreting) B cells ($\mathbf{P_L}$) and antiviral antibody titer (**A**). Note that here we have colored green only those species representing the immune system when activated.

3.12. Hierarchy of Influenza A Virus Models

In order to construct a hierarchical map of all investigated models, we characterize a model by a signature of organizations, which is a set of organization types. For example, the signature of the Handel Model (Figure 7b) is the set {∅, **X**, **X**, **XX**, **XXX**}. An organization type like **XX** means that there is at least one organization that contains uninfected (target) cells (**X**) *and* species of the active immune response (**X**). The deviation of the signatures for all models is shown in Table 1. Note that we ignore species colored black. We include the empty set ∅ because this distinguishes models without any inflow from those that possess an inflow of some species.

Now, we can obtain a partial order among models by defining that a model A is smaller or equal to another model B (A $\leq$ B), if the signature of A is a subset of the signature of model B. For example, the Hernandez Model is smaller than the Lee Model because {**XX**, **XXX**} $\subseteq$ {**X**, **XX**, **XXX**}. This partial order among models leads to a hierarchical map of models, which is visualized by a Hasse-diagram in Figure 15. Note that a model A that is smaller than a model B according to this partial order can possess more species and reactions than B.

In Figure 15, we can see that all models have organizations with uninfected, healthy cells (**X**). There are models that furthermore have infection (**X**) and/or immune response (**X**) in their organizations. There are exactly two models (Hancioglu and Hernandez Model) with immune response in all their organizations which means that these models implicitly assume immune response to be active all the time. Among the models neglecting immune response are those which have infection (Miao Model) or bacteria (**X**) (Smith Model) in their organizations and also those that do not (Baccam and Baccam II Model). For models involving immune response, the situation is more complex. There are those that only have healthy cells in their organizations (Pawelek and Saenz Models). This means that these models implicitly exclude infection and immune response from the long run and thus treat them as transient phenomena a priori. The Bocharov Model is the only one that exhibits only healthy cells and immune response in its organizations but no infection. The remaining five models include all kinds of species (except for bacteria of course) in their organizations.

Table 1. Overview of all models and organization types contained. An organization type like **XX** denotes the type of species contained in an organization, according to our coloring scheme. The set of organization types of a model is called its signature.

Model	Number of Variables	Number of Reactions	Number of Organizations	Organizations	Signature
Baccam [13] 2006	3	4	2	$O_1^{Ba1} = \emptyset$	
				O_2^{Ba1}	X
Miao [14] 2010	3	5	3	$O_1^{M} = \emptyset$	
				O_2^{M}	X
				O_3^{M}	X X
Baccam II [13] 2006	4	5	2	$O_1^{Ba2} = \emptyset$	
				O_2^{Ba2}	X
Pawelek [23] 2012	5	9	2	$O_1^{P} = \emptyset$	
				O_2^{P}	X
Smith [15] 2016	5	12	4	$O_1^{Sm} = \emptyset$	
				O_2^{Sm}	X
				O_3^{Sm}	X
				O_4^{Sm}	X X
Handel [24] 2010	7	12	5	$O_1^{Ha} = \emptyset$	
				O_2^{Ha}	X
				O_3^{Ha}	X
				O_4^{Ha}	X X
				$O_5^{Ha} = \{all\}$	X X X
Hernandez [26] 2012	7	16	2	O_1^{He}	X X
				$O_2^{He} = \{all\}$	X X X
Cao [27] 2015	7	26	3	$O_1^{C} = \emptyset$	
				O_2^{C}	X
				$O_3^{C} = \{all\}$	X X X
Saenz [28] 2010	8	12	4	$O_1^{Sa} = \emptyset$	
				O_2^{Sa}, O_3^{Sa}, O_4^{Sa}	X
Hancioglu [29] 2007	10	44	3	O_1^{Hcg}	X
				O_2^{Hcg}	X X
				$O_3^{Hcg} = \{all\}$	X X X
Bocharov [30] 1994	10	45	2	O_1^{Bo}	X
				O_2^{Bo}	X X
Lee [9] 2009	15	37	8	O_1^{L}	X
				O_2^{L}, O_3^{L}, O_4^{L}, O_5^{L}, O_6^{L}	X X
				O_7^{L}, $O_8^{L} = \{all\}$	X X X

By looking at the hierarchy of models, it becomes evident that there is space for more models. Above the Smith and Handel Model, there could be one in which virus infection as well as bacterial coinfection can be simultaneously persistent ("fully persistent models" denote such hypothetical models in Figure 15). Another extreme case would be a "fully-transient model" in which we have only transient dynamics and all species would finally tend to zero. Such a model would be the smallest one in our partial order of models (Figure 15).

The derived hierarchical map of models might be used to choose the most appropriate model for a particular domain and data set: The model should contain at least one organization for each set of species that were experimentally observed to survive in the long run. If there are several models with such organizations, the one with the smallest organizations might be chosen to provide maximum efficiency in modeling. Table 1 as well as Figure 15 might guide the selection process, complementing established quantitative selection methods, such as those using the area under the viral load curve [31].

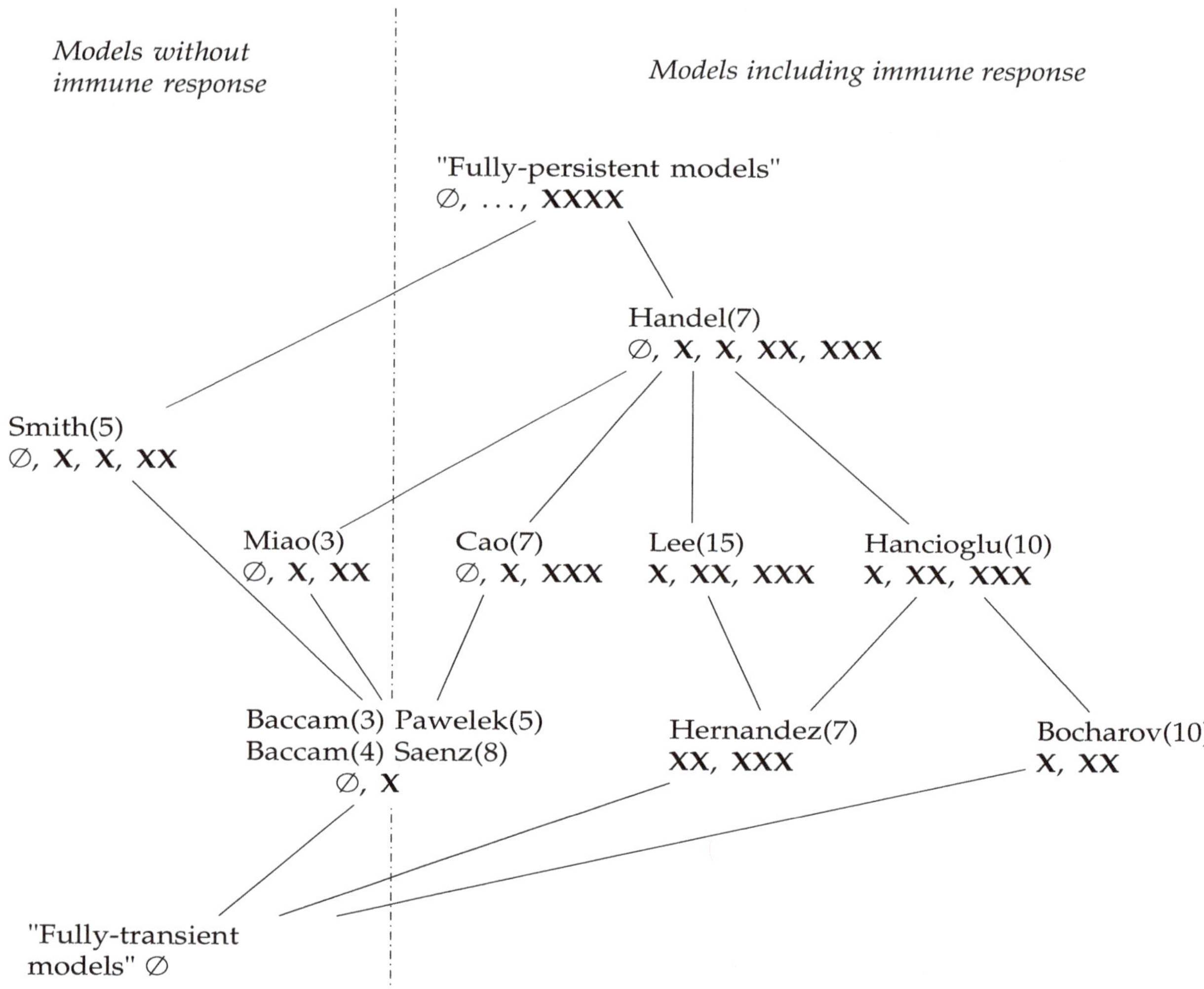

Figure 15. Hasse-diagram of the hierarchy of IAV models with respect to their long-term behaviour. In brackets (), we added the number of species of each model. Underneath (marked by colors) the kinds of species contained in the organizations belonging to each model. The meaning of the four colors is as follows: Species belonging to the healthy state of the organism are colored blue, those belonging to the immune response are colored green, those belonging to infection like infected cells and viruses are colored magenta, and bacteria from bacterial co-infection are colored orange. Horizontally, the diagram consists of four lines. The models in the lowest line contain organizations with exactly two different kinds of species (colors) (including the empty set). In the second line above, there are three different combinations of species (colors) to be found in each model. There is only one model in each of the highest two lines: The Smith model [4] is the only one with bacteria and contains four different combinations of colors. In the Handel Model, there are even five different combinations of colors out of $2^4 = 16$ possible combinations.

4. Conclusions

By analyzing twelve published IAV immune system models, we have shown that we can compute, independently of quantitative kinetic data like rate constants or kinetic laws, all chemical organizations of a typical IAV model, which provides a hierarchical decomposition of the model and an overview of its potential long-term behavior.

It turned out that the derived organizations are meaningful with respect to the model's domain. That is, the composition of species inside an organization can be related to a particular state of the organism, like "healthy", "infected", or "virus controlled by active immune response", and it is possible to annotate organizations accordingly.

By deriving an organizational signature from a model's organizations, we obtained a novel classification scheme and a hierarchy of models with respect to their qualitative long-term behavior. Although the classification via organizational signature is quite coarse-grained, the analysis revealed still a high diversity of models. That is, the models have different potentiality with respect to which variables persist in the long turn and which vanish. Furthermore, the hierarchy map of models contains various empty territories, suggesting space for potential future models.

We envision as a practical use that our method and results can help to select the right model for a particular situation, to relate other models to the present ones, to obtain an overview of the potential long-term dynamics of complex models, and to support model development, for example, by providing a rapid consistency check. Note that measured long-term as well as transient data can be explained with respect to organizations, by defining a projection from a system state to an organization, as demonstrated for the Handel model (Figures 7 and 8).

In addition, our approach is not limited to IAV models, but can be directly applied to other viruses in the same way since their dynamics are similarly modeled by ODEs [32]. Furthermore, our approach is open to include other dynamically relevant components like treatment and vaccination strategies. Another task for future work would be to study the transient virus immune system in more detail, for example, by mapping the basin of attraction of each organization or by systematically analyzing the transition dynamics in-between organizations [33,34].

Author Contributions: S.P. performed the implementation and computations. S.P., P.D. and B.I. analyzed the results with the help of M.H. and K.L. S.P., P.D., and B.I. completed the analysis and final conclusions. S.P., P.D. and B.I. wrote the paper with critical input from M.H., K.L., P.S.d.F., H.A.H., M.M. and S.S. P.D. and B.I. supervised the project and conceived the study. All authors reviewed and approved the final version of the paper.

Appendix A

Here you find the sets of reactions for all example models presented in this paper. Written in brackets behind each reaction is its corresponding term from the differential equations:

$$\begin{aligned} R_1 &: \; T + V \rightarrow I + V \qquad (\beta T V) \\ R_2 &: \; I \rightarrow \varnothing \qquad (\delta I) \\ R_3 &: \; I \rightarrow I + V \qquad (p I) \\ R_4 &: \; V \rightarrow \varnothing \qquad (c V) \end{aligned}$$

Figure A1. Reactions of Baccam model [13].

$$
\begin{aligned}
&R_1: \quad E_P \to 2E_P \qquad (\rho_E E_P)\\
&R_2: \quad E_P + V \to E_P^* + V \qquad (\beta_\alpha E_P V)\\
&R_3: \quad E_P^* \to \varnothing \qquad (\delta_{E^*} E_P^*)\\
&R_4: \quad E_P^* \to E_P^* + V \qquad (\pi_\alpha E_P^*)\\
&R_5: \quad V \to \varnothing \qquad (c_V V)
\end{aligned}
$$

Figure A2. Reactions of Miao model [14].

$$
\begin{aligned}
&R_1: \quad T + V \to I_1 + V \qquad (\beta T V)\\
&R_2: \quad I_1 \to I_2 \qquad (k I_1)\\
&R_3: \quad I_2 \to \varnothing \qquad (\delta I_2)\\
&R_4: \quad I_2 \to I_2 + V \qquad (p I_2)\\
&R_5: \quad V \to \varnothing \qquad (c V)
\end{aligned}
$$

Figure A3. Reactions of Baccam model [14] with delayed virus production.

$$
\begin{aligned}
&R_1: \quad T + V \to I + V \qquad (\beta V T)\\
&R_2: \quad F + T \to F + R \qquad (\phi F T)\\
&R_3: \quad R \to T \qquad (\rho R)\\
&R_4: \quad I \to \varnothing \qquad (\delta I)\\
&R_5: \quad I + F \to F \qquad (\kappa I F)\\
&R_6: \quad I \to I + V \qquad (p I)\\
&R_7: \quad V \to \varnothing \qquad (c V)\\
&R_8: \quad I \to I + F \qquad (q I)\\
&R_9: \quad F \to \varnothing \qquad (d F)
\end{aligned}
$$

Figure A4. Reactions of Pawelek model [23].

$$
\begin{aligned}
&R_1: \quad D \to U \qquad (\lambda D)\\
&R_2: \quad U + V \to E + V \qquad (b U V)\\
&R_3: \quad E \to I \qquad (g E)\\
&R_4: \quad I \to D \qquad (d I)\\
&R_5: \quad I \to I + V \qquad (\frac{p I}{1 + \kappa F})\\
&R_6: \quad V \to \varnothing \qquad (c V)\\
&R_7: \quad U + V \to U \qquad (\gamma b U V)\\
&R_8: \quad V + X \to X \qquad (k V X)\\
&R_9: \quad V \to V + F \qquad (w V)\\
&R_{10}: \quad F \to \varnothing \qquad (\delta F)\\
&R_{11}: \quad V \to V + X \qquad (f V)\\
&R_{12}: \quad X \to 2X \qquad (r X)
\end{aligned}
$$

Figure A5. Reactions of Handel model [24].

$$\begin{aligned}
&R_1: \quad \varnothing \to U_H \qquad (S_H)\\
&R_2: \quad U_H + V \to U_E + V \qquad (k_I U_H V)\\
&R_3: \quad U_H + F \to U_R + F \qquad (k_R U_H F)\\
&R_4: \quad U_H \to \varnothing \qquad (\delta_H U_H)\\
&R_5: \quad U_E \to U_I \qquad (k_E U_E)\\
&R_6: \quad U_E + K \to K \qquad (q_K U_E K)\\
&R_7: \quad U_I \to \varnothing \qquad (\delta_I U_I)\\
&R_8: \quad U_I + K \to K \qquad (q_K U_I K)\\
&R_9: \quad U_R \to \varnothing \qquad (\delta_R U_R)\\
&R_10: \quad U_I \to U_I + V \qquad (\rho_V U_I)\\
&R_11: \quad V \to \varnothing \qquad (\delta_V V)\\
&R_12: \quad U_I \to U_I + F \qquad (a_I U_I)\\
&R_13: \quad F \to \varnothing \qquad (\delta_F F)\\
&R_14: \quad \varnothing \to K \qquad (S_K)\\
&R_15: \quad U_I \to U_I + K \qquad (\Phi_K U_I)\\
&R_16: \quad K \to \varnothing \qquad (\delta_K K)
\end{aligned}$$

Figure A6. Reactions of Hernandez model [26].

$$\begin{aligned}
&R_1: \quad V + T \to V + E_1 \qquad (\beta V T)\\
&R_2: \quad F + T \to F + W \qquad (\phi F T)\\
&R_3: \quad E_1 \to I \qquad (k_1 E_1)\\
&R_4: \quad V + W \to V + E_2 \qquad (m\beta V W)\\
&R_5: \quad W \to R \qquad (aW)\\
&R_6: \quad E_2 \to I \qquad (k_2 E_2)\\
&R_7: \quad I \to \varnothing \qquad (\delta I)\\
&R_8: \quad I \to I + V \qquad (pI)\\
&R_9: \quad V \to \varnothing \qquad (cV)\\
&R_10: \quad E_2 \to E_2 + F \qquad (nqE_2)\\
&R_11: \quad I \to I + F \qquad (qI)\\
&R_12: \quad F \to \varnothing \qquad (dF)
\end{aligned}$$

Figure A7. Reactions of Saenz model [28].

$$
\begin{aligned}
&R_1: \quad I \rightarrow I + V \qquad (\gamma_V I)\\
&R_2: \quad S + A + V \rightarrow S + A \qquad (\gamma_{VA} SAV)\\
&R_3: \quad H + V \rightarrow H \qquad (\gamma_{VH} HV)\\
&R_4: \quad V \rightarrow \varnothing \qquad (\alpha_V V)\\
&R_5: \quad V \rightarrow \varnothing \qquad (\frac{a_{V1} V}{1 + a_{V2} V})\\
&R_6: \quad H \rightarrow 2H \qquad (b_{HD} H)\\
&R_7: \quad R \rightarrow R + H \qquad (b_{HD} R)\\
&R_8: \quad H \rightarrow \varnothing \qquad (-b_{HD} H^2)\\
&R_9: \quad R + H \rightarrow R \qquad (-2b_{HD} HR)\\
&R_{10}: \quad R + H \rightarrow R \qquad (-b_{HD} R^2)\\
&R_{11}: \quad I + H \rightarrow I \qquad (-b_{HD} IH)\\
&R_{12}: \quad I + R + H \rightarrow I + R \qquad (-b_{HD} IR)\\
&R_{13}: \quad R \rightarrow H \qquad (a_R R)\\
&R_{14}: \quad V + H \rightarrow V + I \qquad (\gamma_{HV} VH)\\
&R_{15}: \quad F + H \rightarrow F + R \qquad (b_{HF} FH)\\
&R_{16}: \quad E + I \rightarrow E \qquad (b_{IE} EI)\\
&R_{17}: \quad I \rightarrow \varnothing \qquad (a_I I)\\
&R_{18}: \quad \varnothing \rightarrow M \qquad (b_{MD})\\
&R_{19}: \quad M \rightarrow \varnothing \qquad (-b_{MD} M)\\
&R_{20}: \quad H + M \rightarrow H \qquad (-b_{MD} H)\\
&R_{21}: \quad H + M \rightarrow H + 2M \qquad (b_{MD} HM)\\
&R_{22}: \quad R + M \rightarrow R \qquad (-b_{MD} R)\\
&R_{23}: \quad R + M \rightarrow R + 2M \qquad (b_{MD} RM)\\
&R_{24}: \quad I + M \rightarrow I \qquad (-b_{MD} I)\\
&R_{25}: \quad I + M \rightarrow I + 2M \qquad (b_{MD} IM)\\
&R_{26}: \quad V \rightarrow V + M \qquad (b_{MV} V)\\
&R_{27}: \quad V + M \rightarrow V \qquad (-b_{MV} VM)\\
&R_{28}: \quad M \rightarrow \varnothing \qquad (a_M M)\\
&R_{29}: \quad M \rightarrow M + F \qquad (b_F M)\\
&R_{30}: \quad I \rightarrow I + F \qquad (c_F I)\\
&R_{31}: \quad H + F \rightarrow H \qquad (b_{FH} HF)\\
&R_{32}: \quad F \rightarrow \varnothing \qquad (a_F F)\\
&R_{33}: \quad M + E \rightarrow M + 2E \qquad (b_{EM} ME)\\
&R_{34}: \quad I + E \rightarrow I \qquad (b_{EI} IE)\\
&R_{35}: \quad \varnothing \rightarrow E \qquad (a_E)\\
&R_{36}: \quad E \rightarrow \varnothing \qquad (-a_E E)\\
&R_{37}: \quad M + P \rightarrow M + 2P \qquad (b_{PM} MP)\\
&R_{38}: \quad \varnothing \rightarrow P \qquad (a_P)\\
&R_{39}: \quad P \rightarrow \varnothing \qquad (-a_P P)\\
&R_{40}: \quad P \rightarrow P + A \qquad (b_A P)\\
&R_{41}: \quad S + A + V \rightarrow S + V \qquad (\gamma_{AV} SAV)\\
&R_{42}: \quad A \rightarrow \varnothing \qquad (a_A A)\\
&R_{43}: \quad P \rightarrow P + S \qquad (rP)\\
&R_{44}: \quad P + S \rightarrow P \qquad (-rPS)
\end{aligned}
$$

Figure A8. Reactions of Hancioglu model [29].

$$
\begin{aligned}
R_1 &: C \to C + V_f \quad ()\\
R_2 &: C + E \to C + E + V_f \quad ()\\
R_3 &: V_f + F \to F \quad ()\\
R_4 &: V_f \to \emptyset \quad ()\\
R_5 &: V_f \to \emptyset \quad ()\\
R_6 &: V_f + C \to 2V_f + C \quad ()\\
R_7 &: m + V_f \to m + 2V_f \quad ()\\
R_8 &: V_f + U \to V_f + C \quad ()\\
R_9 &: V_f + C \to V_f + U \quad ()\\
R_{10} &: m + V_f + C \to m + V_f + U \quad ()\\
R_{11} &: C + E \to E + m \quad ()\\
R_{12} &: C \to M_V \quad ()\\
R_{13} &: m \to U\emptyset \quad ()\\
R_{14} &: V_f \to V_f + M_V \quad ()\\
R_{15} &: M_V \to \emptyset \quad ()\\
R_{16} &: M_V + H_E \to M_V + 2H_E \quad ()\\
R_{17} &: m + M_V + H_E \to m + M_V \quad ()\\
R_{18} &: M_V + H_E \to M_V \quad ()\\
R_{19} &: M_V + H_E + E \to M_V + E \quad ()\\
R_{20} &: \emptyset \to H_E \quad ()\\
R_{21} &: H_E \to \emptyset \quad ()\\
R_{22} &: M_V + H_B \to M_V + 2H_B \quad ()\\
R_{23} &: m + M_V + H_B \to m + M_V \quad ()\\
R_{24} &: M_V + H_B \to M_V \quad ()\\
R_{25} &: M_V + H_B + B \to M_V + B \quad ()\\
R_{26} &: \emptyset \to H_B \quad ()\\
R_{27} &: H_B \to \emptyset \quad ()\\
R_{28} &: M_V + H_E + E \to M_V + H_E + 2E \quad ()\\
R_{29} &: m + M_V + H_E + E \to m + M_V + H_E \quad ()\\
R_{30} &: M_V + H_E + E \to M_V + H_E \quad ()\\
R_{31} &: C + E \to C \quad ()\\
R_{32} &: \emptyset \to E \quad ()\\
R_{33} &: E \to \emptyset \quad ()\\
R_{34} &: M_V + H_B + B \to M_V + H_B + 2B \quad ()\\
R_{35} &: m + M_V + H_B + B \to m + M_V + H_B \quad ()\\
R_{36} &: M_V + H_B + B \to M_V + H_B \quad ()\\
R_{37} &: \emptyset \to B \quad ()\\
R_{38} &: B \to \emptyset \quad ()\\
R_{39} &: M_V + H_B + B \to M_V + H_B + B + P \quad ()\\
R_{40} &: m + M_V + H_B + B + P \to m + M_V + H_B + B \quad ()\\
R_{41} &: \emptyset \to P \quad ()\\
R_{42} &: P \to \emptyset \quad ()\\
R_{43} &: P \to P + F \quad ()\\
R_{44} &: V_f + F \to V_f \quad ()\\
R_{45} &: F \to \emptyset \quad ()
\end{aligned}
$$

Figure A9. Reactions of Bocharov model [30].

$$
\begin{aligned}
R_1 &: \varnothing \rightarrow E \quad ()\\
R_2 &: E \rightarrow \varnothing \quad ()\\
R_3 &: E + V \rightarrow V + E_p^* \quad ()\\
R_4 &: E_p^* + T_E \rightarrow T_E \quad ()\\
R_5 &: E_p^* \rightarrow \varnothing \quad ()\\
R_6 &: E_p^* \rightarrow E_p^* + V \quad ()\\
R_7 &: V \rightarrow \varnothing \quad ()\\
R_8 &: V + A \rightarrow A \quad ()\\
R_9 &: \varnothing \rightarrow D \quad ()\\
R_{10} &: D \rightarrow \varnothing \quad ()\\
R_{11} &: D + V \rightarrow V + D^* \quad ()\\
R_{12} &: D^* \rightarrow D^* + D_M \quad ()\\
R_{13} &: D^* + D_M \rightarrow D^* \quad ()\\
R_{14} &: D_M \rightarrow \varnothing \quad ()\\
R_{15} &: \varnothing \rightarrow H_N \quad ()\\
R_{16} &: H_N \rightarrow \varnothing \quad ()\\
R_{17} &: D_M + H_N \rightarrow D_M + H_E \quad ()\\
R_{18} &: D_M + H_E \rightarrow D_M + 2H_E \quad ()\\
R_{19} &: D_M + H_E \rightarrow D_M \quad ()\\
R_{20} &: \varnothing \rightarrow T_N \quad ()\\
R_{21} &: T_N \rightarrow \varnothing \quad ()\\
R_{22} &: D_M + T_N \rightarrow D_M + T_E \quad ()\\
R_{23} &: D_M + T_E \rightarrow D_M + 2T_E \quad ()\\
R_{24} &: D_M + T_E \rightarrow D_M \quad ()\\
R_{25} &: \varnothing \rightarrow B_N \quad ()\\
R_{26} &: B_N \rightarrow \varnothing \quad ()\\
R_{27} &: D_M + B_N \rightarrow D_M + B_A \quad ()\\
R_{28} &: D_M + B_A \rightarrow D_M + 2B_A \quad ()\\
R_{29} &: H_E + B_A \rightarrow H_E + 2B_A \quad ()\\
R_{30} &: B_A \rightarrow \varnothing \quad ()\\
R_{31} &: B_A \rightarrow P_S \quad ()\\
R_{32} &: H_E + B_A \rightarrow H_E + P_L \quad ()\\
R_{33} &: P_S \rightarrow \varnothing \quad ()\\
R_{34} &: P_L \rightarrow \varnothing \quad ()\\
R_{35} &: P_S \rightarrow P_S + A \quad ()\\
R_{36} &: P_L \rightarrow P_L + A \quad ()\\
R_{37} &: A \rightarrow \varnothing \quad ()
\end{aligned}
$$

Figure A10. Reactions of Lee model [9].

References

1. Stöhr, K. Influenza—WHO cares. *Lancet Infect. Dis.* **2002**, *2*, 517. [CrossRef]
2. Petrova, V.N.; Russell, C.A. The evolution of seasonal influenza viruses. *Nat. Rev. Microbiol.* **2018**, *16*, 47–60. [CrossRef] [PubMed]

3. Krammer, F.; Smith, G.J.; Fouchier, R.A.; Peiris, M.; Kedzierska, K.; Doherty, P.C.; Palese, P.; Shaw, M.L.; Treanor, J.; Webster, R.G.; et al. Influenza. *Nat. Rev. Dis. Prim.* **2018**, *4*. [CrossRef] [PubMed]
4. Smith, A.M.; Perelson, A.S. Influenza A virus infection kinetics: Quantitative data and models. *Wiley Interdiscip. Rev. Syst. Biol. Med.* **2011**, *3*, 429–445. [CrossRef] [PubMed]
5. Beauchemin, C.A.; Handel, A. A review of mathematical models of influenza A infections within a host or cell culture: Lessons learned and challenges ahead. *BMC Public Health* **2011**, *11* (Suppl. 1), S7. [CrossRef] [PubMed]
6. Dobrovolny, H.M.; Reddy, M.B.; Kamal, M.A.; Rayner, C.R.; Beauchemin, C.A. Assessing mathematical models of influenza infections using features of the immune response. *PLoS ONE* **2013**, *8*, e57088. [CrossRef] [PubMed]
7. Boianelli, A.; Nguyen, V.K.; Ebensen, T.; Schulze, K.; Wilk, E.; Sharma, N.; Stegemann-Koniszewski, S.; Bruder, D.; Toapanta, F.R.; Guzman, C.A.; et al. Modeling Influenza Virus Infection: A Roadmap for Influenza Research. *Viruses* **2015**, *7*, 5274–5304. [CrossRef] [PubMed]
8. Handel, A.; Liao, L.E.; Beauchemin, C.A. Progress and trends in mathematical modelling of influenza A virus infections. *Curr. Opin. Syst. Biol.* **2018**, *12*, 30–36. [CrossRef]
9. Lee, H.Y.; Topham, D.J.; Park, S.Y.; Hollenbaugh, J.; Treanor, J.; Mosmann, T.R.; Jin, X.; Ward, B.M.; Miao, H.; Holden-Wiltse, J.; et al. Simulation and prediction of the adaptive immune response to influenza A virus infection. *J. Virol.* **2009**, *83*, 7151–7165. [CrossRef]
10. Dittrich, P.; Speroni di Fenizio, P. Chemical Organization Theory. *Bull. Math. Biol.* **2007**, *69*, 1199–1231. [CrossRef]
11. Matsumaru, N.; Centler, F.; di Fenizio, P.S.; Dittrich, P. Chemical organization theory applied to virus dynamics. *IT-Inf. Technol.* **2006**, *48*, 154–160.
12. Peter, S.; Dittrich, P. On the Relation between Organizations and Limit Sets in Chemical Reaction Systems. *Adv. Complex Syst.* **2011**, *14*, 77–96. [CrossRef]
13. Baccam, P.; Beauchemin, C.; Macken, C.A.; Hayden, F.G.; Perelson, A.S. Kinetics of influenza A virus infection in humans. *J. Virol.* **2006**, *80*, 7590–7599. [CrossRef]
14. Miao, H.; Hollenbaugh, J.A.; Zand, M.S.; Holden-Wiltse, J.; Mosmann, T.R.; Perelson, A.S.; Wu, H.; Topham, D.J. Quantifying the early immune response and adaptive immune response kinetics in mice infected with influenza A virus. *J. Virol.* **2010**, *84*, 6687–6698. [CrossRef]
15. Smith, A.M.; Smith, A.P. A Critical, Nonlinear Threshold Dictates Bacterial Invasion and Initial Kinetics During Influenza. *Sci. Rep.* **2016**, *6*, 38703. [CrossRef]
16. Soliman, S.; Heiner, M. A unique transformation from ordinary differential equations to reaction networks. *PLoS ONE* **2010**, *5*, e14284. [CrossRef] [PubMed]
17. Heinrich, R.; Schuster, S. *The Regulation of Cellular Systems*; Springer Science & Business Media: Berlin, Germany, 2012.
18. Fontana, W.; Buss, L.W. "The arrival of the fittest": Toward a theory of biological organization. *Bull. Math. Biol.* **1994**, *56*, 1–64.
19. Kreyssig, P.; Wozar, C.; Peter, S.; Veloz, T.; Ibrahim, B.; Dittrich, P. Effects of small particle numbers on long-term behaviour in discrete biochemical systems. *Bioinformatics* **2014**, *30*, 475–481. [CrossRef] [PubMed]
20. Ibrahim, B. Toward a systems-level view of mitotic checkpoints. *Prog. Biophys. Mol. Biol.* **2015**, *117*, 217–224. [CrossRef]
21. Kreyssig, P.; Escuela, G.; Reynaert, B.; Veloz, T.; Ibrahim, B.; Dittrich, P. Cycles and the qualitative evolution of chemical systems. *PLoS ONE* **2012**, *7*, e45772. [CrossRef]
22. Smith, A.P.; Moquin, D.J.; Bernhauerova, V.; Smith, A.M. Influenza virus infection model with density dependence supports biphasic viral decay. *Front. Microbiol.* **2018**, *9*, 1554. [CrossRef]
23. Pawelek, K.A.; Huynh, G.T.; Quinlivan, M.; Cullinane, A.; Rong, L.; Perelson, A.S. Modeling within-host dynamics of influenza virus infection including immune responses. *PLoS Comput. Biol.* **2012**, *8*, e1002588. [CrossRef] [PubMed]
24. Handel, A.; Longini, I.M., Jr.; Antia, R. Towards a quantitative understanding of the within-host dynamics of influenza A infections. *J. R. Soc. Interface* **2009**, *7*, 35–47. [CrossRef] [PubMed]
25. Handel, A.; Longini, I.M., Jr; Antia, R. Neuraminidase inhibitor resistance in influenza: Assessing the danger of its generation and spread. *PLoS Comput. Biol.* **2007**, *3*, e240. [CrossRef]

26. Hernandez-Vargas, A.E.; Meyer-Hermann, M. Innate immune system dynamics to influenza virus. *IFAC Proc. Vol.* **2012**, *45*, 260–265. [CrossRef]
27. Cao, P.; Yan, A.W.; Heffernan, J.M.; Petrie, S.; Moss, R.G.; Carolan, L.A.; Guarnaccia, T.A.; Kelso, A.; Barr, I.G.; McVernon, J.; et al. Innate Immunity and the Inter-exposure Interval Determine the Dynamics of Secondary Influenza Virus Infection and Explain Observed Viral Hierarchies. *PLoS Comput. Biol.* **2015**, *11*, e1004334. [CrossRef] [PubMed]
28. Saenz, R.A.; Quinlivan, M.; Elton, D.; Macrae, S.; Blunden, A.S.; Mumford, J.A.; Daly, J.M.; Digard, P.; Cullinane, A.; Grenfell, B.T.; et al. Dynamics of influenza virus infection and pathology. *J. Virol.* **2010**, *84*, 3974–3983. [CrossRef] [PubMed]
29. Hancioglu, B.; Swigon, D.; Clermont, G. A dynamical model of human immune response to influenza A virus infection. *J. Theor. Biol.* **2007**, *246*, 70–86. [CrossRef] [PubMed]
30. Bocharov, G.A.; Romanyukha, A.A. Mathematical model of antiviral immune response. III. Influenza A virus infection. *J. Theor. Biol.* **1994**, *167*, 323–360. [CrossRef]
31. Cao, P.; McCaw, J. The mechanisms for within-host influenza virus control affect model-based assessment and prediction of antiviral treatment. *Viruses* **2017**, *9*, 197. [CrossRef]
32. Zitzmann, C.; Kaderali, L. Mathematical Analysis of Viral Replication Dynamics and Antiviral Treatment Strategies: From Basic Models to Age-Based Multi-Scale Modeling. *Front. Microbiol.* **2018**, *9*, 1546. [CrossRef] [PubMed]
33. Mu, C.; Dittrich, P.; Parker, D.; Rowe, J.E. Organisation-Oriented Coarse Graining and Refinement of Stochastic Reaction Networks. *IEEE/ACM Trans. Comput. Biol. Bioinform.* **2018**, *15*, 1152–1166. [CrossRef] [PubMed]
34. Henze, R.; Mu, C.; Puljiz, M.; Kamaleson, N.; Huwald, J.; Haslegrave, J.; di Fenizio, P.S.; Parker, D.; Good, C.; Rowe, J.E.; et al. Multi-scale stochastic organization-oriented coarse-graining exemplified on the human mitotic checkpoint. *Sci. Rep.* **2019**, *9*, 3902. [CrossRef] [PubMed]

3

The Third Annual Meeting of the European Virus Bioinformatics Center

Franziska Hufsky [1,2,*,†], Bashar Ibrahim [1,3,†], Sejal Modha [4,†], Martha R. J. Clokie [5], Stefanie Deinhardt-Emmer [1,6,7,8], Bas E. Dutilh [1,9,10], Samantha Lycett [11], Peter Simmonds [12], Volker Thiel [1,13,14], Aare Abroi [15], Evelien M. Adriaenssens [1,16], Marina Escalera-Zamudio [17], Jenna Nicole Kelly [13,14], Kevin Lamkiewicz [1,2], Lu Lu [18], Julian Susat [19], Thomas Sicheritz [20], David L. Robertson [1,4,†] and Manja Marz [1,2,*,†]

1 European Virus Bioinformatics Center, 07743 Jena, Germany; bashar.ibrahim@uni-jena.de (B.I.); Stefanie.Deinhardt-Emmer@med.uni-jena.de (S.D.-E.); bedutilh@gmail.com (B.E.D.); volker.thiel@vetsuisse.unibe.ch (V.T.); Evelien.Adriaenssens@quadram.ac.uk (E.M.A.); kevin.lamkiewicz@uni-jena.de (K.L.); David.L.Robertson@glasgow.ac.uk (D.L.R.)

2 RNA Bioinformatics and High-Throughput Analysis, Friedrich Schiller University Jena, 07743 Jena, Germany

3 Chair of Bioinformatics, Matthias-Schleiden-Institute, Friedrich Schiller University Jena, 07743 Jena, Germany

4 MRC-University of Glasgow Centre for Virus Research, Glasgow G61 1QH, UK; s.modha.1@research.gla.ac.uk

5 Department of Genetics and Genome Biology, University of Leicester, Leicester LE1 7RH, UK; mrjc1@leicester.ac.uk

6 Institute of Medical Microbiology, Jena University Hospital, Am Klinikum 1, D-07747 Jena, Germany

7 Section for Experimental Virology, Jena University Hospital, Hans-Knöll-Straße 2, D-07745 Jena, Germany

8 Center for Sepsis Control and Care, Jena University Hospital, D-07747 Jena, Germany

9 Theoretical Biology and Bioinformatics, Science4Life, Utrecht University, Padualaan 8, Utrecht 3584 CH, The Netherlands

10 Centre for Molecular and Biomolecular Informatics, Radboud Institute for Molecular Life Sciences, Radboud University Medical Centre, Geert Grooteplein 26, Nijmegen 6525 GA, The Netherlands

11 Infection & Immunity Division, Roslin Institute, University of Edinburgh, Midlothian EH25 9RG, UK; samantha.lycett@ed.ac.uk

12 Nuffield Department of Medicine, University of Oxford, Peter Medawar Building, South Parks Road, Oxford OX1 3SY, UK; peter.simmonds@ndm.ox.ac.uk

13 Institute of Virology and Immunology, 3012 Bern, Switzerland; jenna.kelly@vetsuisse.unibe.ch

14 Department of Infectious Diseases and Pathobiology, Vetsuisse Facility, University of Bern, 3012 Bern, Switzerland

15 University of Tartu, Institute of Technology, 50411 Tartu, Estonia; aare.abroi@ut.ee

16 Quadram Institute Bioscience, Norwich Research Park, Norwich NR4 7UQ, UK

17 Department of Zoology, University of Oxford, Parks Rd, Oxford OX1 3PS, UK; marina.escalerazamudio@zoo.ox.ac.uk

18 Usher Institute of Population Health Sciences & Informatics, Ashworth Laboratories, Kings Buildings, University of Edinburgh, Charlotte Auerbach Road, Edinburgh EH9 3FL, UK; lu.lu@ed.ac.uk

19 Institute of Clinical Molecular Biology, Kiel University, 24118 Kiel, Germany; jsusat@ikmb.uni-kiel.de

20 Natural History Museum of Denmark, University of Copenhagen, DK-1123 Copenhagen, Denmark; thomassp@snm.ku.dk

* Correspondence: franziska.hufsky@uni-jena.de (F.H.); manja@uni-jena.de (M.M.)

† These authors organized the conference.

Abstract: The Third Annual Meeting of the European Virus Bioinformatics Center (EVBC) took place in Glasgow, United Kingdom, 28–29 March 2019. Virus bioinformatics has become central to virology research, and advances in bioinformatics have led to improved approaches to investigate viral infections and outbreaks, being successfully used to detect, control, and treat infections of humans and animals. This active field of research has attracted approximately 110 experts in virology and bioinformatics/computational biology from Europe and other parts of the world to attend the two-day meeting in Glasgow to increase scientific exchange between laboratory- and computer-based researchers. The meeting was held at the McIntyre Building of the University of Glasgow; a perfect location, as it was originally built to be a place for "rubbing your brains with those of other people", as Rector Stanley Baldwin described it. The goal of the meeting was to provide a meaningful and interactive scientific environment to promote discussion and collaboration and to inspire and suggest new research directions and questions. The meeting featured eight invited and twelve contributed talks, on the four main topics: (1) systems virology, (2) virus-host interactions and the virome, (3) virus classification and evolution and (4) epidemiology, surveillance and evolution. Further, the meeting featured 34 oral poster presentations, all of which focused on specific areas of virus bioinformatics. This report summarizes the main research findings and highlights presented at the meeting.

Keywords: virology; virus bioinformatics; software; systems virology; metagenomics; virome; viral taxonomy; virus classification; genome evolution; bacteriophage; virosphere

1. Introduction

The European Virus Bioinformatics Center (EVBC) was conceived of in 2017 to bring together experts in virology and virus bioinformatics in Europe [1,2]. EVBC's member numbers have increased steadily since then with currently 151 members from 78 research institutions distributed over 26 countries across Europe and internationally. This spring, the Annual Meeting of the EVBC was held for the third time (Table 1). The Third Annual Meeting of the EVBC attracted experts at all career stages to attend the two-day meeting in Glasgow in an inspiring and interactive scientific environment to promote discussion, exchange of ideas and collaboration and to inspire and suggest new research directions and opportunities.

Table 1. History of the Annual Meeting of the European Virus Bioinformatics Center(EVBC).

Date	Location	# of Participants	Key outcomes
6–8 March 2017	Friedrich Schiller University Jena, Germany	~100	Founding of the Center; Discussion of the role of EVBC; Election of the first Board of Directors; Insights into EU policy and funding opportunities.
9–10 April 2018	Utrecht University, Netherlands	~120	Extending of the EVBC network to include America and Asia; Discussion and design of joint projects; Insights on first applied European fund among EVBC members [3].
28–29 March 2019	University of Glasgow, United Kingdom	~110	Inclusion of contributed talks in themed sections in the scientific programme; Establishment of travel, poster and best contributed talk awards for junior scientists; Need for greater coordination and communication within the European virology community.

2. Sessions and Oral Presentations

During the two-day conference, about 110 participants from 20 countries contributed in productive discussion on the four topics: (1) systems virology, (2) virus-host interactions and the virome, (3) virus classification and evolution and (4) epidemiology, surveillance and evolution. A number of high quality presentations were given by leading virologists and junior scientists. In addition to the eight invited speakers, we had twelve talks selected from the contributed submissions (see http://evbc.uni-jena.de/events/3rd-evbc-meeting). It was clear that the distinction between laboratory and computer researchers is often blurred. That collaborating teams of individuals with different skill sets are often a road to success, while individuals working alone can still make massive contributions. Data-driven research is now mainstream, and the scale and complexity of datasets is ever increasing. Discussions highlighted how virology, like all of biology, is now a data science, exploiting methods from dimensionality reduction of large datasets to data visualisation. We took from this that virus bioinformatics is evolving and succeeding as an area of research in its own right at the interface of virology and computer science and that there are many ways to be a successful researcher.

2.1. *Systems Virology*

This session was chaired by Philippe Le Mercier (University of Geneva Medical School, Switzerland), board member of the EVBC. Two speakers have been invited on this topic. Volker Thiel (University of Bern, Switzerland), board member of the EVBC, presented about host proteins composing the microenvironment of coronavirus replicase complexes. EVBC member Stefanie Deinhardt-Emmer (Jena University Hospital, Germany) presented about co-infection between *Staphylococcus aureus* and influenza virus. From the submitted abstracts, we selected talks by Jenna Nicole Kelly (University of Bern, Switzerland) on single-cell analysis of influenza virus infection, Florian Erhard (University of Würzburg, Germany) on tools revealing core features of CMV-induced regulation in single cells and Daniel Blanco Melo (Icahn School of Medicine at Mount Sinai, New York, USA) on in-depth transcriptomic analysis in influenza A virus infection. Studying virus infections at the molecular level is as complex as studying the host systems they infect.

2.1.1. Determination of Host Proteins Composing the Microenvironment of Coronavirus Replicase Complexes, by Volker Thiel

Coronaviruses are positive-sense RNA viruses that infect a variety of mammalian and avian species and are mainly associated with respiratory and enteric diseases. In humans, there are four coronaviruses known to cause rather mild respiratory symptoms; however, the appearance of zoonotic viruses, such as the Severe Acute Respiratory Syndrome (SARS) and Middle East Respiratory Syndrome (MERS) coronaviruses, exemplified that coronaviruses can also cause severe and lethal diseases in humans. Within their target cells, coronaviruses replicate their RNA genome at host-derived membranes in the host cell cytoplasm. The Replicase Complex (RC) that is synthesizing the viral RNA is encoded on the genomic RNA and comprises a set of 15–16 non-structural proteins (nsps). Besides canonical functions associated with RNA synthesis, such as RNA-dependent RNA polymerase, helicase and methyltransferases, a wealth of additional enzymatic activities, such as endoribonuclease, ADP-ribosylation and de-ubiquitination, are included within the coronaviral RC, suggesting that various virus–host interactions are taking place at the site of viral RNA synthesis. However, our knowledge about host factors at the interface between the RC and the host cell cytoplasm is rudimentary. To identify the composition of the viral RC and adjacent host cell proteins composing the RC-microenvironment, we engineered a biotin ligase into a coronaviral RC. This allowed us to biotinylate, affinity-purify and identify specifically all viral components constituting the coronavirus RC and host cell proteins that are in close proximity (Figure 1). Amongst the >500 host proteins constituting the RC-microenvironment, we identified numerous proteins associated with vesicular trafficking pathways, ubiquitin-dependent and autophagy-related processes and translation initiation. Notably, following the detection of translation initiation factors at the RC, we were able to visualize and demonstrate active translation proximal to the site of viral RNA synthesis of several coronaviruses. Collectively, our work established a spatial link between viral RNA synthesis and diverse host factors of unprecedented breadth. Many of the coronavirus RC-proximal host proteins and pathways have also documented roles in the life cycle of other positive-stranded RNA viruses, suggesting considerable commonalities and conserved virus-host interactions at the RCs of a broad range of RNA viruses.

Our data may thus serve as a paradigm for other RNA viruses and provide a starting point for a comprehensive analysis of critical virus-host interactions that represent targets for therapeutic intervention [4].

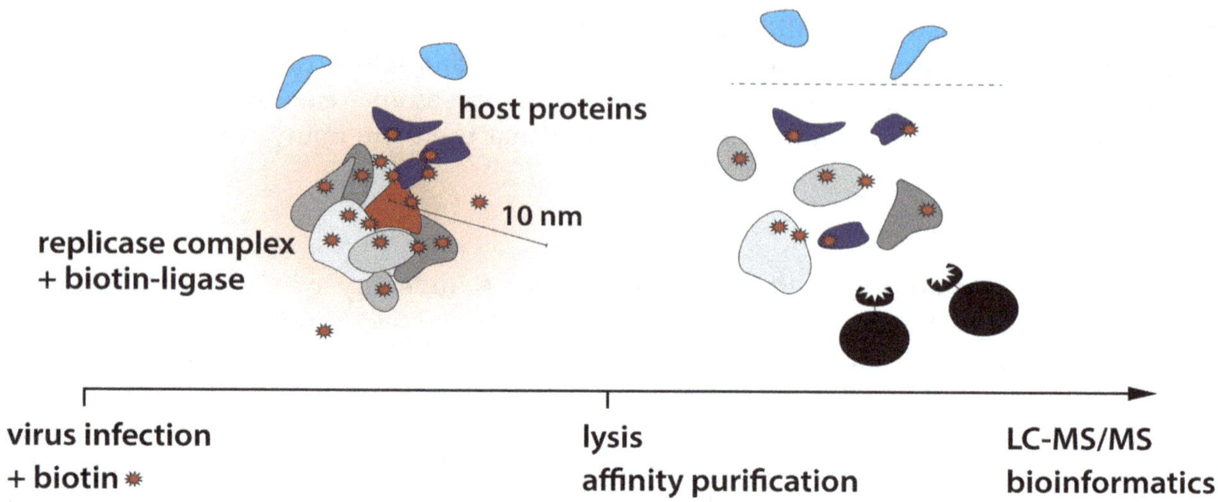

Figure 1. Illustration of the experimental design to determine the microenvironment of coronavirus Replicase Complexes (RCs) (adapted from V'kovski et al. [4]).

2.1.2. Co-Infection between *Staphylococcus aureus* and Influenza Virus Reduces Endothelial Barrier Function, by Stefanie Deinhardt-Emmer

Pneumonia is the most serious inflammatory disease of the respiratory tract and also the most common infectious disease. The classification of pneumonia into Hospital-Acquired Pneumonia (HAP), Community-Acquired Pneumonia (CAP) and Ventilated-Acquired Pneumonia (VAP) indicates the source of disease by a wide variety of microorganisms including bacteria, viruses and fungi. Respiratory tract infections and in particular pneumonia represent the most common cause of sepsis [5]. Long-time associated with bacterial infection, sepsis definition became more in focus as a multifaceted host response to an infecting pathogen, which leads to organ failure [6]. However, Influenza Virus (IV) as a pneumotropic virus can lead to lung failure and systemic host reaction with subsequent multiple organ failure. IV circulates worldwide and causes highly contagious respiratory diseases characterized by mild to severe symptoms. The seasonal IV-associated bronchopneumonia is one of these infectious diseases with the highest population-based mortality rates [7]. Besides virulence factors, the sudden increase of pathogenicity is the most striking problem of influenza accompanied by bacterial co-infection. In a single-centre study conducted at the Jena University Hospital during the winter season 2017/2018, we detected 1197 influenza-virus-positive samples and 89 *S. aureus*-positive respiratory specimens. However, the diagnosis of a co-infection was significantly lower with 17 samples. Interestingly, the mortality rate increased dramatically from single infection (approximately 20%) to co-infection (approximately 80%). Even larger studies indicating similarly dates and also the Spanish flu of 1918 showed that co-infection results in high mortality rates [8]. While the pathogen–host interaction-induced severe dysregulations of the immune response is under investigation in many studies, the regulatory effects between the different pathogens and the subsequent impact on the host are barely understood. In a multifactorial process, a wide range of pathogen factors and pathogen-regulated signalling events are involved in co-pathogenesis. This process is associated with elevated host-response, changed repair-processes, and modifications in the cellular immune response [9]. It is shown that primary IV-infection inhibits the apoptosis mechanism and the following infection with *S. aureus* inhibits IV-induced apoptosis by procaspase-8 activation [10]. Various models are available for studying the mechanisms of the viral–bacterial interference. However, the use of murine models is adversely regarded because of obvious discrepancies between men and mice despite the attempts of humanized murine models to fill the gaps. New methods enable investigations with cost-saving and efficient cell culture models as an excellent supplement to animal experiments. Organ-on-a-chip technology allows species-specific investigations for different cell types and also immune cells. Using this method, viral-bacterial interference can be investigated in a human-specific manner.

2.1.3. Single Cell Analysis of iNfluenza Virus Infection in Its Natural Target Cells Reveals Cell Type-Specific Host Responses and Disparate Viral Burden, by Jenna Nicole Kelly

The human respiratory epithelium is a pseudostratified epithelium that constitutes the first line of defence against invading respiratory pathogens, including influenza viruses. Although several studies have now shown that both viral transcript production and the innate immune response to infection vary widely among single influenza-infected cells, the cause of this extreme heterogeneity remains unclear [11,12]. More specifically, it remains unknown how key innate immune components are distributed among the different cell populations found in the respiratory epithelium and how the latter may influence the host response to infection. To determine the distribution of these innate immune components and to examine how specific cell types respond to influenza infection, we used single-cell RNA sequencing to acquire transcriptomes from primary human Airway Epithelial Cells (hAEC) infected with Influenza A Virus (IAV) (Figure 2) [13]. A low MOI was used to infect hAECs with either Wild-Type (WT) pandemic IAV or an NS1mutated form of the virus (NS1R38A) that impairs its ability to counteract Interferon (IFN) and produces an amplified innate immune response. We then annotated both host and viral transcriptomes of more than 19,000 single cells across the five major hAEC cell types for mock, WT, and NS1R38A conditions. We observed a large heterogeneity in viral burden; however, in contrast to what was found in previous studies, no absence of viral genes was detected. Interestingly, in both WT- and NS1R38A-infected cultures, there was a significant decrease in the fraction of ciliated and goblet cells compared to mock hAECs. We also identified a number of cell-type-specific innate immune responses, including the expression of type I and III IFNs in all major cell types. Collectively, our results represent the first comprehensive report on how individual cells contribute to the antiviral response during IAV infection in the context of the human respiratory epithelium.

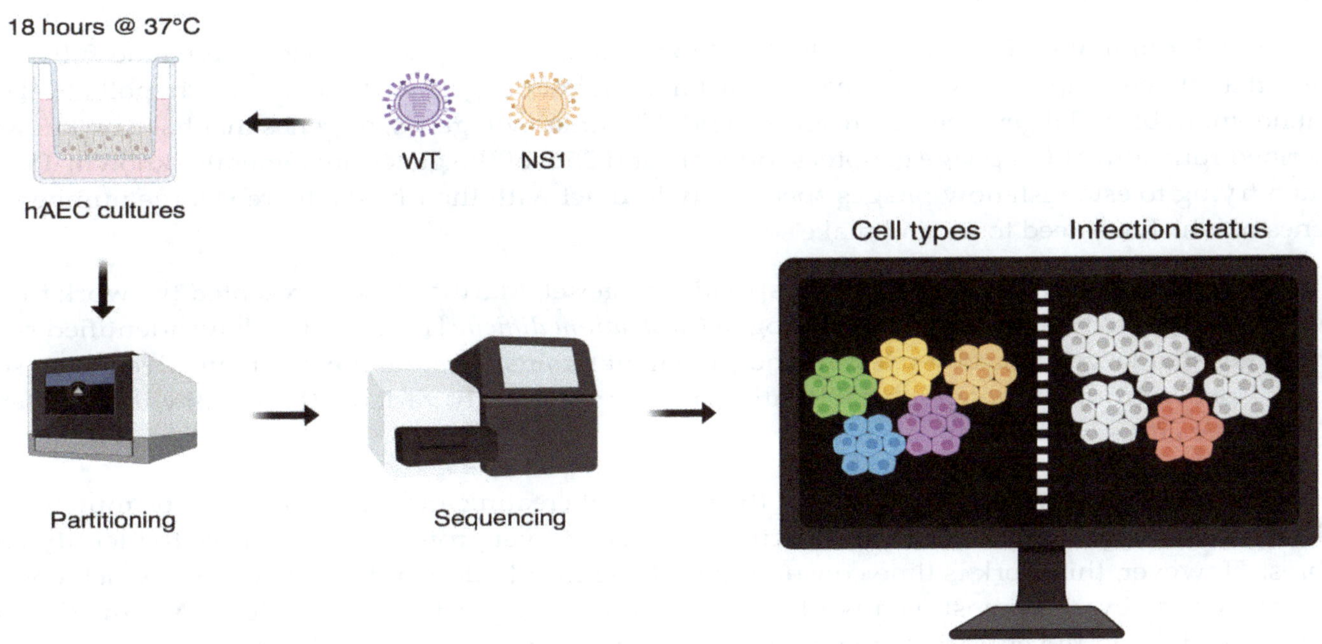

Figure 2. Sequencing and annotation workflow for single influenza-infected cells in the human respiratory epithelium. hAEC, human Airway Epithelial Cells.

2.2. *Virus–Host Interactions and the Virome*

This session was chaired by David Robertson (MRC-University of Glasgow Centre for Virus Research, United Kingdom), the local organizer of the meeting. Bacteriophages, the viruses of bacteria, are an important and usually neglected component of microbiome studies. Two speakers have been invited on this topic. Martha Clokie (University of Leicester, United Kingdom) presented about the roles of phages in impacting infectious diseases in human microbiomes. EVBC member Bas Dutilh (Utrecht University, Netherlands) presented about global phylogeography and the ancient evolution of the widespread human gut virus crAssphage. From the submitted abstracts, we selected talks by Katherine Brown (University of Cambridge, United Kingdom) on viral transcripts in RNA-seq datasets from bees, mites, and ants, Evelien Adriaenssens (Quadram Institute Bioscience, Norwich, United Kingdom) on genome-resolved metaviromics for the detection of pathogenic viruses in the environment and Josquin Daron (Centre National de la Recherche Scientifique, Montpellier, France) on codon usage preference similarity among human-infecting viruses and their hosts.

2.2.1. Roles of Phages in Impacting Infectious Diseases in Human Microbiomes, by R. J. Martha Clokie

Most of the roles of phages in human health and disease are yet to be unravelled. However, phages in all environments including the human microbiome are increasingly acknowledged to be the puppeteers of their bacterial hosts, shaping their structure and evolution and physiology. Phages associated with bacterial pathogens have multiple, often complex interactions with their bacterial hosts, forcing them to interact differently with other bacterial and human cells. Besides being the ultimate bacterial killers, phages can change bacterial surfaces to prevent recognition by the human immune system. In cystic fibrosis, they can allow their hosts to cope with anaerobic conditions found in mucus-laden lungs, and in many bacteria, they encode potent toxins [14]. There is indeed a plethora of unknown phage-mediated bacterial phenotypes that could be critical for our understanding of disease. Their ability to be developed as targeted removers of pathogenic bacteria is likely to be critical to solving the antimicrobial resistance crisis.

A major limitation for our ability to develop therapeutic phages and also understand fully the ways that phages impact bacteria is that the vast majority of phage gene functions are hypothetical or unknown. In bacterial genomes, there are around 25% unknown genes, or genes that have no known ascribed function, but in phage genomes, only around 25% of the genes are generally known! Thus, when trying to establish how phages specifically interact with their hosts, there is large number of genes of which we need to try and make sense.

To illustrate the diversity within one specific phage set, Martha Clokie presented the work from her lab on phages that infect the gut pathogen *Clostridium difficile* [15–17]. They have identified sets of phages that target clinically-relevant and prevalent strains. Despite the most effective phage set being isolated from one geographical location, they are strikingly variable (Figure 3) with very few identifiable genes in common.

Martha Clokie's group is currently in the process of creating and examining genetic mutants to identify phenotypes and conducting structural work on novel proteins, for example to identify tail fibres. However, this work is time consuming and technically demanding. Choosing which genes to focus on is key, as downstream work is key to unravelling critical phenotypes. Martha Clokie presented data on the efficacy of this phage set to treat disease along with a framework for their ongoing work to use different machine learning approaches to examine the genomes of these phages and their associated bacteria robustly in order to identify hard-to-identify features, for example shared and unique genes of interest. These approaches will direct work to unravel the mechanics of phage efficacy for virulent phages and modes of action for lysogens.

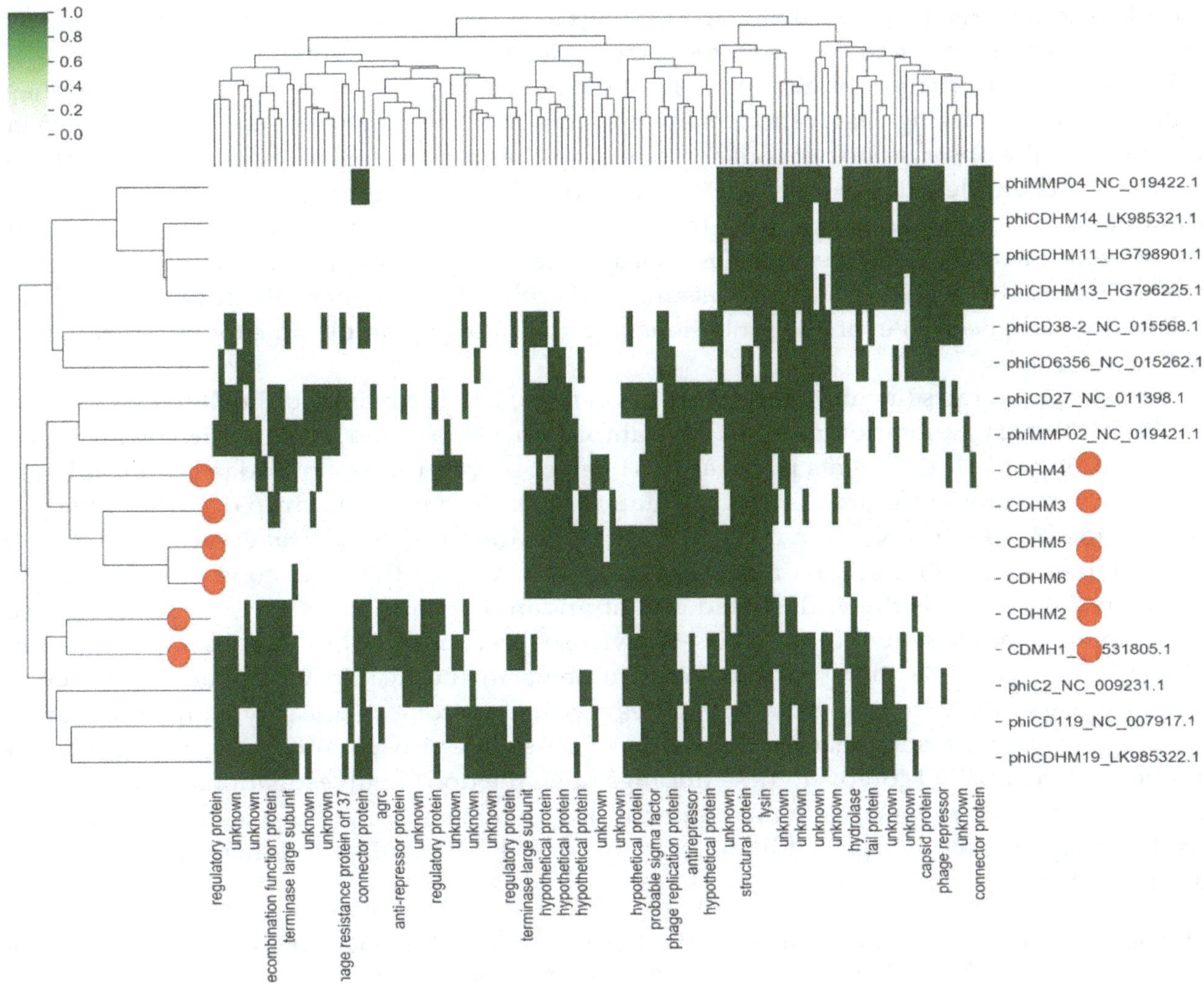

Figure 3. Set of *Clostridium difficile* phages on the vertical axis, which includes six well-characterised myoviruses from Martha Clokies' laboratory (red dots). The genes commonly identified in *C. difficile* phages are shown on the horizontal axis and homologous genes represented by a green line. It is clear that these phages do not share a large common gene set.

2.2.2. Global Phylogeography and Ancient Evolution of the Widespread Human Gut Virus crAssphage, by Bas E. Dutilh

While viruses are vastly abundant and ubiquitous throughout the biosphere, they have remained a relatively unexplored superkingdom of life. Early findings of genomic mosaicism [18] and enhanced mutation rates of especially RNA viruses [19] have led to the conception of viruses as genomically highly variable entities. This was further supported as metagenomics unveiled the extent of genetic diversity of viruses, initially in marine water and human faeces [20], and in many different biomes since. Images of an unparalleled diversity that is dominated by unknown sequences has been the common theme of viral metagenomic explorations. However, while the virosphere is undoubtedly diverse, ubiquitous viruses are increasingly being discovered by metagenomic analysis of globally-distributed, ecologically-stable ecosystems, including once again the global oceans [21,22] and the human gut [23–25].

Moreover, the genome sequence in individual viral lineages may be more conserved than could previously be recognized. Recently, large-scale comparisons of gene order in the genome sequences of dsDNA bacteriophages revealed a surprisingly conserved genomic structure [26,27]. A possible mechanism at play is the genomic encoding of different transcriptional regions with promoters that govern the expression of early, middle and late specific genes, such as known from the well-studied case of the T4 bacteriophage [28]. Together, these findings suggest a highly-optimized genomic encoding of gene expression regulation that is consistent across globally-diverse viral populations.

While the conservation of genomic architecture between distantly-related bacteriophages as outlined above is a striking observation, many open questions remain. For example, it remains unclear to what extent the observations of conserved genomic architecture described above reflect a biased sampling, for example of temperate, dsDNA and/or tailed bacteriophages that have been observed to dominate, e.g., marine systems [29]. Indeed, the modes of genome evolution differ for viruses with different lifestyles [30]. Nevertheless, viruses have vast global population sizes that result in highly-efficient evolutionary selection pressures and optimized genomes. Moreover, viruses and their cellular hosts have been co-evolving for billions of years, allowing ample time for optimization of their genome structures.

Viral mutation rates (including recombination rates) have remained difficult to quantify due to a lack of evolutionary calibration points. For example, on a short time scale of thirty years, a constant recombination rate of five events per year has been observed for Siphoviridae bacteriophages [31], but when longer timespans are assessed, mutation rate estimates may drop dramatically by orders of magnitude [32]. One way of obtaining ancient calibration points in viral evolution in the absence of fossil data is by exploiting their association to hosts. One of the most conserved constituents of the human gut virome is the widespread and abundant bacteriophage crAssphage [23]. Recently, near-complete genome sequences of crAss-like viruses were detected in faecal samples of a range of wild non-human primates living on different continents, including Old-World monkeys, New-World monkeys and apes [33]. Strikingly, these genomes revealed a strong collinearity with human-associated crAss-like viruses, suggesting that the association of crAss-like viruses with the primate gut biome may be millions of years old. Moreover, these findings open the door to investigations into viral mutation rates at long time-scales, once again illustrating how viral metagenomics opens up a treasure trove for virus discovery [34], as well as evolutionary analyses of these smallest and most abundant biological entities on Earth.

2.2.3. Genome-Resolved Metaviromics for the Detection of Pathogenic Viruses in the Environment: Will Eating Shellfish Make You Ill?, by Evelien M. Adriaenssens

Viromics or viral metagenomics has been proposed as an alternative method to qPCR-based approaches for the detection of pathogenic viruses linked to food- and water-borne illness in the aquatic environment [35,36]. The main advantage is that viral communities can be investigated without prior knowledge of the genome sequences or genotypes of the viruses present in the sample. There are, however, several drawbacks associated with viromics, such as laboratory and computational costs, scalability and the issue of viral dark matter in which sequence data are classified as "unknown". In her presentation, Evelien Adriaenssens focused on the latter aspect and showed that reconstruction of Uncultivated Virus Genomes (UViGs) [37] and classification into families reduced the fraction of completely unknown sequences, particularly for RNA viruses. Using read mapping approaches followed by visualisation and analysis with Anvi'o [38], she showed that they can identify pathogenic virus genomes present in the Conwy River catchment area, mainly found in wastewater [39], and showed changing abundance patterns between sample sites and types. Using species-level clustering and differential read mapping, comparative genomics and phylogenetics, she could gradually descend from the bigger picture of viral diversity to strain-level resolution, identifying the genotype of potentially pathogenic viruses. This workflow is ideally suited to find new pathogenic viral species and identify markers for wastewater contamination of the environment.

Evelien M. Adriaenssens was funded by the Biotechnology and Biological Sciences Research Council (BBSRC) under the BBSRC Institute Strategic Programme Gut Microbes and Health BB/R012490/1.

2.3. Virus Classification and Evolution

This session was chaired by Darren Obbard (University of Edinburgh, United Kingdom). Two speakers have been invited on this topic. Peter Simmonds (University of Oxford, United Kingdom) presented about classification of viruses in metagenomic datasets. Unfortunately, Olga Kalinina (Max Planck Institute for Informatics, Saarbrücken, Germany) was unable to make it to the meeting. Instead, Manja Marz (Friedrich Schiller University Jena, Germany), Managing director of the EVBC, presented about machine learning applied to virus data. From the submitted abstracts, we selected talks by Julian Susat (Institute of Clinical Molecular Biology, Kiel, Germany) on the detection of viruses in ancient human remains, Aare Abroi (University of Tartu, Estonia) on the relation between virosphere and biosphere, and Kevin Lamkiewicz (Friedrich Schiller University Jena, Germany) on RNA secondary structures in whole genome alignments of viruses. Based on this submission, Kevin was competitively awarded the PhD travel award.

2.3.1. The Classification of Viruses in Metagenomic Datasets—Where Do You Draw the Line?, by Peter Simmonds

Methodological advances, such as High-Throughput Sequencing (HTS), and new capabilities to recover and assemble genome sequences has unearthed vast numbers of previously-undescribed viruses from environmental, human clinical, veterinary and plant samples. How such viruses can be incorporated into the current virus taxonomy is a major challenge, especially at the family and species levels, which have been historically based largely on descriptive taxon definitions of phenotypic properties that "sequence-only" viruses often lack. These assignments typically encapsulate descriptions of replication strategies, virion structure, and clinical and epidemiological features, such as host range, geographical distribution and disease outcomes. If "sequence-only" viruses are to be formally placed into the classification maintained by the International Committee on the Taxonomy of Viruses (ICTV) as recently proposed [40], then their assignments will have to be based largely or entirely on metrics of genetic relatedness and any other features that might be inferred from their genome sequences. However, there are no published guidelines in the ICTV code on how similar or how divergent viruses must be in order to be considered as new species or new families (https://talk.ictvonline.org/information/w/ictv-information/383/ictv-code).

Peter Simmonds described their investigations of the extent to which the existing virus taxonomy could be reproduced by the recoverable genetic relationships between sequences of viruses currently classified by the ICTV. Comparisons of viruses were based on extraction of protein coding gene signatures and genome organisational features from virus sequences and using these to construct a metric of genetic relatedness through computation of Composite Generalised Jaccard (CGJ) distances between each pair of viruses [41]. For eukaryotic viruses, there was large-scale consistency between such genetic relationships and their current family- and genus-level taxonomic assignments, irrespective of genome configurations and genome sizes. The analysis pipeline, "Genome Relationships Applied to Virus Taxonomy" (GRAViTy), diagrammatically summarised in Figure 4, predicted family membership of eukaryotic viruses with close to 100% accuracy and specificity; this method should therefore enable the vast collection of metagenomic sequences to be classified in a manner consistent with the current ICTV taxonomy. Preliminary analysis of such datasets revealed that over one half (460/921) of (near)-complete genome sequences from recently-generated eukaryotic virus datasets could be assigned to 127 novel family-level groupings, more than double the number of eukaryotic virus families in the ICTV taxonomy.

The taxonomy of the 20 currently-classified prokaryotic virus families differs substantially [42]. Members of three families in particular (*Podoviridae*, *Siphoviridae* and *Myoviridae*) were far more divergent from each other than observed within eukaryotic and archaeal virus families. Applying a CGJ distance threshold of 0.8, prokaryotic viruses form over 100 groupings equivalent to eukaryotic virus families. The use of a common benchmark with which to compare taxonomies of eukaryotic and prokaryotic viruses supports ongoing efforts by the ICTV to revise thoroughly the phage taxonomy so that assignment criteria are consistent across all virus groups. Developing a consistent classification of viruses in which assignments at family and other taxonomic levels extending the current framework, but which will be underpinned both by metrics of genomic relatedness, is essential for future, evidence-based classification of metagenomic viruses.

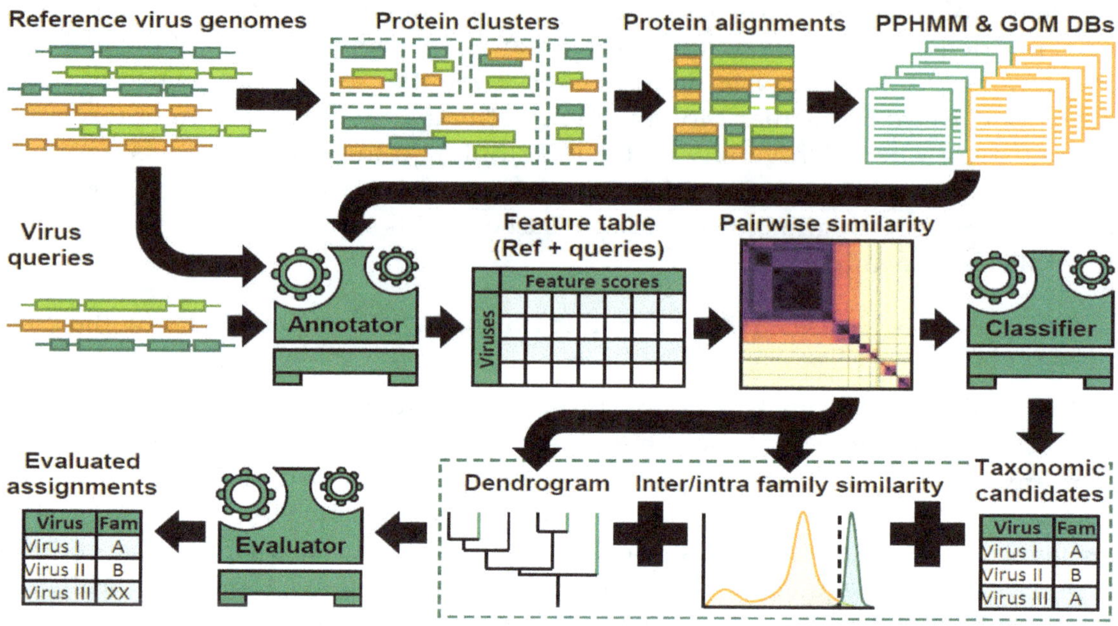

Figure 4. Overview of virus taxonomy prediction by "Genome Relationships Applied to Virus Taxonomy" (GRAViTy). A simplified diagram of the steps used to construct profile tables from sequences of viruses with assigned taxonomic status (reference virus genomes). It further illustrates the steps to classify viruses of of undetermined taxonomic relationships. The method is based on extraction of protein sequences from reference virus genomes and their clustering using pairwise BLASTp bit scores. Sequences in each cluster are then aligned and turned into a Protein Profile Hidden Markov Model (PPHMM). Reference genomes are subsequently scanned against the database of PPHMMs to determine the locations of their genes, and Genomic Organisation Models (GOMs) for each virus family are constructed. These models form the core of the genome annotator (Annotator), which is used to annotate query sequences with information on the presence of genes and the degree of similarity of their genomic organisation to reference virus sequences. From this, genome relationships can be extracted by computation of various genetic distance metrics, including composite generalised Jaccard similarity, which forms the basis for heat maps and dendrograms that depict the relationships of query sequences to the dataset of classified viruses (Classifier) and recommendations for their taxonomic assignments (Evaluator).

2.3.2. Detecting Viruses in Ancient Human Remains, by Julian Susat

The field of ancient DNA covers a wide range of research topics, spanning from human evolution, megafauna to pathogen evolution. Despite the recent advantages in ancient DNA techniques and modern metagenomic screening tools, the identification of authentic viral sequences from ancient material is still challenging. The materials that are mainly used in ancient DNA research, teeth and petrous bones, already limit the number of detectable viruses by their nature. Only viruses that are present in the bloodstream can be detected. The fast evolution of viral pathogens and therefore the comparability to modern variability in viruses makes it even more difficult to identify their ancestors reliably. The highly-fragmented and degraded nature of ancient genetic material and the high risk of modern contamination are causing further problems in the analysis. For the detection of viruses, a wide variety of software utilizing different approaches like HMMs, dedicated marker genes and complete genome references are available to screen these ancient samples for the presence of pathogens. Each of these approaches has its own characteristic strengths and weaknesses. In a competitive alignment approach using all complete virus genomes as a reference, we were able to detect three Hepatitis-B Viruses (HBV) during our regular screening. All three samples originated in Germany and dated to the mediaeval times (1000 BP) and the Neolithic (5000 and 7000 BP). After sequencing and competitive mapping against 16 HBV references, complete HBV genomes could be recovered from all three samples. This resulted in the oldest human pathogenic viral genome that is known up to know. Phylogenetic analysis revealed that the medieval strain was genotype D and surprisingly conserved. The ancient Neolithic strains were closer together than to any other modern and closest to strains from Old-World monkeys. These findings might suggest reciprocal cross-species transmission between human and ape. Furthermore, we could show that the genomic structure of ancient strains closely resembles the structure of modern HBV strains. Since publishing these results, we and others detected more HBV-positive samples, supporting the notion that viruses will become more important for the aDNAcommunity (Figure 5). The new HBV genomes we reconstructed support our earlier findings. A bigger number of HBV cases spanning over longer time frames opens the door for reliable diachronic analysis and maybe even epidemiological analysis. Besides the recent findings of ancient viruses (e.g., Parvovirus), an open question still remains how we could detect and reconstruct extinct or highly-altered virus genomes. Bioinformatic protocols for the detection of unknown viral protein families based on long sequencing reads and high coverage data are published and available, but due to the above-described nature of aDNA, applying these methods is not straightforward, and strong optimization needs to be carried out. Still, these HBV and other findings have opened a new door within the aDNA community and blazed a trail for upcoming viral ancient DNA studies. This work was done by a team composed of Ben Krause-Kyora, Julian Susat, Felix M. Key, Denise Kühnert, Alexander Immel, Alexander Herbig, Almut Nebel and Johannes Krause.

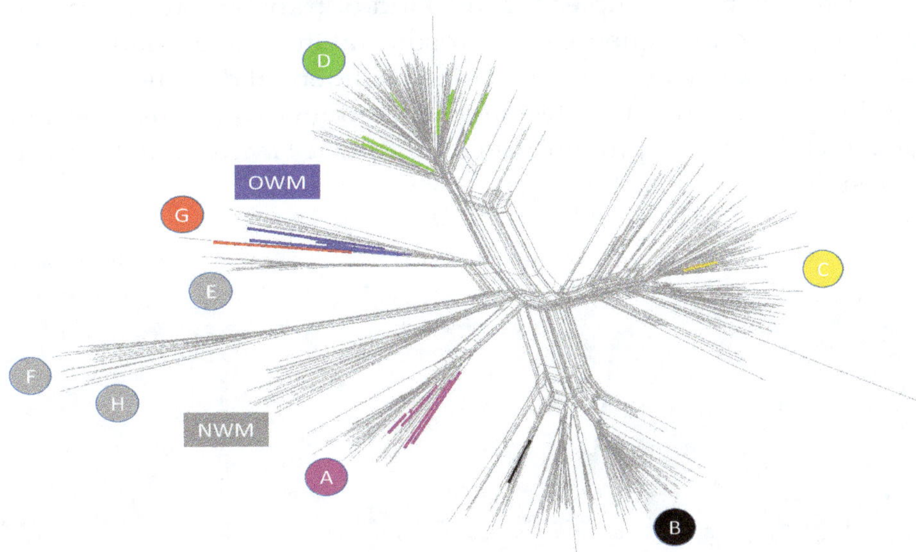

Figure 5. Network of 493 modern genomes, 15 published ancient strains and 12 newly-discovered ancient strains. Single letters indicate HBV genotypes (A–H); coloured strains are of ancient origin; OWM = Old-World Monkey HBV strains, NWM = New-World Monkey HBV strains. D: five new ancient strains, six ancient strains [43–45]; C: one ancient strain [46]; B: one ancient strain [45]; A: two new ancient strains, three ancient strains [45]; G: one new ancient strain; OWM: three new ancient strains, four ancient strains [44,45].

2.3.3. Virosphere and Biosphere—How Related They Are? A Protein (Domain) Based View, by Aare Abroi

Viruses are not always pathogens, and they are also an important and inseparable part of the biosphere and should be studied as such. Unfortunately, the wider functional and evolutionary role of viruses in the biosphere is not yet widely accepted in most disciplines, a good exception being marine biology/ecology, where viruses are already accepted as important players. How the virosphere is related to the rest of the biosphere can be examined in several different ways. One of these ways is a protein domain-based view. We analysed how virosphere protein domain occurrence is related to the occurrence of protein domains in all (sequenced) organisms (we called the last the phylogenomic space of protein domains). This is based on the distribution of protein domains in viruses and in organisms (by superkingdom), i.e., which protein domains are found in viruses (or a specific set of viruses) and to what extent and where these domains are found elsewhere in organisms. In our analysis, we used predefined protein domain databases Pfam, Superfamily and Gene3D. Domains found in the virosphere can be found in a different number of organisms, starting from a few organisms for some viral domains up to all organisms in the others. However, if we specify a narrower set of viruses (Baltimore class, viral family or host range), differences between viral taxons appear. Therefore, the heterogeneity of viruses is also very clearly expressed by where in the phylogenomic space the domains that are found in different viral taxons are located. A few examples are shown in Figure 6. An important conclusion from our analysis is the existence of virosphere-specific protein domains (domains not found in cellular organisms), even at the level of structural homology. Several evolutionary routes that may lead to virosphere specificity (absence in cellular organisms) will be discussed. Considering the new knowledge on virus-to-host gene transfers in eukaryotes during the last ten years, it is clear that the virosphere is a source of functional and structural novelties also for this superkingdom. A possible route for the genesis of novel domains in viruses (as well as in organisms) is double coding or overprinted genes. We have developed a web-tool cRegions (http://bioinfo.ut.ee/cRegions/), which helps to find potential double coding regions (and other embedded functional elements) in coding sequences [47,48]. Of course, there exist many domains that are shared by viruses and organisms. Beside others, virus-to-host gene transfer is one process leading to shared domains. A number of examples for this kind of transfer have been described; however, they are all based on sequence-to-sequence comparison. Taking into account the very fast evolution of viruses, the sequence similarity may fall below the confidential detection limit relatively fast. We applied structure-guided information to detect more ancestral virus-to-host transfers. Our data show that "as a proof of principle", using protein structure-guided HMM models, it is possible to detect V2Htransfers not "visible"; with BLAST analysis.

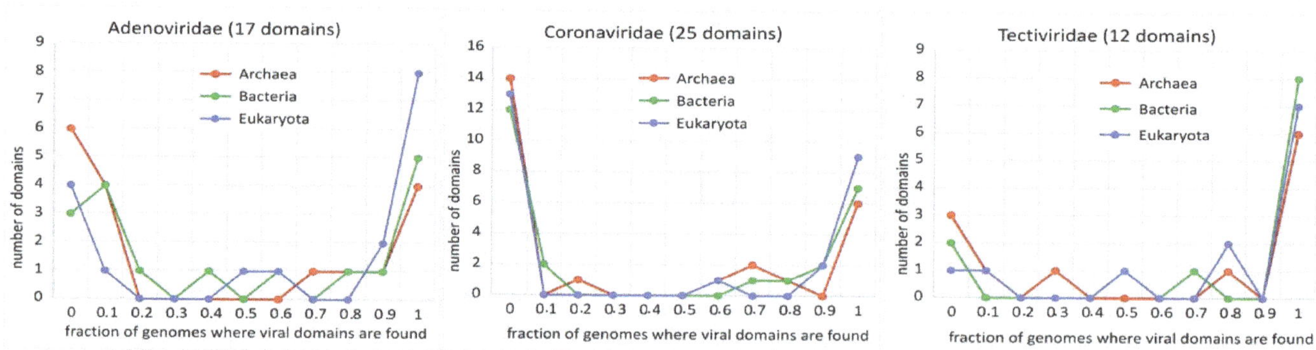

Figure 6. Distribution of the protein domains found in three viral families according to their occurrence in different superkingdoms. Protein domains as they are defined in SCOPat the superfamily level and the occurrence of these domains according to Superfamily assignment (www.supfam.org). For example, Coronaviridae encodes 13 protein domains not found in eukaryotic genomes and nine domains found in more than 90% of eukaryotic genomes.

2.3.4. RNA Secondary Structures in Whole Genome Alignments of Viruses, by Kevin Lamkiewicz

RNA secondary structures are known to play important roles in viruses, and especially in RNA viruses, since they can initiate and facilitate transcription, translation and replication. Several studies indicate that structures are cis-acting regulators for transcription. However, only looking at local structures is not sufficient to capture all RNA–RNA interactions of one molecule. Long-Range Interactions (LRI) are described in a few RNA virus families [49], but are computationally intensive to predict. Further, studies show that a single nucleotide changing can disrupt the replication of a coronavirus completely [50]. Thus, a deep understanding of conserved RNA structures is necessary to develop anti-viral therapies.

In order to increase the confidence of predictions, Multiple Sequence Alignments (MSA) are needed, since they provide conservation information between viruses. Identifying conserved secondary structures in whole genomes of viruses is computationally challenging, as the whole genome has to be considered for possible structures and interactions.

Here, we give an overview of the landscape of RNA secondary structures in viruses and provide a pipeline that generates whole genome alignments with structure annotation for downstream analyses. Our pipeline distinguishes itself from other tools by considering both the sequence and structure of input genomes for the final alignment. Therefore, for the first time, the generation of structure-annotated whole genome alignments for viruses enables sophisticated and comprehensive downstream analysis for RNA structures and RNA functions. This is achieved with an iterative combination of the sequence-based aligner MAFFT [51] and the structure-based aligner LocARNA [52]. For our example case, we were able to predict structures in the genus *Flavivirus* [53] that are consistent with described structures in the literature (Figure 7). Further, we predicted novel structural elements in coding regions of genomes.

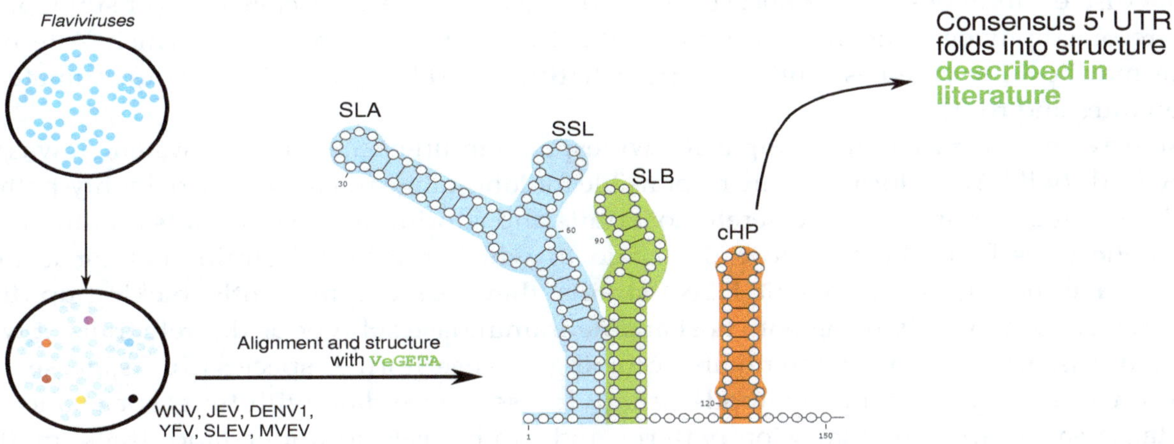

Figure 7. First results of our VeGETApipeline on an example input set consisting of flaviviruses. We were able to identify the West-Nile Virus (WNV), Dengue Virus 1 (DENV1), Japanese Encephalitis Virus (JEV), Yellow Fever Virus (YFW), Saint Louis Encephalitis Virus (SLEV) and Murray Valley Encephalitis Virus (MVEV) as representative viruses from downloaded virus genomes [53]. The resulting alignment calculated by VeGETA has structure annotations for the complete genomes, including 5′ UTR, coding regions and 3′ UTR. Here, we extracted the 5′ UTR from the alignment and visualized the annotated structure elements. These elements agree with the literature [54], as we were able to reconstruct the SLA, SLL, SLBand cHPelements accurately. The first two elements were recognized by the viral replication mechanism (NS5) [55]. The sequence embedded in the SLB structure is known to play a role in the genome circularization of flaviviruses [56], whereas the cHP facilitates the translation of the coding region by pausing the translation machinery and finding the correct starting triplet [57].

2.4. Epidemiology, Surveillance and Evolution

This session was chaired by Edward Hutchinson (MRC-University of Glasgow Centre for Virus Research, United Kingdom). Two speakers have been invited on this topic. Samantha Lycett (University of Edinburgh, United Kingdom) presented about phylodynamics for tracking epidemic, endemic and evolving viral strains. Roman Biek (University of Glasgow, United Kingdom) presented about leveraging pathogen genomics to reveal and control the spread of rabies virus. From the submitted abstracts, we selected talks by Marina Escalera-Zamudio (University of Oxford, United Kingdom) on parallel evolution and the emergence of highly-pathogenic avian influenza A viruses, David Bauer

(University of Oxford, United Kingdom) on the structure of the influenza A virus genome and Lu Lu (University of Edinburgh, United Kingdom) on the evolutionary origins of the epidemic potential among human RNA viruses.

2.4.1. Phylodynamics for Tracking Epidemic, Endemic and Evolving Viral Strains, by Samantha Lycett

Infectious diseases caused by viral pathogens in animal and livestock populations can have important economic and health consequences globally. The ability to foresee where, in which host species and under what conditions outbreaks could occur is key to developing prevention and control strategies. Sequencing pathogens from infected animals has become much more affordable and widespread in recent years, especially during outbreaks and in endemic disease settings with targeted surveillance programmes. Consequently, there are growing collections of animal virus sequences from around the globe. In this talk, the use of viral sequence data together with phylodynamic methodologies for understanding the transmission patterns in animal populations was discussed, using Avian Influenza (AI), Foot-and-Mouth Disease (FMD) and Porcine Reproductive and Respiratory Syndrome (PRRS) as examples [58–60].

Since RNA viruses have fast mutation rates and variable sequences, transmission routes between places and host species can be inferred [59,60]. One approach is to group sequences from individual hosts into discrete locations and/or host species and consider these as discrete traits or subpopulations on time-resolved phylogenetic trees, with the goal to infer which group infected which. Alternatively, locations may be represented as continuous traits (latitude and longitude) in order to estimate spatial diffusion rates and routes.

Using avian influenza as an example of a widespread multi-species disease system, it was shown that wild birds (wild Anseriformes) were responsible for long-range transmissions of highly-pathogenic H5N8, by using a combination of discrete host traits and continuous spatial traits on time-resolved phylogenetic trees [58]. Furthermore the clade to which the H5N8 strains belong is unusual because unlike the highly-pathogenic H5N1 strains, they reassort frequently, picking up different neuraminidase subtypes. By using both host and neuraminidase subtype as discrete traits, it was also shown that reassortment was preferentially occurring in Anseriformes species (ducks, geese, etc.).

To conclude, phylodynamic methods using viral sequence data with time, space and species metadata reveal complex transmission patterns and can be used to understand, track, model and ultimately inform disease control measures.

2.4.2. Parallel Evolution and the Emergence of Highly-Pathogenic Avian Influenza A Viruses, by Marina Escalera-Zamudio

Avian Influenza A Viruses (AIVs) circulate among wild and domestic bird populations worldwide. While some strains only cause mild to asymptomatic infections, known as Low Pathogenicity avian influenza viruses (LP), High Pathogenicity avian influenza viruses (HP) can have an extremely high mortality rate in both domestic and wild bird populations, leading to huge economic loses (Figure 8A) [61]. Thus, surveillance of AIVs is crucial for early detection of outbreaks. Although virulence is a polygenic trait, molecular determinants of virulence have been well characterised for AIVs, such as a polybasic proteolytic cleavage site within the hemagglutinin protein, which enables a systemic viral spread within the host [62]. We hypothesise that the parallel evolution of HP lineages from LP ancestors may have been facilitated by permissive or compensatory secondary mutations occurring anywhere in the viral genome, preceding or following the appearance of a polybasic proteolytic cleavage site. We used a comparative phylogenetic and structural approach to detect shared mutations evolving under positive selection across the whole genome of HP AIVs of the H7NX and H5NX subtypes and developed a model that statistically assesses genotype-phenotype associations. We present cumulative evolutionary and structural evidence that supports the association between parallel mutations and the evolution of the HP phenotype. Parallel mutations occur frequently among HP lineages of the same viral subtype (Figure 8B). Many of the mutations have been previously determined to increase viral fitness in terms of their biological properties, whilst most of these are ranked as stabilising to protein structure, supporting that these are rather permissive/compensatory. The mutational panel provided here may function as an early detection system for transitional virulence stages. Circulating AIVs that do not have a polybasic cleavage site yet, but show all or some of the amino acid changes ranked, should remain under surveillance.

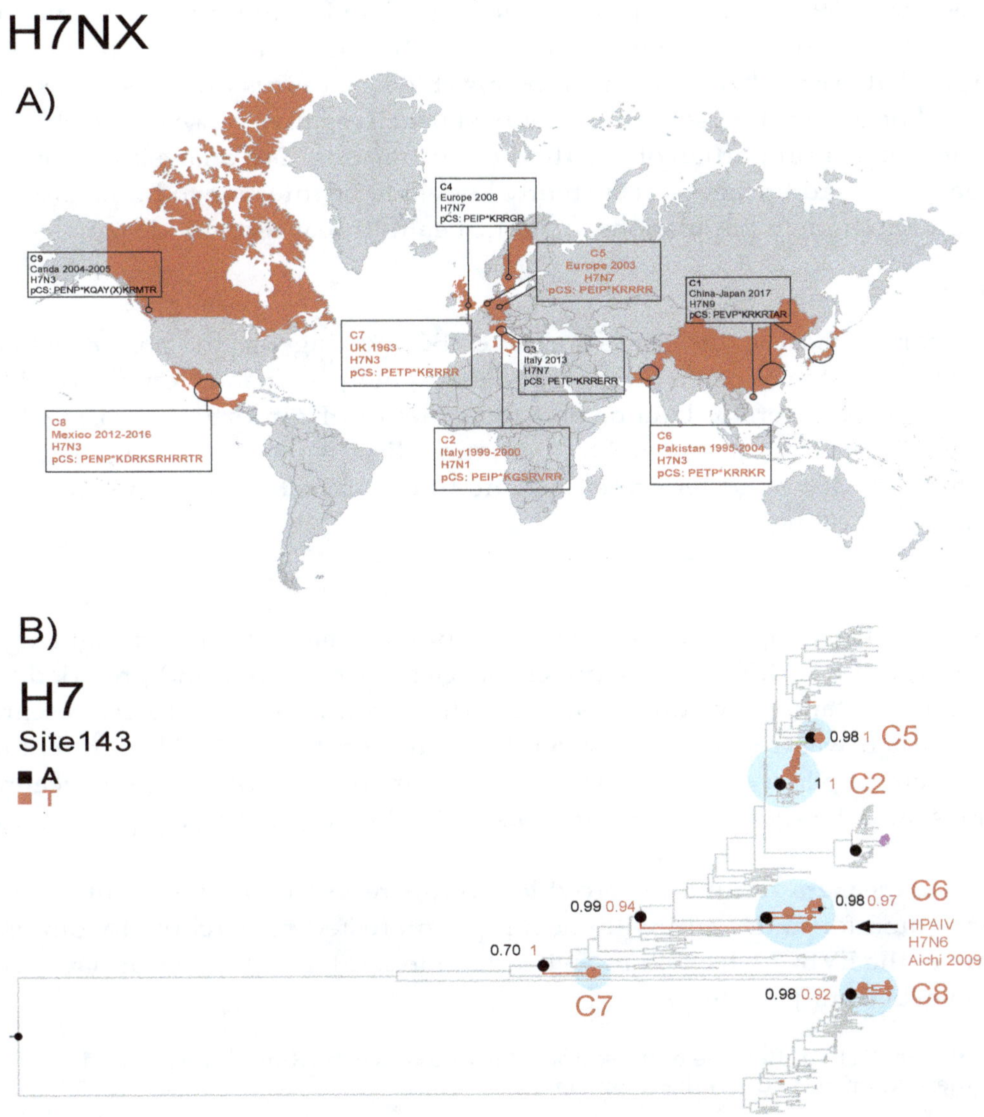

Figure 8. (**A**) Geographical occurrence of historic Highly-Pathogenic (HP) outbreaks for the H7NX viruses. Countries of emergence are highlighted in red. Year of circulation, virus subtype and consensus sequence for the polybasic Cleavage Site (pCS) within the Hemagglutinin (HA) protein are indicated for the selected outbreaks used in this work (C1-C9). Each outbreak corresponds to a distinct genotype, defined as well- supported clusters within all viral genome segment trees (data not shown). (**B**) MCCtree for the HA protein with reconstruction of ancestral states for site 143, as mutation A143T was found to be evolving under parallel evolution and to be associated with the HP phenotype, occurring in 4/9 of the HP clusters analysed. This mutation is a non-conservative amino acid change located within an antigenic pocket site. Branches within the trees are coloured according to the corresponding amino acid states in nodes (tip states not shown). Ancestral nodes preceding the emergence of a mutation associated with the HP lineages are represented with coloured circles. The probabilities of a given amino acid state occurring within ancestral/descending nodes are indicated. The HP clusters of interest are highlighted with blue circles. Mutations strongly associated with an HP phenotype may function as an early detection system for transitional virulence stages.

2.4.3. Evolutionary Origins of Epidemic Potential among Human RNA Viruses, by Lu Lu

For a virus to have epidemic potential in human populations, an infected individual must be capable of transmitting the infection to other individuals. However, for the majority of human RNA virus species, human infections are acquired only from non-human reservoirs. The evolution of human transmissibility is poorly understood. Through parallel analyses of 1755 RNA viruses, we identified at least 90 nodes across 39 genus-level phylogenies associated with transitions involving the gain of human infectivity and/or transmissibility. Human-infective and human-transmissible viruses evolve independently, and at least 73% of human-transmissible RNA virus lineages emerged directly from non-human virus lineages in diverse mammal or bird taxa. Negative sense single-stranded RNA virus lineages generate a higher proportion of strictly zoonotic viruses. Our analysis demonstrates that RNA viruses from mammal/bird lineages not currently known to be infective to humans are a likely source of future epidemics in human populations, a public health threat recently designated "Disease X".

3. Poster Session

Another important facet of this year's annual EVBC meeting was the poster session on Thursday evening. The standard of the research presented was extremely high and, combined with a networking event in the Glasgow University Union, provided plenty of opportunity to meet the presenters. The relaxed atmosphere was instrumental to promoting discussions and developing new interactions between attendees. The list of poster presenters and titles can be found online (http://evbc.uni-jena.de/events/3rd-evbc-meeting).

4. Conclusions

The Third Annual Meeting of the European Virus Bioinformatics Center brought together scientists in the field with expertise in different disciplines for scientific exchange and provided the opportunity for discussing ongoing and new collaborations. The meeting attracted new researchers to virus bioinformatics, which was reflected by several first-time attendees. The presentations strongly underlined the interdisciplinary "virology meets bioinformatics" character of the meeting. We enjoyed lively discussions after the speakers' presentations, in the breaks, during the poster session and at the social events.

We hope that speakers summaries provided in this report will give an interesting insight into the field of virus bioinformatics and will encourage interested researchers to join us at the Fourth Annual Meeting of the EVBC to be held in Switzerland in 2020. For more information, do not hesitate to contact us via evbc@uni-jena.de.

Author Contributions: F.H. drafted the manuscript. The speakers contributed the paragraphs on their own talks. All authors commented on and edited the manuscript.

Acknowledgments: The conference organisers acknowledge sponsorship from Viruses, an open access journal from MDPI, MSD, Biomex, Theolytics and the MRC-University of Glasgow Centre for Virus Research. Their support paid for the refreshments, room and poster board hire.

References

1. Hufsky, F.; Ibrahim, B.; Beer, M.; Deng, L.; Le Mercier, P.; McMahon, D.P.; Palmarini, M.; Thiel, V.; Marz, M. Virologists—Heroes need weapons. *PLoS Pathog.* **2018**, *14*, e1006771. [CrossRef] [PubMed]
2. Ibrahim, B.; McMahon, D.P.; Hufsky, F.; Beer, M.; Deng, L.; Le Mercier, P.; Palmarini, M.; Thiel, V.; Marz, M. A new era of virus bioinformatics. *Virus Res.* **2018**, *251*, 86–90. [CrossRef]
3. Ibrahim, B.; Arkhipova, K.; Andeweg, A.; Posada-Céspedes, S.; Enault, F.; Gruber, A.; Koonin, E.; Kupczok, A.; Lemey, P.; McHardy, A.; et al. Bioinformatics Meets Virology: The European Virus Bioinformatics Center's Second Annual Meeting. *Viruses* **2018**, *10*, 256. [CrossRef] [PubMed]
4. V'kovski, P.; Gerber, M.; Kelly, J.; Pfaender, S.; Ebert, N.; Lagache, S.B.; Simillion, C.; Portmann, J.; Stalder, H.; Gaschen, V.; et al. Determination of host proteins composing the microenvironment of coronavirus replicase complexes by proximity-labeling. *eLife* **2019**, *8*, e42037. [CrossRef] [PubMed]

5. Mayr, F.B.; Yende, S.; Angus, D.C. Epidemiology of severe sepsis. *Virulence* **2014**, *5*, 4–11. [CrossRef] [PubMed]
6. Singer, M.; Deutschman, C.S.; Seymour, C.W.; Shankar-Hari, M.; Annane, D.; Bauer, M.; Bellomo, R.; Bernard, G.R.; Chiche, J.D.; Coopersmith, C.M.; et al. The Third International Consensus Definitions for Sepsis and Septic Shock (Sepsis-3). *JAMA* **2016**, *315*, 801–810. [CrossRef] [PubMed]
7. Iuliano, A.D.; Roguski, K.M.; Chang, H.H.; Muscatello, D.J.; Palekar, R.; Tempia, S.; Cohen, C.; Gran, J.M.; Schanzer, D.; Cowling, B.J.; et al. Estimates of global seasonal influenza-associated respiratory mortality: A modelling study. *Lancet (London, England)* **2018**, *391*, 1285–1300. [CrossRef]
8. Papanicolaou, G.A. Severe influenza and S. aureus pneumonia: For whom the bell tolls? *Virulence* **2013**, *4*, 666–668. [CrossRef] [PubMed]
9. Klemm, C.; Bruchhagen, C.; van Krüchten, A.; Niemann, S.; Löffler, B.; Peters, G.; Ludwig, S.; Ehrhardt, C. Mitogen-activated protein kinases (MAPKs) regulate IL-6 over-production during concomitant influenza virus and *Staphylococcus aureus* infection. *Sci. Rep.* **2017**, *7*, 42473. [CrossRef]
10. van Krüchten, A.; Wilden, J.J.; Niemann, S.; Peters, G.; Löffler, B.; Ludwig, S.; Ehrhardt, C. *Staphylococcus aureus* triggers a shift from influenza virus-induced apoptosis to necrotic cell death. *FASEB J.* **2018**, *32*, 2779–2793. [CrossRef] [PubMed]
11. Russell, A.B.; Trapnell, C.; Bloom, J.D. Extreme heterogeneity of influenza virus infection in single cells. *eLife* **2018**, *7*, e32303. [CrossRef] [PubMed]
12. Steuerman, Y.; Cohen, M.; Peshes-Yaloz, N.; Valadarsky, L.; Cohn, O.; David, E.; Frishberg, A.; Mayo, L.; Bacharach, E.; Amit, I.; et al. Dissection of Influenza Infection In Vivo by Single-Cell RNA Sequencing. *Cell Syst.* **2018**, *6*, 679–691. [CrossRef] [PubMed]
13. Jonsdottir, H.R.; Dijkman, R. Characterization of Human Coronaviruses on Well-Differentiated Human Airway Epithelial Cell Cultures. In *Coronaviruses*; Springer: New York, NY, USA, 2015; pp. 73–87. [CrossRef]
14. James, C.E.; Davies, E.V.; Fothergill, J.L.; Walshaw, M.J.; Beale, C.M.; Brockhurst, M.A.; Winstanley, C. Lytic activity by temperate phages of *Pseudomonas aeruginosa* in long-term cystic fibrosis chronic lung infections. *ISME J.* **2015**, *9*, 1391–1398. [CrossRef] [PubMed]
15. Shan, J.; Ramachandran, A.; Thanki, A.M.; Vukusic, F.B.I.; Barylski, J.; Clokie, M.R.J. Bacteriophages are more virulent to bacteria with human cells than they are in bacterial culture; insights from HT-29 cells. *Sci. Rep.* **2018**, *8*, 5091. [CrossRef]
16. Nale, J.Y.; Redgwell, T.A.; Millard, A.; Clokie, M.R.J. Efficacy of an Optimised Bacteriophage Cocktail to Clear *Clostridium difficile* in a Batch Fermentation Model. *Antibiotics* **2018**, *7*, 13. [CrossRef]
17. Nale, J.Y.; Spencer, J.; Hargreaves, K.R.; Buckley, A.M.; Trzepiński, P.; Douce, G.R.; Clokie, M.R.J. Bacteriophage Combinations Significantly Reduce *Clostridium difficile* Growth In Vitro and Proliferation In Vivo. *Antimicrob. Agents Chemother.* **2016**, *60*, 968–981. [CrossRef]
18. Hendrix, R.W.; Smith, M.C.; Burns, R.N.; Ford, M.E.; Hatfull, G.F. Evolutionary relationships among diverse bacteriophages and prophages: All the world's a phage. *Proc. Natl. Acad. Sci. USA* **1999**, *96*, 2192–2197. [CrossRef]
19. Sanjuán, R.; Nebot, M.R.; Chirico, N.; Mansky, L.M.; Belshaw, R. Viral mutation rates. *J. Virol.* **2010**, *84*, 9733–9748. [CrossRef]
20. Breitbart, M.; Salamon, P.; Andresen, B.; Mahaffy, J.M.; Segall, A.M.; Mead, D.; Azam, F.; Rohwer, F. Genomic analysis of uncultured marine viral communities. *Proc. Natl. Acad. Sci. USA* **2002**, *99*, 14250–14255. [CrossRef]
21. Roux, S.; Brum, J.R.; Dutilh, B.E.; Sunagawa, S.; Duhaime, M.B.; Loy, A.; Poulos, B.T.; Solonenko, N.; Lara, E.; Poulain, J.; et al. Ecogenomics and potential biogeochemical impacts of globally abundant ocean viruses. *Nature* **2016**, *537*, 689. [CrossRef]
22. Breitbart, M.; Rohwer, F. Here a virus, there a virus, everywhere the same virus? *Trends Microbiol.* **2005**, *13*, 278–284. [CrossRef] [PubMed]
23. Dutilh, B.E.; Cassman, N.; McNair, K.; Sanchez, S.E.; Silva, G.G.Z.; Boling, L.; Barr, J.J.; Speth, D.R.; Seguritan, V.; Aziz, R.K.; et al. A highly abundant bacteriophage discovered in the unknown sequences of human faecal metagenomes. *Nat. Commun.* **2014**, *5*, 4498. [CrossRef] [PubMed]
24. Stern, A.; Mick, E.; Tirosh, I.; Sagy, O.; Sorek, R. CRISPR targeting reveals a reservoir of common phages associated with the human gut microbiome. *Genome Res.* **2012**, *22*, 1985–1994. [CrossRef] [PubMed]
25. Manrique, P.; Bolduc, B.; Walk, S.T.; van der Oost, J.; de Vos, W.M.; Young, M.J. Healthy human gut phageome. *Proc. Natl. Acad. Sci. USA* **2016**, *113*, 10400–10405. [CrossRef] [PubMed]

26. Mahmoudabadi, G.; Phillips, R. A comprehensive and quantitative exploration of thousands of viral genomes. *eLife* **2018**, *7*, e31955. [CrossRef] [PubMed]
27. Kang, H.S.; McNair, K.; Cuevas, D.; Bailey, B.; Segall, A.; Edwards, R.A. Prophage genomics reveals patterns in phage genome organization and replication. *bioRxiv* **2017**, 114819. [CrossRef]
28. Miller, E.S.; Kutter, E.; Mosig, G.; Arisaka, F.; Kunisawa, T.; Ruger, W. Bacteriophage T4 Genome. *Microbiol. Mol. Biol. Rev.* **2003**, *67*, 86–156. [CrossRef]
29. Brum, J.R.; Schenck, R.O.; Sullivan, M.B. Global morphological analysis of marine viruses shows minimal regional variation and dominance of non-tailed viruses. *ISME J.* **2013**, *7*, 1738–1751. [CrossRef] [PubMed]
30. Mavrich, T.N.; Hatfull, G.F. Bacteriophage evolution differs by host, lifestyle and genome. *Nat. Microbiol.* **2017**, *2*. [CrossRef]
31. Kupczok, A.; Neve, H.; Huang, K.D.; Hoeppner, M.P.; Heller, K.J.; Franz, C.M.A.P.; Dagan, T. Rates of Mutation and Recombination in Siphoviridae Phage Genome Evolution over Three Decades. *Mol. Biol. Evol.* **2018**, *35*, 1147–1159. [CrossRef]
32. Simmonds, P.; Aiewsakun, P.; Katzourakis, A. Prisoners of war — host adaptation and its constraints on virus evolution. *Nat. Rev. Microbiol.* **2018**, 17, 321–328. [CrossRef] [PubMed]
33. Edwards, R.; Vega, A.; Norman, H.; Ohaeri, M.C.; Levi, K.; Dinsdale, E.; Cinek, O.; Aziz, R.; McNair, K.; Barr, J.; et al. Global phylogeography and ancient evolution of the widespread human gut virus crAssphage. *bioRxiv* **2019**, 527796. [CrossRef]
34. Mokili, J.L.; Rohwer, F.; Dutilh, B.E. Metagenomics and future perspectives in virus discovery. *Curr. Opin. Virol.* **2012**, *2*, 63–77. [CrossRef] [PubMed]
35. Symonds, E.M.; Breitbart, M. Affordable Enteric Virus Detection Techniques Are Needed to Support Changing Paradigms in Water Quality Management. *Clean* **2014**, *43*, 8–12. [CrossRef]
36. Bibby, K. Metagenomic identification of viral pathogens. *Trends Biotechnol.* **2013**, *31*, 275–279. [CrossRef] [PubMed]
37. Roux, S.; Adriaenssens, E.M.; Dutilh, B.E.; Koonin, E.V.; Kropinski, A.M.; Krupovic, M.; Kuhn, J.H.; Lavigne, R.; Brister, J.R.; Varsani, A.; et al. Minimum Information about an Uncultivated Virus Genome (MIUViG). *Nat. Biotechnol.* **2018**. [CrossRef]
38. Eren, A.M.; Esen, Ö.C.; Quince, C.; Vineis, J.H.; Morrison, H.G.; Sogin, M.L.; Delmont, T.O. Anvi'o: An advanced analysis and visualization platform for 'omics data. *PeerJ* **2015**, *3*, e1319. [CrossRef]
39. Adriaenssens, E.; Farkas, K.; Harrison, C.; Jones, D.; Allison, H.E.; McCarthy, A.J. Viromic analysis of wastewater input to a river catchment reveals a diverse assemblage of RNA viruses. *bioRxiv* **2018**, 248203. [CrossRef]
40. Simmonds, P.; Adams, M.J.; Benkő, M.; Breitbart, M.; Brister, J.R.; Carstens, E.B.; Davison, A.J.; Delwart, E.; Gorbalenya, A.E.; Harrach, B.; et al. Consensus statement: Virus taxonomy in the age of metagenomics. *Nat. Rev. Microbiol.* **2017**, *15*, 161–168. [CrossRef]
41. Aiewsakun, P.; Adriaenssens, E.M.; Lavigne, R.; Kropinski, A.M.; Simmonds, P. Evaluation of the genomic diversity of viruses infecting bacteria, archaea and eukaryotes using a common bioinformatic platform: Steps towards a unified taxonomy. *J Gen Virol* **2018**, *99*, 1331–1343. [CrossRef]
42. Aiewsakun, P.; Simmonds, P. The genomic underpinnings of eukaryotic virus taxonomy: Creating a sequence-based framework for family-level virus classification. *Microbiome* **2018**, *6*, 38. [CrossRef] [PubMed]
43. Patterson Ross, Z.; Klunk, J.; Fornaciari, G.; Giuffra, V.; Duchêne, S.; Duggan, A.T.; Poinar, D.; Douglas, M.W.; Eden, J.S.; Holmes, E.C.; et al. The paradox of HBV evolution as revealed from a 16th century mummy. *PLoS Pathog.* **2018**, *14*, e1006750. [CrossRef]
44. Krause-Kyora, B.; Susat, J.; Key, F.M.; Kühnert, D.; Bosse, E.; Immel, A.; Rinne, C.; Kornell, S.C.; Yepes, D.; Franzenburg, S.; et al. Neolithic and medieval virus genomes reveal complex evolution of hepatitis B. *eLife* **2018**, *7*, e36666. [CrossRef] [PubMed]
45. Mühlemann, B.; Jones, T.C.; Damgaard, P.d.B.; Allentoft, M.E.; Shevnina, I.; Logvin, A.; Usmanova, E.; Panyushkina, I.P.; Boldgiv, B.; Bazartseren, T.; et al. Ancient hepatitis B viruses from the Bronze Age to the Medieval period. *Nature* **2018**, *557*, 418–423. [CrossRef] [PubMed]
46. Bar-Gal, G.K.; Kim, M.J.; Klein, A.; Shin, D.H.; Oh, C.S.; Kim, J.W.; Kim, T.H.; Kim, S.B.; Grant, P.R.; Pappo, O.; et al. Tracing hepatitis B virus to the 16th century in a Korean mummy. *Hepatology* **2012**, *56*, 1671–1680. [CrossRef]
47. Puustusmaa, M.; Abroi, A. Conservation of the E8 CDS of the $E8^{E2}$ protein among mammalian papillomaviruses. *J. Gen. Virol.* **2016**, *97*, 2333–2345. [CrossRef] [PubMed]

48. Puustusmaa, M.; Abroi, A. cRegions—a tool for detecting conserved cis-elements in multiple sequence alignment of diverged coding sequences. *PeerJ* **2019**, *6*, e6176. [CrossRef]
49. Nicholson, B.L.; White, K.A. Functional long-range RNA-RNA interactions in positive-strand RNA viruses. *Nat. Rev. Microbiol.* **2014**, *12*, 493–504. [CrossRef] [PubMed]
50. Madhugiri, R.; Karl, N.; Petersen, D.; Lamkiewicz, K.; Fricke, M.; Wend, U.; Scheuer, R.; Marz, M.; Ziebuhr, J. Structural and functional conservation of cis-acting RNA elements in coronavirus 5′-terminal genome regions. *Virology* **2018**, *517*, 44–55. [CrossRef] [PubMed]
51. Kuraku, S.; Zmasek, C.M.; Nishimura, O.; Katoh, K. aLeaves facilitates on-demand exploration of metazoan gene family trees on MAFFT sequence alignment server with enhanced interactivity. *Nucleic Acids Res.* **2013**, *41*, W22–W28. [CrossRef] [PubMed]
52. Will, S.; Reiche, K.; Hofacker, I.L.; Stadler, P.F.; Backofen, R. Inferring noncoding RNA families and classes by means of genome-scale structure-based clustering. *PLoS Comput. Biol.* **2007**, *3*, e65. [CrossRef]
53. Pickett, B.E.; Sadat, E.L.; Zhang, Y.; Noronha, J.M.; Squires, R.B.; Hunt, V.; Liu, M.; Kumar, S.; Zaremba, S.; Gu, Z.; et al. ViPR: An open bioinformatics database and analysis resource for virology research. *Nucleic Acids Res.* **2012**, *40*, D593–D598. [CrossRef] [PubMed]
54. Fernández-Sanlés, A.; Ríos-Marco, P.; Romero-López, C.; Berzal-Herranz, A. Functional Information Stored in the Conserved Structural RNA Domains of Flavivirus Genomes. *Front. Microbiol.* **2017**, *8*, 546. [CrossRef]
55. Filomatori, C.V.; Lodeiro, M.F.; Alvarez, D.E.; Samsa, M.M.; Pietrasanta, L.; Gamarnik, A.V. A 5′ RNA element promotes dengue virus RNA synthesis on a circular genome. *Genes Dev.* **2006**, *20*, 2238–2249. [CrossRef] [PubMed]
56. Clyde, K.; Barrera, J.; Harris, E. The capsid-coding region hairpin element (cHP) is a critical determinant of dengue virus and West Nile virus RNA synthesis. *Virology* **2008**, *379*, 314–323. [CrossRef] [PubMed]
57. Kozak, M. Downstream secondary structure facilitates recognition of initiator codons by eukaryotic ribosomes. *Proc. Natl. Acad. Sci. USA* **1990**, *87*, 8301–8305. [CrossRef]
58. Global Consortium for H5N8 and Related Influenza Viruses. Role for migratory wild birds in the global spread of avian influenza H5N8. *Science* **2016**, *354*, 213–217. [CrossRef]
59. Lycett, S.; Tanya, V.N.; Hall, M.; King, D.; Mazeri, S.; Mioulet, V.; Knowles, N.; Wadsworth, J.; Bachanek-Bankowska, K.; Victor, N.N.; et al. The evolution and phylodynamics of serotype A and SAT2 foot-and-mouth disease viruses in endemic regions of Africa. *bioRxiv* **2019**, 572198. [CrossRef]
60. Duchatel, F.; Bronsvoort, M.; Lycett, S. Phylogeographic analysis and identification of factors impacting the diffusion of Foot-and-Mouth disease virus in Africa. *bioRxiv* **2018**, 358044. [CrossRef]
61. Dhingra, M.S.; Artois, J.; Dellicour, S.; Lemey, P.; Dauphin, G.; Von Dobschuetz, S.; Van Boeckel, T.P.; Castellan, D.M.; Morzaria, S.; Gilbert, M. Geographical and Historical Patterns in the Emergences of Novel Highly Pathogenic Avian Influenza (HPAI) H5 and H7 Viruses in Poultry. *Front. Vet. Sci.* **2018**, *5*, 84. [CrossRef]
62. Abdelwhab, E.M.; Veits, J.; Ulrich, R.; Kasbohm, E.; Teifke, J.P.; Mettenleiter, T.C. Composition of the Hemagglutinin Polybasic Proteolytic Cleavage Motif Mediates Variable Virulence of H7N7 Avian Influenza Viruses. *Sci. Rep.* **2016**, *6*, 3950. [CrossRef] [PubMed]

4

The Transcriptional Landscape of Marek's Disease Virus in Primary Chicken B Cells Reveals Novel Splice Variants and Genes

Luca D. Bertzbach [1,†], Florian Pfaff [2,*,†], Viktoria I. Pauker [3], Ahmed M. Kheimar [1,4], Dirk Höper [2], Sonja Härtle [5], Axel Karger [3,*] and Benedikt B. Kaufer [1,*]

1 Institute of Virology, Freie Universität Berlin, Robert von Ostertag-Straße 7-13, 14163 Berlin, Germany; luca.bertzbach@fu-berlin.de (L.D.B.); ahmed.kheimar@fu-berlin.de (A.M.K.)
2 Institute of Diagnostic Virology, Friedrich-Loeffler-Institut, Federal Research Institute for Animal Health, Südufer 10, 17493 Greifswald-Insel Riems, Germany; dirk.hoeper@fli.de
3 Institute of Molecular Virology and Cell Biology, Friedrich-Loeffler-Institut, Federal Research Institute for Animal Health, Südufer 10, 17493 Greifswald-Insel Riems, Germany; viktoria.pauker@uni-greifswald.de
4 Department of Poultry Diseases, Faculty of Veterinary Medicine, Sohag University, Sohag 82424, Egypt
5 Department of Veterinary Sciences, Institute for Animal Physiology, Ludwig-Maximilians-Universität München, 80539 Munich, Germany; sonja.haertle@tiph.vetmed.uni-muenchen.de
* Correspondence: florian.pfaff@fli.de (F.P.); axel.karger@fli.de (A.K.); benedikt.kaufer@fu-berlin.de (B.B.K.)
† These authors contributed equally to this work.

Abstract: Marek's disease virus (MDV) is an oncogenic alphaherpesvirus that infects chickens and poses a serious threat to poultry health. In infected animals, MDV efficiently replicates in B cells in various lymphoid organs. Despite many years of research, the viral transcriptome in primary target cells of MDV remained unknown. In this study, we uncovered the transcriptional landscape of the very virulent RB1B strain and the attenuated CVI988/Rispens vaccine strain in primary chicken B cells using high-throughput RNA-sequencing. Our data confirmed the expression of known genes, but also identified a novel spliced MDV gene in the unique short region of the genome. Furthermore, *de novo* transcriptome assembly revealed extensive splicing of viral genes resulting in coding and non-coding RNA transcripts. A novel splicing isoform of MDV UL15 could also be confirmed by mass spectrometry and RT-PCR. In addition, we could demonstrate that the associated transcriptional motifs are highly conserved and closely resembled those of the host transcriptional machinery. Taken together, our data allow a comprehensive re-annotation of the MDV genome with novel genes and splice variants that could be targeted in further research on MDV replication and tumorigenesis.

Keywords: Marek's disease virus (MDV); RNA-seq; transcriptome; splicing; polycistronic viral transcripts; primary B cells; RB1B; CVI988/Rispens; ICP0

---

1. Introduction

Marek's disease virus (MDV), also known as Gallid alphaherpesvirus 2, causes a deadly lymphoproliferative disease in chickens. Typical clinical symptoms include immunosuppression, paralysis and polyneuritis, acute brain edema, and lymphoma that develop as early as 3 weeks post infection [1,2]. MDV has a big economic impact on the poultry industry worldwide due to animal losses, reduced growth, decreased egg production, and cost of vaccination [3]. Vaccines are crucial for the protection against MDV, as very virulent strains can cause mortalities of up to 100% in susceptible unvaccinated chickens [4]. Live attenuated MDV vaccines such as the gold standard Rispens strain (CVI988) are highly effective in preventing tumor formation [3,5], but do not provide sterilizing immunity.

MDV infection is initiated by the inhalation of the virus from a contaminated environment. Macrophages and other phagocytic cells are thought to take up the virus and transfer it to lymphatic tissues, where B and T cells are infected [6]. B cells are efficiently infected during the initial lytic virus replication, whereas T cell subsets play key roles in MDV pathogenesis [7–11]. MDV establishes latency predominantly in CD4+ T cells, which can also transport the virus to the feather follicle epithelium. These cells efficiently produce the infectious virus and shed it into the environment [3]. Latently infected T cells can also be transformed, resulting in deadly lymphomas.

MDV has a 180 kilo base pairs class E genome consisting of a unique long (U_L) and a unique short (U_S) sequence that are flanked by terminal (TR_L and TR_S) and internal (IR_L and IR_S) inverted repeat regions [12,13]. MDV encodes about 100 proteins that orchestrate the virus life cycle and/or contribute to pathogenesis [12,14]. Until now, analyses of the viral transcriptome has been limited to chicken fibroblasts that are not infected in chickens, ex vivo samples [15,16] and tumor cells [17]; however, the mRNA expression in the primary target cells of lytic replication in vivo remained elusive. This is mainly due to the short lifespan of B and T cells in culture and the low quantity of infected cells in lymphoid organs of chickens [7,18]. To overcome this obstacle, we recently developed an in vitro infection system for primary B and T cells that allows for a prolonged survival and efficient infection of these cells [19]. We used this system to analyze the MDV transcriptome in the most frequent lytically infected cell type in vivo, the B cells.

In this study, we performed next generation RNA-sequencing and protein profiling in primary B cells infected with the very virulent RB1B strain or the vaccine strain CVI988. Our data reveal that the coding capacity of the MDV genome is larger than expected. We identified novel MDV genes and splice variants, and confirmed them either on the protein level or by RT-PCR. This comprehensive approach provides novel insights into the transcriptome of MDV in the natural target cells and supply a basis for future research on MDV pathogenesis.

2. Materials and Methods

2.1. Ethics Statement

Valo specific-pathogen-free (SPF) chickens (VALO BioMedia GmbH, Osterholz-Scharmbeck, Germany) were housed for 6 to 11 weeks of age and humanely euthanized prior to the isolation of the bursa of Fabricius. The animal work was approved by the governmental agency, the Landesamt für Gesundheit und Soziales (LAGeSo) in Berlin, Germany (approval number T0245/14, approval date 23 October 2014).

2.2. Cells

Embryonated SPF Valo chicken eggs (VALO BioMedia GmbH,) were used for the preparation of chicken embryo cells (CEC). CEC were maintained in minimal essential medium (MEM, PAN Biotech; Aidenbach, Germany) supplemented with 1-10% fetal bovine serum (FBS) and penicillin/streptomycin as previously described [20]. B cells were obtained from the bursa of Fabricius by dissociation of the organ and subsequent isolation of the cells by density gradient centrifugation as previously described [21]. Briefly, the bursa of Fabricius was homogenized through a 40 μm cell filter to obtain a uniform single cell suspension. Suspension cells were carefully applied on a Biocoll separating solution (Biochrom; Berlin, Germany), centrifuged for 12 min at 650 × g with slow acceleration, and deactivated deceleration. Lymphocytes at the interphase were carefully transferred to a new tube, washed with PBS, and maintained in RPMI 1640 (PAN Biotech) supplemented with 10% FBS and penicillin [100 U/mL]/streptomycin [100 μg/mL] at 41 °C under a 5% CO_2 atmosphere. B cells were activated using recombinant soluble chicken CD40 ligand (chCD40L) [22], which was expressed in HEK293 cells and purified using a Vivacell 250 ultrafiltration concentrator (Sartorius; Göttingen, Germany).

2.3. Viruses

All viruses were reconstituted by calcium phosphate transfection of CEC with purified bacterial artificial chromosome (BAC) DNA as previously described [23]. The very virulent RB1B strain and the vaccine strain CVI988 both express a green fluorescent protein (GFP) under the control of the early thymidine kinase promotor. The viruses were propagated on CEC for up to six passages, and infected cells were stored in liquid nitrogen until further use. All virus stocks were titrated on fresh CEC.

2.4. Infection of Primary Chicken B Cells

Primary chicken B cells were infected by co-cultivation with infected CEC due to the strictly cell-associated nature of MDV. One million CEC were infected with 30,000 plaque-forming units (PFU) of CVI988, RB1B, or mock infected. After 4 days, one million B cells were seeded on the infected 6-well-plates in the presence of CD40L for 16 h at 41 °C. All B cells were then carefully removed from the CEC monolayer, washed with phosphate buffered saline (PBS), and prepared for fluorescence-activated cell sorting.

2.5. Flow Cytometry

Viable bursal B cells were detected using the eFluor780 fixable viability dye at a dilution of 1:1000 (Affymetrix eBioscience; San Diego, CA, USA) as previously described [19]. Cells were sorted using a FACS Aria III sorter and the FACSDiva software (Becton Dickinson; Franklin Lakes, NJ, USA). For each sample, approximately 10^5 to 10^6 infected B cells were sorted at 4 °C and stored at −80 °C until further analysis. The purity of GFP+ sorted fractions was determined by FACS reanalysis and yielded 99.73% (±0.46 SD) for mock-infected cells, 95.33% (±1.29 SD) for RB1B infected cells and 97.47% (±1.33 SD) for CVI988 infected cells.

2.6. High-Throughput RNA-Sequencing

RNA was isolated from three independent experiments of CVI988 or RB1B infected chicken B cell cultures and sequenced as described [24]. Briefly, total RNA was extracted using TRIzol reagent (Life Technologies; Carlsbad, CA, USA) in combination with the RNeasy Mini Kit (Qiagen; Hilden, Germany) following the manufacturer's instructions. Additionally, RNA was treated with DNase using the RNase-Free DNase Set (Qiagen). Subsequently, ERCC ExFold RNA Spike-In mix 1 (Invitrogen; Carlsbad, CA, USA) was added to the total RNA as an internal control and the polyadenylated (poly(A)) RNA fraction was extracted using the Dynabeads mRNA DIRECT Micro kit (Invitrogen). Whole transcriptome libraries were prepared using the Ion Total RNA-Seq Kit v2 (Life Technologies) following the manufacturer's instructions. Quality and quantity of the nucleic acids was controlled at each step using the NanoDrop 1000 spectrophotometer (Peqlab) or Agilent 2100 Bioanalyzer (Agilent Technologies; Böblingen, Germany) in combination with appropriate chips, respectively. The resulting libraries were finally quantified using the KAPA Library Quantification Kit for Ion Torrent (Kapa Biosystems; Wilmington, MA, USA) on a CFX96 Real-Time PCR Detection System (BioRad Laboratories) and sequenced on an Ion S5XL system (Life Technologies) using the Ion 540 OT2 and Chip kit (Life Technologies).

2.7. Sequence and Bioinformatic Analyses

Reads from separate sequencing runs (technical replicates) of the same library (biological replicates) were combined and quality-trimmed using the 454 Sequencing System Software (v. 3.0; Roche; Mannheim, Germany) along with appropriate Ion Torrent specific adapter sequences. Each quality-trimmed data set was then mapped to a non-redundant version of the MDV reference NC_002229.3 [12] (only segments U_L-IR_L-IR_S-U_S) using STAR (version 2.6.1a; [25]), running in basic two-pass mode. In this manuscript, the MDV genes were designated according to the current gene nomenclature used for the prototype alphaherpesvirus herpes simplex virus 1 (HSV-1) [26,27].

As the mRNA libraries were amplified for several rounds, the sequence duplicates were removed prior to *de-novo* assembly and coverage analysis. Therefore, the duplicated reads were marked and removed in each mapped dataset using Picard (version 2.18.20; http://broadinstitute.github.io/picard). Subsequently, the unique aligned reads were directionally sorted using samtools (version 1.9; [28]) and the sequence depth was deduced from each dataset using bedtools (version 2.15.0; [29]) and samtools. The directionally sorted reads were then used for *de-novo* assembly using the 454 Sequencing System Software (v. 3.0; Roche) running in "-cdna" mode. Assembly was done for each biological replicate separately and with the combined read data from all replicates. Deduplicated and directionally sorted reads were also used as basis for the coverage plots. In order to receive high quality splice junctions, all assembled "isotigs" (transcript variants) were then re-mapped to the non-redundant version of MDV reference NC_002229.3 [12] using STARlong (version 2.6.1a; [25]) in basic two-pass mode. Options were set to allow a single mapping isotig to yield a splice junction with a maximum intron length of 10,000 bp. Only positions corresponding to the high-quality splice junctions were then selected from the splice junctions of the initial mapping approach for further analysis. The frequency of spliced reads was calculated by dividing the number of reads with splice junction by the total number of reads at the respective donor site. The overall splice frequency at a single donor site for CVI988 and RB1B was then averaged from the individual splice frequencies in the replicates. Based on the deduced splice junctions, the positions up- and downstream of these were extracted and visualized using the R (version 3.4.1; [30]) package "ggseqlogo" (version 0.1; [31]) in combination with RStudio (version 1.0.153) in order to receive information on donor and acceptor motif sequences. Polyadenylation cleavage clusters were determined with ContextMap (version 2.7.9; [32]) in combination with bowtie2 (version 2.2.9; [33]) using the complete trimmed dataset as input and the "–polyA" parameter. The resulting poly(A) cleavage sites were then combined into clusters, as the exact position of mRNA cleavage downstream of a cleavage signal can be heterogeneous [34]. Starting with the first poly(A) cleavage site, all other sites on the same strand within a window of the next 30 nt were combined into a single cluster. The window was then moved to the next cleavage site that was not within the last cluster. This was repeated for all cleavage sites. To scan for enriched regulatory motifs within the three prime untranslated region (3′-UTR) of MDV transcripts, the sequences 50 nt up- and 20 nt downstream of the start position of each identified poly(A) cleavage cluster were extracted and analyzed using DREME as part of the MEME suite (version 4.9.0; [35]) using default settings. Subsequently, the identified enriched motif (AWTAAA) was searched in the non-redundant version of MDV reference NC_002229.3 [12] using FIMO (version 4.9.0; [36]) with default settings. Poly(A) cleavage clusters and regulatory motifs were then grouped into relevant pairs based on the FIMO *p*-value and the their distance, allowing a maximum distance of 50 nt. Differential gene expression between RB1B and the vaccine strain CVI988 was conducted using Salmon (version 0.12.0; [37]) in combination with DESeq2 (version 1.18.1; [38]) as described earlier [24]. All relevant MDV CDS sequences were used as transcript reference and genes with an adjusted p-value > 0.01 were considered significant. Potential phosphorylation sites in novel protein SORF6 were predicted using the NetPhos 3.1 Server [39] and its DNA-protein binding probability using the DNABIND server [40].

2.8. LC-MALDI TOF/TOF Mass Spectrometry

Infected FACS sorted primary chicken B cells and mock infected primary chicken B cells were lysed in batches of 1.5×10^6 cells using 150 µl of a lysis buffer containing 0.1 M DL-Dithothreitol (DTT) and 2% SDS in 0.1 M Tris-HCl (pH 8.0) at 99 °C for 5 min. Protein contents were determined by densitometry of Coomassie stained SDS gels [41,42]. After cell lysis, 20 µg aliquots were digested using the FASP protocol as described [43]. Samples were differentially labeled by dimethylation [44] using unlabeled and ^{13}C-labeled formaldehyde, respectively, and subjected to nano-LC MALDI-TOF/TOF mass spectrometry as described previously [45]. Briefly, peptides were separated by nano reversed-phase liquid chromatography (EASY-nLC II, Bruker; Bremen, Germany), spotted to a MALDI target (Proteineer fcII, Bruker), and analyzed with an UltrafleXtreme MALDI-TOF/TOF mass spectrometer

(Bruker) as described previously [46]. Peptide spectra were acquired in the m/z range 700 to 3.500 Da with a minimum signal-to-noise (S/N) ratio of 7. Proteins were identified with a Mascot server (version 2.4.1; Matrix Science Ltd; London, UK) and analyzed using ProteinScape software (version 3; Bruker). Oxidation of methionine, acetylation of protein N-termini, and dimethylation of lysine and peptide N-termini with either isotopomer were set as variable modifications, whereas the carbamydomethylation of cysteine residues was set as a fixed modification. Two independent experiments were performed with inverted labeling. As database for the protein identification with the MASCOT search engine (Matrix Science Ltd), the *Gallus gallus* proteome was downloaded from the ENSEMBL website [47] and the viral sequences were added to the FASTA file. Viral protein content was calculated in mol% using the exponentially modified protein abundance index (emPAI) [48]. To identify peptides covering the potential new splicing sites discovered by RNA-sequencing, a database with sequence fragments covering a 67 amino acid region centered on the splicing site was constructed and used for the database search with MASCOT.

2.9. *Reverse Transcription and PCR over Splice Junctions*

RNA was isolated as described above. cDNA was synthesized after DNAse treatment (Promega; Fitchburg, WI, USA) with the Applied Biosystems High-Capacity cDNA Reverse Transcription Kit (Thermo Fischer; Waltham, MA, USA). Conventional Taq-PCR was performed with primers specific to the respective viral gene (Table S2). Amplification of BAC DNA was used as a positive control. Mock-infected cells and samples without reverse transcriptase to exclude a contamination with genomic DNA were included as negative controls.

2.10. *Data Availability*

The RNA-seq raw data were deposited in the ArrayExpress database at EMBL-EBI (www.ebi.ac.uk/arrayexpress) under accession number E-MTAB-7772. A supplementary GFF file for the reference sequence NC_002229.3 containing annotations for all detected introns, poly(A) cleavage sites and associated motifs, as well as the novel CDS for SORF6 can be found in the Supplementary File S1.

3. Results and Discussion

3.1. *The MDV Transcriptional Landscape*

To assess the transcriptional landscape of MDV in the primary target cells of lytic replication in vivo, we used a previously established in vitro infection system for primary chicken B cells [19]. B cells were infected for 16 h with the very virulent MDV strain RB1B and the vaccine strain CVI988 and analyzed by high-throughput RNA sequencing.

The overall RNA-seq dataset consisted of 82.6 million reads from three biological replicates of CVI988 (48.2 million reads) and two biological replicates of RB1B (34.4 million reads). A third replicate of RB1B did not yield sufficient amount of reads and was therefore excluded from our analysis. An average of 10.5% and 11.2% of the CVI and RB1B datasets respectively could be mapped to the MDV reference sequence (see Table S3).

The position and direction of mapped reads fitted very well to the previously annotated MDV genes (Figure 1A). Highly abundant genes like the immediate early gene SORF1 (ICP4) or the UL49 tegument protein (VP22) correlate well with previously published data [49]. Surprisingly, only minor differences were detected between the transcriptome of CVI988 and RB1B. Comparing the 94 detected MDV genes, two variants of MDV075 encoding the 14-kDa polypeptides (pp14), were significantly higher expressed in RB1B infected primary B cells (Figure 1B). These phosphorylated cytoplasmic proteins arise from splice variants of the same gene and are thought to be involved in transcriptional regulation and increased neurovirulence [50–52]. Furthermore, the hypothetical gene MDV082 [53] that is located on the same transcript as the ICP4 gene, was significantly higher expressed in CVI988 infected B cells. However, the transcriptome of RB1B and CVI988 only shows subtle differences in primary chicken B cells, suggesting that the differences in their pathogenesis might be due to sequence changes on the protein level and/or functional differences of virulence factors. Similarly, we only detected marginal quantitative differences between the expression levels of viral proteins in a proteome analysis using LC-MALDI TOF/TOF mass spectrometry (Figure S1 and Table S1). Taken together, our data indicate that there are only minor differences in the mRNA and protein expression levels after B cell infections with the very virulent RB1B or the CVI988 vaccine strain.

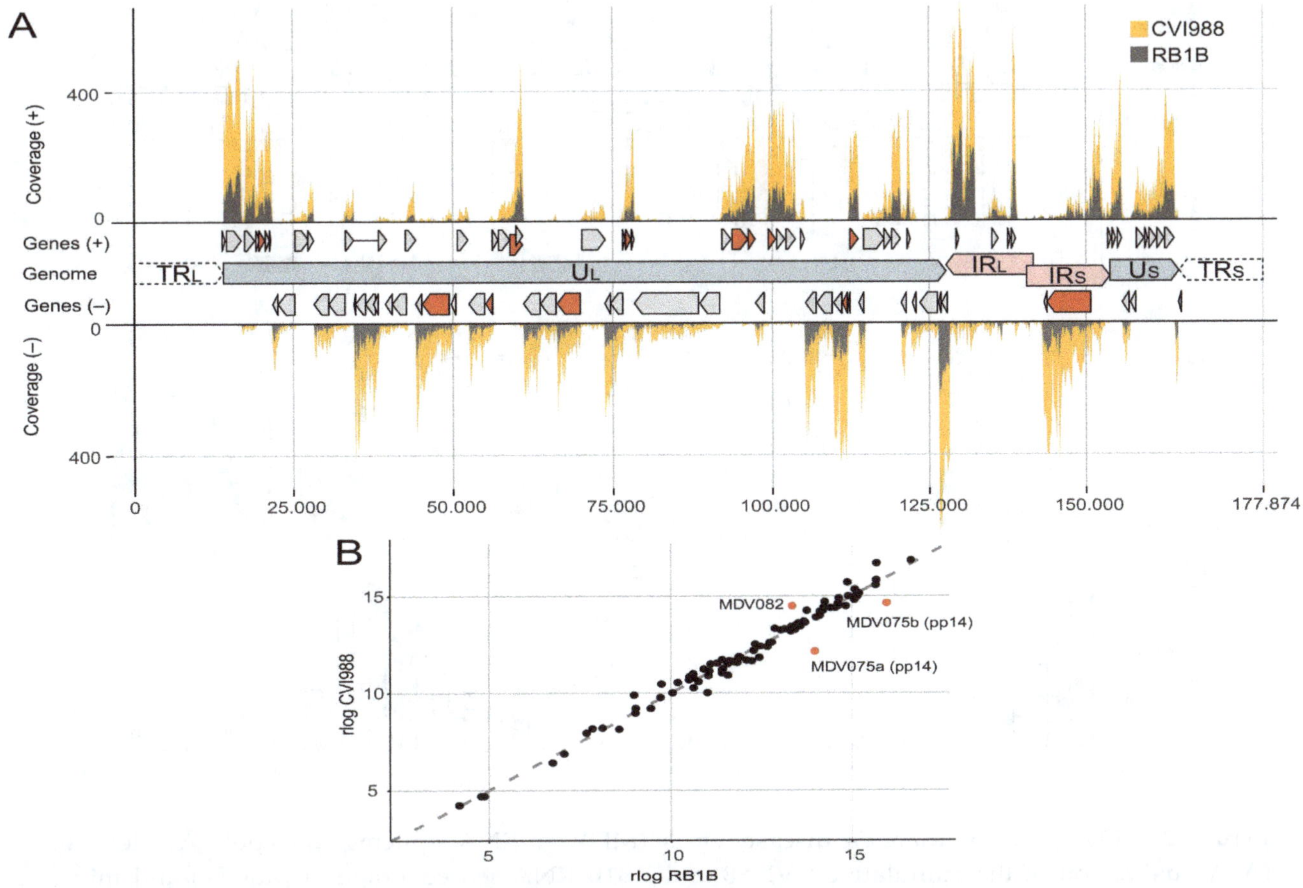

Figure 1. The Marek's disease virus (MDV) transcriptome and proteome in in vitro infected primary chicken B cells. (**A**) Visualization of the deduplicated and strand-specific RNA-seq read coverage of plus (+) and minus (-) strand-encoded genes across the MDV genome. Orange curves indicate CVI988 reads and grey curves indicate reads for RB1B (with the respective annotated genes as grey arrow bars). Red bars depict proteins identified by MS. The two unique regions, unique long (U_L) and short (U_S) are flanked by terminal (TR_L and TR_S) and internal (IR_L and IR_S) inverted repeat regions. Nucleotide position numbers are derived from [12]. (**B**) Gene expression scatterplot comparing normalized expression levels (rlog) in RB1B and CVI988 infected primary chicken B cells. Red dots indicate significantly differentially expressed genes.

3.2. Splicing of Polycistronic MDV Transcripts

In addition to the transcriptional profile, we could readily identify 71 introns that were represented by at least one *de novo* transcript (Figure 2A). Some of the introns and associated spliced genes have been previously described such as the viral lipase (vLIP) [54], LORF2 (MDV012) [55], UL15 [12], UL44 (glycoprotein C) [56], vIL8 [57] and pp14 [51]. However, analysis of the MDV transcriptome revealed a number of novel splice forms (Table S4). These results are in line with previous RNA-seq analysis for other alphaherpesviruses [58] that also revealed a plethora of novel splice products. The detected splice variants could contribute to viral proteomic diversity and could prevent viral mRNA degradation through the virion host shutoff UL41 endoribonuclease [59]. For HSV-1 it has been shown that UL41 not only targets many cellular but also viral mRNAs. Spliced mRNAs are protected from UL41-mediated degradation by bound exon junction complexes (EJCs) [60].

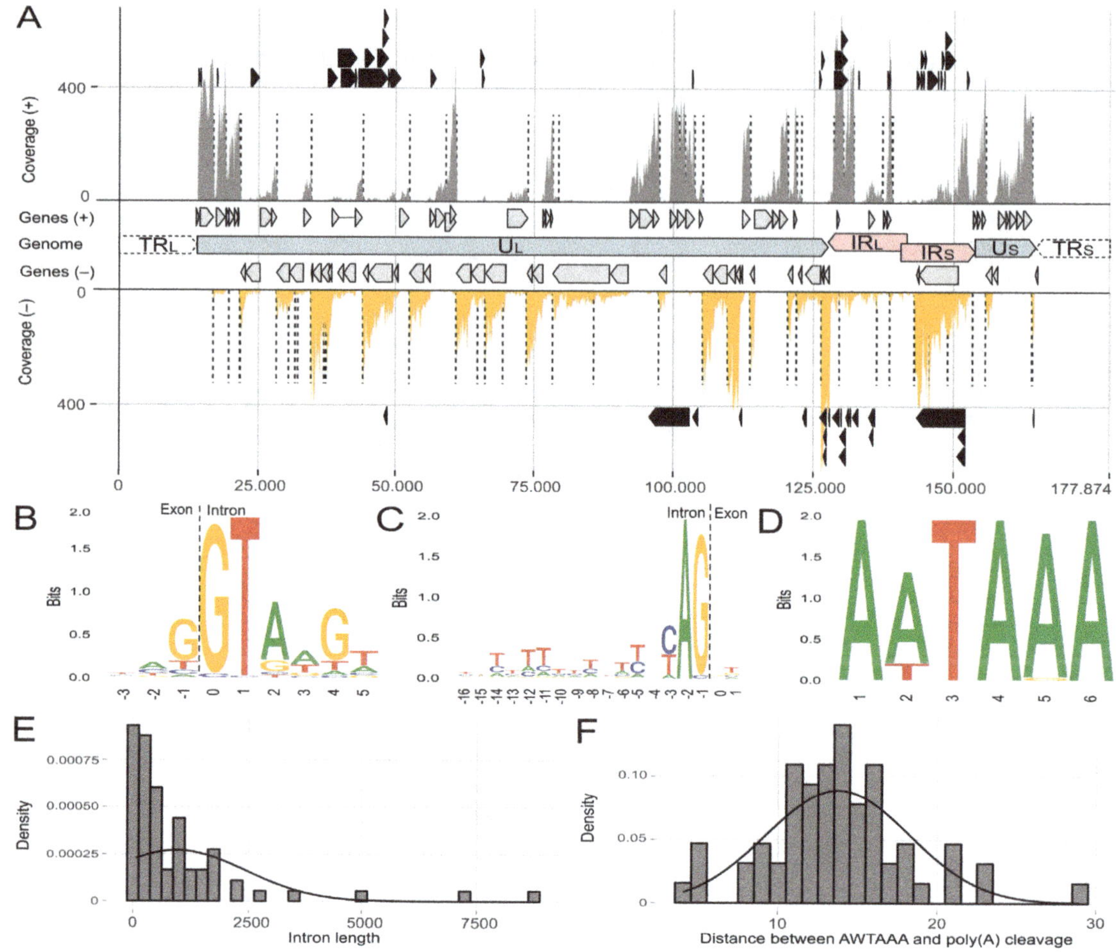

Figure 2. Overview of Marek's disease virus (MDV) mRNA splicing and poly(A) cleavage. (**A**) Visualization of the cumulative CVI988 and RB1B RNA-seq coverage of plus (+) and minus (-) strand-encoded genes across the MDV genome. Black arrow bars indicate introns and dashed lines indicate poly(A) cleavage sites. More detailed information is shown in Table S4 and Table S5. (**B**) Nucleotide frequency maps (sequence logo) of splice donor sites in MDV-encoded mRNAs. The relative heights of letters correspond to frequencies of bases at each position. (**C**) Sequence logo of splice acceptor sites in MDV-encoded mRNAs. (**D**) Sequence logo of polyadenylation signals in MDV-encoded three prime untranslated regions (3′-UTR). (**E**) Histogram depicting MDV intron length distributions. (**F**) Histogram depicting the distance from AWTAAA-like motifs to the poly(A) cleavage site.

The identified splice site sequences mostly represent canonical splicing motifs, containing the GT at the donor and AG at the acceptor sites (Figure 2B,C). The intron length varied between 70 and 8651 nt (Figure 2E). Intriguingly, the intron frequencies differed between RB1B and CVI988 in infected primary chicken B cells (Figures 3–5). By matching the intron positions with our MS data, we identified a peptide that spans the exon-exon junction of UL15 (Figure S2).

The analysis of poly(A) cleavage signals within the RNA-seq data revealed abundant bicistronic and polycistronic MDV transcripts (Table S5). These transcripts encode for two or more proteins and were characterized as regions of high coverage that were not separated by a poly(A) cleavage site. Here we found that the canonical AATAAA motif is the most frequent and functional polyadenylation signal in MDV, followed by ATTAAA (Figure 2D). Interestingly, we also found evidence for alternative non-canonical polyadenylation signals in MDV mRNA 3′ UTRs (Figure 2 and Table S5). The distances between the detected AWTAAA polyadenylation signal motif and the actual poly(A) cleavage site was 13.8 nt (±4.4 SD) and confirmed that not only the polyadenylation signal sequence, but also its distance from the poly(A) is highly conserved (Figure 2D,F) [61,62].

3.3. The Transcriptional Makeup of the MDV Unique Long Region (U_L).

The MDV unique regions mainly harbor genes that are conserved among alphaherpesviruses and are involved in DNA replication and production of progeny virus [26]. The U_L spans over roughly 113,000 base pairs and harbors the majority of the MDV-encoded genes [26]. Within the U_L, we could detect high transcription rates of nearly all annotated genes. Splicing was identified in multiple genes including RLORF14 (pp24), vLIP [63], and LORF2 (MDV012) [55] and in a transcript antisense to UL5 (MDV017) (Figure 3). Only minor differences were observed in the intron frequencies between RB1B and CVI988. To confirm the splice events and frequencies detected by RNA-seq, we performed RT-PCR analyses on several randomly selected genes (Figure 3C, Figure 4B, Figure 5C, Figures S2 and S3). All analyzed genes showed a comparable splice pattern in both RNA-seq and RT-PCR. In addition, we confirmed a novel splice site of UL15 by MS and RT-PCR (Figure S2). UL15 encodes the tripartite terminase subunit that is involved in DNA packaging into the viral capsid. Splicing of UL15 mRNA has already been observed in herpes simplex virus type 1 (HSV-1) [64] and duck enteritis virus (DEV) [65]. However, the observed UL15 isoforms in MDV are to our knowledge unknown and expand the number of potential proteins encoded by UL15 to at least five.

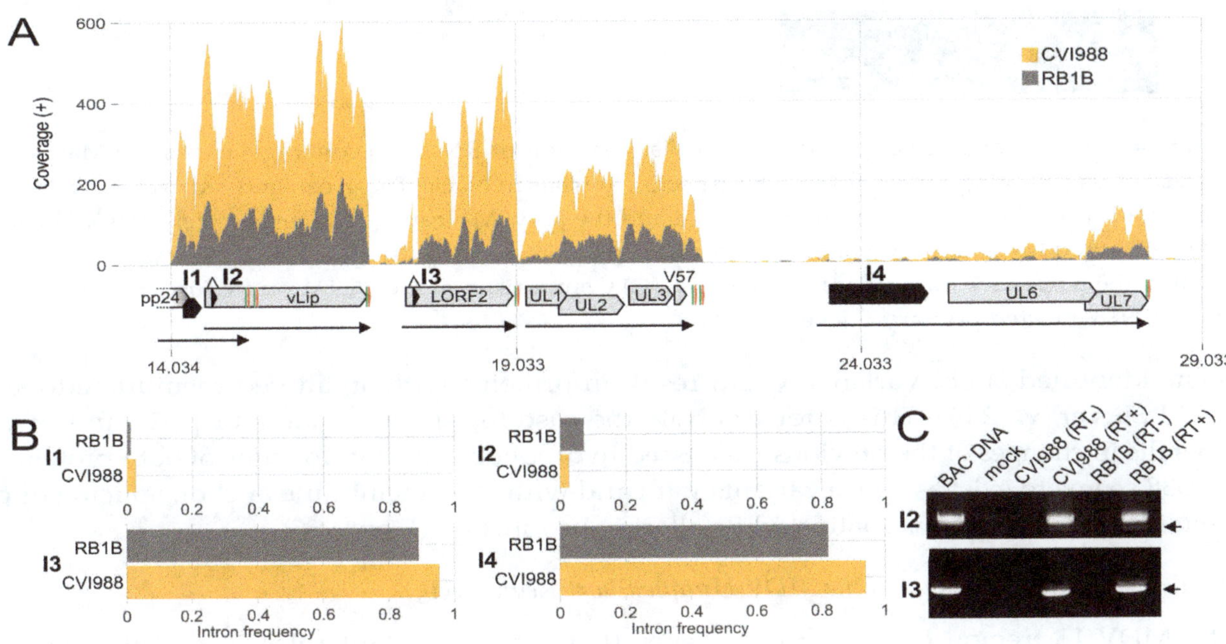

Figure 3. The Marek's disease virus (MDV) unique long (U_L) region. (**A**) Visualization of RNA-Seq coverage across parts of the MDV U_L region with respective introns in black. Green and red arrows indicate the polyadenylation signal and the poly(A) cleavage site respectively. Underlying black arrows suggest unspliced (mono-, bi-, or polycistronic) mRNAs. (**B**) Comparison of intron frequencies in RB1B and CVI988 infected primary chicken B cells. I1: pp24, I2: vLIP, I3: LORF2 (MDV012), I4: transcript antisense to UL5 (MDV017). (**C**) RT-PCR was performed to validate the splicing event. PCR products were derived using forward/reverse primers to amplify the respective intron-flanking regions. The representative gel images illustrate the results of RT-PCR analysis. The black arrows indicate the spliced form of the respective gene.

We could also confirm known splice sites in UL44 (gC) in our analysis (Figure 4) [56]. These splice variants lead to a gC protein that lacks the transmembrane domain and is secreted into the supernatant [56]. Beyond that, we also confirmed novel splice sites like in RLORF14a (pp38) (Figure S3). Several capsid and tegument protein-encoding genes encoded in the U_L and U_S regions, like UL18 (triplex capsid protein 2), UL19 (VP5), UL49 (VP22), UL49.5 (gN), US7 (gI) and US8 (gE) also undergo different splicing events (Table S4), contradicting the long-standing paradigm that splicing is a rare phenomenon throughout the alphaherpesvirus family [13].

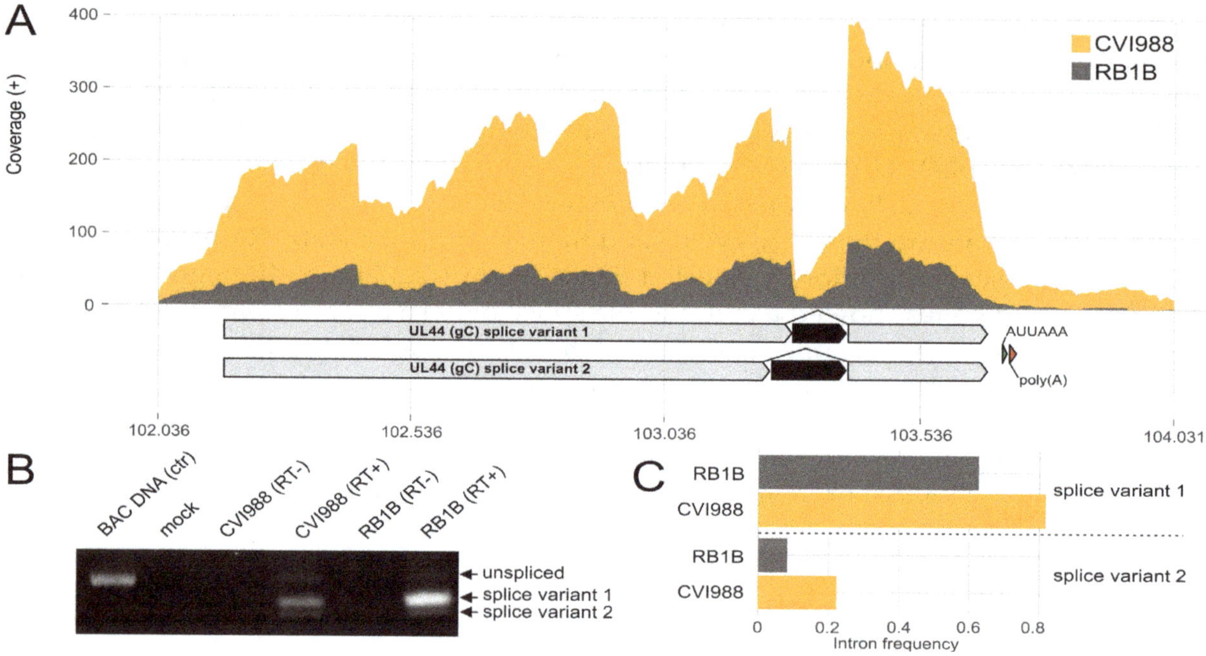

Figure 4. Glycoprotein C (gC) splicing. (**A**) Visualization of RNA-seq coverage across the Marek's disease virus (MDV) gC gene UL44 with respective introns in black. The green and red arrow indicate the canonical ATTAAA polyadenylation signal and the poly(A) cleavage site respectively. (**B**) RT-PCR was performed to validate gC splicing. PCR products were derived using forward/reverse primers to amplify the respective intron-flanking regions. (**C**) Comparison of the gC intron frequencies in RB1B and CVI988 infected primary chicken B cells.

Some identified splice variants would result in proteins with an altered membrane topology (TMHMM Server, v. 2.0). This is for example the case for splice variants of pp24 and of pp38, which results in changes of the previously assessed hydrophobic anchor domains of both proteins [66]. While pp38 seems to exist as splice variants with and without a membrane anchor, splicing of pp28 could retain its membrane association while altering its function (Figure S5).

3.4. *The Transcriptional Makeup of the MDV Unique Short Region (U_S).*

The MDV U_S region contains many genes that play important roles in the viral life cycle. Intriguingly, we detected splice variants of several envelope glycoproteins as described above. In addition, we identified a hitherto uncharacterized spliced transcript of a gene located downstream of SORF2A, termed SORF6 (Figure 5). This novel gene possesses an upstream TATA box in the transcriptional regulatory region, an intron and exon with respective donor and acceptor site and a downstream polyadenylation signal with the poly(A) cleavage cluster (as described in Figure 2). Furthermore, the resulting protein is predicted to be 85 amino acids in size, shows several predicted phosphorylation sites (Figure S6) and may act as a DNA-binding protein (DNABIND server [40]). The region containing this novel gene (Figure 5) has previously been associated with the virulence of the virus [67]. However, more work needs to be done to understand the contribution of this novel gene and the region in MDV virulence.

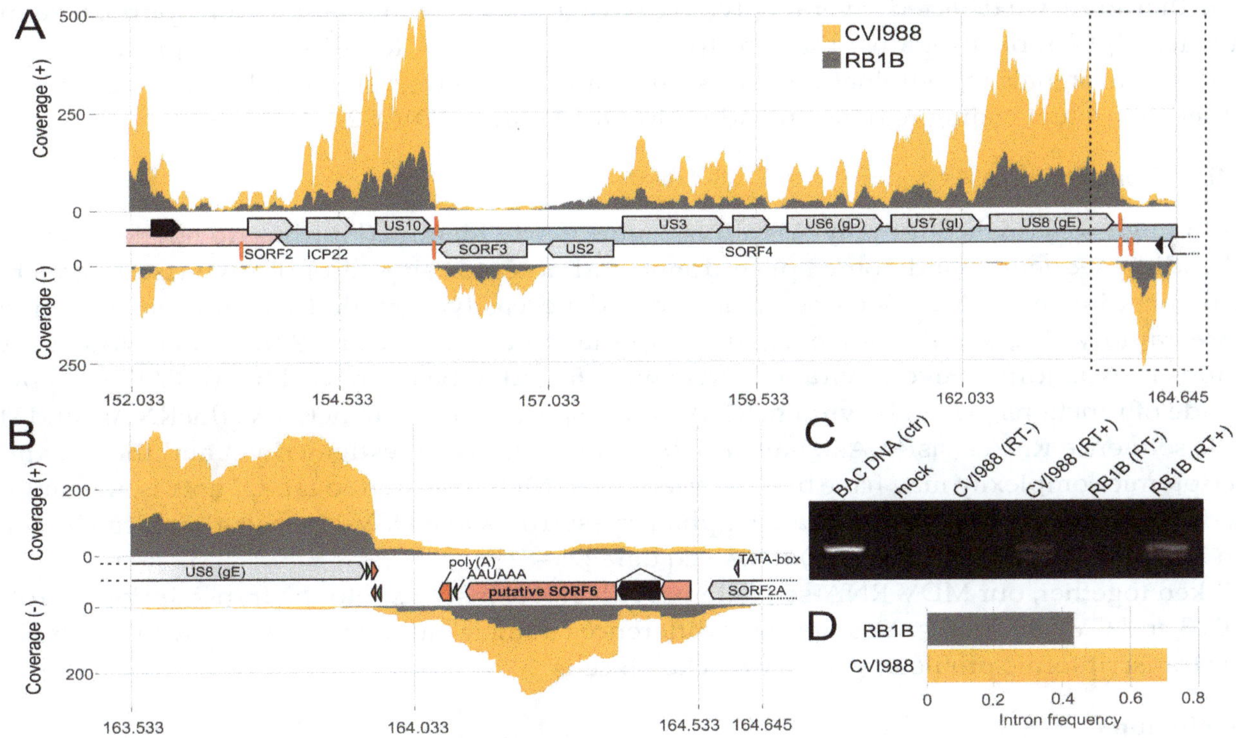

Figure 5. The Marek's disease virus (MDV) unique short (U_S) region. (**A**) Visualization of RNA-seq coverage across the MDV U_S region with respective introns in black. (**B**) Zoom into the far-right region of the MDV US with depiction of the novel U_S gene SORF6. Green and red arrows indicate the polyadenylation signals and the poly(A) cleavage sites, respectively. (**C**) RT-PCR was performed to validate the splicing event in the novel gene SORF6. PCR products were derived using forward/reverse primers to amplify the respective intron-flanking regions as full-length (upper band) and spliced (lower band). (**D**) Comparison of the novel U_S gene (SORF6) intron frequencies in RB1B and CVI988 infected primary chicken B cells.

3.5. The Transcriptional Makeup of the MDV Repeat Regions

The repeat regions mostly contain MDV-specific genes encoding for proteins or RNA that play a role in the cell tropism, MDV pathogenesis, latency, and transformation [14]. Here, we observed excessive splicing antisense to ICP4. These transcripts are part of the latency associated transcript (LAT) region, have a complex splice pattern, and their functions remain largely unknown [68,69]. Some of these RNAs function as MDV-encoded micro RNAs and are described elsewhere [70,71].

Only moderate RLORF7 (Meq) and vIL8 splicing was detected in infected primary B cells 16 hpi although more extensive splicing activity has been observed in this region of the MDV genome in vitro and in vivo [72–74]. These splice variants are likely higher expressed in latently infected and transformed cells. The splice variants of the neurovirulence factor pp14 encoded by MDV075 were efficiently detected as published previously [15,44].

In the RNA-seq data, we did not detect any reads complementary to vTR. However, this region is annotated as a hypothetical MDV gene termed RLORF1 (an arginine-rich protein/ICP0-like protein) [75]. RLORF1 is discussed as a potential positional orthologue of alphaherpesviral ICP0 proteins; however, it does not contain typical ICP0 features such as a C3HC4 zinc RING finger at the N-terminus or a nuclear localization signal (NLS). To assess if ICP0 protein is expressed and if it plays a role in replication, we generated recombinant MDV mutants harboring an HA-tagged ICP0 (RB1B_ICP0-HA) or an ICP0 knockout (RB1B_ΔMetICP0). The knockout did not affect MDV replication and cell to cell spread in vitro (Method S1 and Figure S4A) and no ICP0 was detected by western blotting (Method S1 and Figure S4B), suggesting that ICP0 is not expressed and therefore does not play a role in the virus life cycle.

Furthermore, we detected several poly(A) cleavage sites in combination with appropriate motifs within the MDV repeat regions, that are to our knowledge undescribed. The presence of these transcriptional signals in combination with sufficient read coverage suggest the existence of hitherto hypothetical protein coding regions, including RLORF11 and MDV082.

3.6. MDV Noncoding RNAs

Although we enriched for poly(A) mRNA, some newly identified introns do not lie in annotated MDV coding sequences and some splice donor and acceptor sites do not give rise to conclusive protein-encoding mRNAs. Such sequences could be easily regarded as 'nonsense' transcripts that are rapidly degraded. However, the importance of noncoding RNAs (ncRNAs) in MDV infections is expanding. Several viral ncRNAs were found to be expressed by MDV [75–78] but the multitude of functions played by viral ncRNAs, and especially by long ncRNAs (lncRNAs) and stable intronic sequence RNAs (sisRNAs), have not been thoroughly investigated yet and the unexpected transcriptomic complexity may have been overlooked in MDV research so far. Of note is, that antisense transcription was also observed in related alphaherpesviruses like HSV-1 or pseudorabies virus (PRV), and in human herpesvirus 6 (HHV-6) RNA-seq data [79–81].

Taken together, our MDV RNA-seq data provide novel insights into the transcriptional profile of the RB1B and CVI988 strains. Despite stark differences in their pathogenicity, the two viruses show a similar transcriptomic profile in primary chicken B cells.

4. Conclusions

B cells are a major target for lytic MDV replication in vivo [8,10]; however, it remained impossible to assess the MDV transcriptome in primary B cells, due to the short-lived nature of these cell. The aim of this study was to evaluate the gene expression profiles of the very virulent RB1B strain and the commercial MDV vaccine CVI988 in primary chicken B cells by RNA-seq using our recently established in vitro infection system [19]. We developed a bioinformatics pipeline that can be easily transferred to other herpesviruses or large DNA viruses to identify unknown transcript isoforms and associated motifs.

The RNA-seq revealed the expression of 94 MDV transcripts and the presence of 71 introns that lead to mostly novel splice forms and antisense transcripts. In addition, we could identify a novel gene in the U_S region of the MDV genome that we will characterize in future studies. While some of the detected splice sites were previously published, we identified several novel splice variants and confirmed some of them by RT-PCR and/or MS. However, more work is certainly required to dissect their relevance in the MDV life cycle.

We found that MDV produces bicistronic and polycistronic transcripts as a mechanism to maximize its coding capacities. Poly(A) cleavage after the upstream AATAAA motif seems to be the most frequent and functional polyadenylation signal in MDV. The identification of possible alternative transcript termination (ATT) needs further experimental evidence (Figure 2 and Table S5). ATT is a strong regulatory factor in eukaryotes [82], but there is only limited data for ATT in herpesvirus transcription.

The comparison of the transcriptome between the very virulent RB1B strain and the CVI988 vaccine revealed differences in only a few transcripts (Figure 1). However, more work needs to be done to unravel significant differences that could possibly point towards a mechanism of attenuation or provide valuable information for the development of diagnostic tools.

In summary, our data demonstrate that the MDV genome is more complex than previously assumed. It provides a source of reference for MDV transcripts expressed in primary chicken B cells and lays the foundation for future research on MDV-encoded gene products and splice variants.

Supplementary Materials:
Method S1: Generation and in vitro characterization of RB1BΔMetICP0, Figure S1: Protein concentration (mol%) scatterplot comparing levels of detected viral proteins in RB1B and CVI988 infected primary chicken B cells,

Figure S2: Detected splice variants of UL15, Figure S3: RT-PCR confirmation of splicing events in pp38, Figure S4: Plaques size assay and western blot analysis for the ICP0 mutant RB1BΔMetICP0, Figure S5: Transmembrane helix prediction for spliced and unspliced pp24 and pp38, Figure S6: Prediction of serine, threonine or tyrosine phosphorylation sites in the hypothetical MDV protein SORF6 encoded on the U_S segment, Table S1: Marek's disease virus (MDV) proteins detected by mass spectrometry, Table S2: Primers used in this study, Table S3: Summary of RNA-seq read mapping, Table S4: Introns identified from RNA-seq data, Table S5: Poly(A) cleavage sites, polyadenylation signals and polycistronic transcripts identified in MDV transcriptomes, Supplementary File S1: Annotations for NC_002229.3 in GFF format.

Author Contributions: Conceptualization: L.D.B., F.P., A.M.K., A.K. and B.B.K.; methodology: L.D.B., F.P., V.I.P., A.K. and B.B.K.; formal analysis: L.D.B., F.P., A.K. and B.B.K..; investigation: L.D.B., F.P., V.I.P. and A.M.K.; resources: S.H.; writing—original draft preparation: L.D.B. and F.P.; writing—review and editing: L.D.B., F.P., A.K. and B.B.K.; visualization: L.D.B., F.P., A.K. and B.B.K.; supervision: D.H., A.K. and B.B.K.; project administration: A.K. and B.B.K.; funding acquisition: A.K. and B.B.K.; all authors read and approved the final manuscript.

Acknowledgments: We thank Ann Reum and Yu You (Institut of Virology, Freie Universität Berlin), Marina Kohn (Institute for Animal Physiology, Ludwig-Maximilians-Universität München) and Patrick Zitzow (Institute of Diagnostic Virology, Friedrich-Loeffler-Institut) for outstanding technical assistance.

References

1. Nair, V. Evolution of Marek's disease—A paradigm for incessant race between the pathogen and the host. *Vet. J.* **2005**, *170*, 175–183. [CrossRef]
2. Davison, T.F.; Nair, V. *Marek's Disease: An Evolving Problem*; Elsevier: Amsterdam, The Netherlands, 2004.
3. Davison, F.; Nair, V. Use of Marek's disease vaccines: Could they be driving the virus to increasing virulence? *Expert Rev. Vaccines* **2005**, *4*, 77–88. [CrossRef]
4. Witter, R.L. Increased virulence of Marek's disease virus field isolates. *Avian Dis.* **1997**, *41*, 149–163. [CrossRef]
5. Baigent, S.J.; Nair, V.K.; Le Galludec, H. Real-time PCR for differential quantification of CVI988 vaccine virus and virulent strains of Marek's disease virus. *J. Virol. Methods* **2016**, *233*, 23–36. [CrossRef]
6. Jarosinski, K.W.; Tischer, B.K.; Trapp, S.; Osterrieder, N. Marek's disease virus: Lytic replication, oncogenesis and control. *Expert Rev. Vaccines* **2006**, *5*, 761–772. [CrossRef]
7. Baigent, S.J.; Ross, L.J.; Davison, T.F. Differential susceptibility to Marek's disease is associated with differences in number, but not phenotype or location, of pp38+ lymphocytes. *J. Gen. Virol.* **1998**, *79 Pt 11*, 2795–2802. [CrossRef]
8. Baaten, B.J.; Staines, K.A.; Smith, L.P.; Skinner, H.; Davison, T.F.; Butter, C. Early replication in pulmonary B cells after infection with Marek's disease herpesvirus by the respiratory route. *Viral Immunol.* **2009**, *22*, 431–444. [CrossRef]
9. Shek, W.; Calnek, B.; Schat, K.; Chen, C. Characterization of Marek's disease virus-infected lymphocytes: Discrimination between cytolytically and latently infected cells. *J. Natl. Cancer Inst.* **1983**, *70*, 485–491.
10. Bertzbach, L.D.; Laparidou, M.; Hartle, S.; Etches, R.J.; Kaspers, B.; Schusser, B.; Kaufer, B.B. Unraveling the role of B cells in the pathogenesis of an oncogenic avian herpesvirus. *Proc. Natl. Acad. Sci. USA* **2018**, *115*, 11603–11607. [CrossRef]
11. Calnek, B.W.; Schat, K.A.; Ross, L.J.; Shek, W.R.; Chen, C.L. Further characterization of Marek's disease virus-infected lymphocytes. I. In vivo infection. *Int. J. Cancer* **1984**, *33*, 389–398. [CrossRef]
12. Tulman, E.R.; Afonso, C.L.; Lu, Z.; Zsak, L.; Rock, D.L.; Kutish, G.F. The genome of a very virulent Marek's disease virus. *J. Virol.* **2000**, *74*, 7980–7988. [CrossRef]
13. Davison, A.J. Comparative analysis of the genomes. In *Human Herpesviruses: Biology, Therapy, and Immunoprophylaxis*; Arvin, A., Campadelli-Fiume, G., Mocarski, E., Moore, P.S., Roizman, B., Whitley, R., Yamanishi, K., Eds.; Cambridge University Press: Cambridge, UK, 2007.
14. Bertzbach, L.D.; Kheimar, A.; Ali, F.A.Z.; Kaufer, B.B. Viral Factors Involved in Marek's Disease Virus (MDV) Pathogenesis. *Curr. Clin. Microbiol. Rep.* **2018**, *5*, 238–244. [CrossRef]
15. Heidari, M.; Huebner, M.; Kireev, D.; Silva, R.F. Transcriptional profiling of Marek's disease virus genes during cytolytic and latent infection. *Virus Genes* **2008**, *36*, 383–392. [CrossRef]
16. Neerukonda, S.N.; Tavlarides-Hontz, P.; McCarthy, F.; Pendarvis, K.; Parcells, M.S. Comparison of the Transcriptomes and Proteomes of Serum Exosomes from Marek's Disease Virus-Vaccinated and Protected and Lymphoma-Bearing Chickens. *Genes* **2019**, *10*, 116. [CrossRef]
17. Mwangi, W.N.; Vasoya, D.; Kgosana, L.B.; Watson, M.; Nair, V. Differentially expressed genes during spontaneous lytic switch of Marek's disease virus in lymphoblastoid cell lines determined by global gene expression profiling. *J. Gen. Virol.* **2017**, *98*, 779–790. [CrossRef]

18. Baigent, S.J.; Ross, L.J.; Davison, T.F. A flow cytometric method for identifying Marek's disease virus pp38 expression in lymphocyte subpopulations. *Avian Pathol.* **1996**, *25*, 255–267. [CrossRef]
19. Schermuly, J.; Greco, A.; Hartle, S.; Osterrieder, N.; Kaufer, B.B.; Kaspers, B. In vitro model for lytic replication, latency, and transformation of an oncogenic alphaherpesvirus. *Proc. Natl. Acad. Sci. USA* **2015**, *112*, 7279–7284. [CrossRef]
20. Schat, K.; Purchase, H. *Cell-Culture Methods. A Laboratory Manual for the Isolation and Identification of Avian Pathogens*; American Association of Avian Pathologists: Kennett Square, PA, USA, 1998.
21. Martin, A.; Lillehoj, H.S.; Kaspers, B.; Bacon, L.D. Antigen-specific T cell proliferation following coccidia infection. *Poult Sci.* **1993**, *72*, 2084–2094. [CrossRef]
22. Kothlow, S.; Morgenroth, I.; Tregaskes, C.A.; Kaspers, B.; Young, J.R. CD40 ligand supports the long-term maintenance and differentiation of chicken B cells in culture. *Dev. Comp. Immunol.* **2008**, *32*, 1015–1026. [CrossRef]
23. Engel, A.T.; Selvaraj, R.K.; Kamil, J.P.; Osterrieder, N.; Kaufer, B.B. Marek's disease viral interleukin-8 promotes lymphoma formation through targeted recruitment of B cells and CD4+ CD25+ T cells. *J. Virol.* **2012**, *86*, 8536–8545. [CrossRef]
24. Pfaff, F.; Hägglund, S.; Zoli, M.; Blaise-Boisseau, S.; Laloy, E.; Koethe, S.; Zühlke, D.; Riedel, K.; Zientara, S.; Bakkali-Kassimi, L.; et al. Proteogenomics Uncovers Critical Elements of Host Response in Bovine Soft Palate Epithelial Cells Following In Vitro Infection with Foot-And-Mouth Disease Virus. *Viruses* **2019**, *11*, 53. [CrossRef]
25. Dobin, A.; Davis, C.A.; Schlesinger, F.; Drenkow, J.; Zaleski, C.; Jha, S.; Batut, P.; Chaisson, M.; Gingeras, T.R. STAR: Ultrafast universal RNA-seq aligner. *Bioinformatics* **2013**, *29*, 15–21. [CrossRef]
26. Lee, L.F.; Wu, P.; Sui, D.; Ren, D.; Kamil, J.; Kung, H.J.; Witter, R.L. The complete unique long sequence and the overall genomic organization of the GA strain of Marek's disease virus. *Proc. Natl. Acad. Sci. USA* **2000**, *97*, 6091–6096. [CrossRef]
27. Davison, A.J. Herpesvirus systematics. *Vet. Microbiol.* **2010**, *143*, 52–69. [CrossRef]
28. Li, H.; Handsaker, B.; Wysoker, A.; Fennell, T.; Ruan, J.; Homer, N.; Marth, G.; Abecasis, G.; Durbin, R. 1000 Genome Project Data Processing Subgroup. The Sequence Alignment/Map format and SAMtools. *Bioinformatics* **2009**, *25*, 2078–2079. [CrossRef]
29. Quinlan, A.R.; Hall, I.M. BEDTools: A flexible suite of utilities for comparing genomic features. *Bioinformatics* **2010**, *26*, 841–842. [CrossRef]
30. R Core Team. *R: A Language and Environment for Statistical Computing*; R Core Team: Vienna, Austria, 2017.
31. Wagih, O. ggseqlogo: A versatile R package for drawing sequence logos. *Bioinformatics* **2017**, *33*, 3645–3647. [CrossRef]
32. Bonfert, T.; Kirner, E.; Csaba, G.; Zimmer, R.; Friedel, C.C. ContextMap 2: Fast and accurate context-based RNA-seq mapping. *BMC Bioinform.* **2015**, *16*, 122. [CrossRef]
33. Langmead, B.; Salzberg, S.L. Fast gapped-read alignment with Bowtie 2. *Nat. Methods* **2012**, *9*, 357–359. [CrossRef]
34. Pauws, E.; van Kampen, A.H.; van de Graaf, S.A.; de Vijlder, J.J.; Ris-Stalpers, C. Heterogeneity in polyadenylation cleavage sites in mammalian mRNA sequences: Implications for SAGE analysis. *Nucl. Acids Res.* **2001**, *29*, 1690–1694. [CrossRef]
35. Bailey, T.L.; Boden, M.; Buske, F.A.; Frith, M.; Grant, C.E.; Clementi, L.; Ren, J.; Li, W.W.; Noble, W.S. MEME SUITE: Tools for motif discovery and searching. *Nucl. Acids Res.* **2009**, *37*, W202–W208. [CrossRef]
36. Grant, C.E.; Bailey, T.L.; Noble, W.S. FIMO: Scanning for occurrences of a given motif. *Bioinformatics* **2011**, *27*, 1017–1018. [CrossRef]
37. Patro, R.; Duggal, G.; Love, M.I.; Irizarry, R.A.; Kingsford, C. Salmon provides fast and bias-aware quantification of transcript expression. *Nat. Methods* **2017**, *14*, 417–419. [CrossRef]
38. Love, M.I.; Huber, W.; Anders, S. Moderated estimation of fold change and dispersion for RNA-seq data with DESeq2. *Genome Biol.* **2014**, *15*, 550. [CrossRef]
39. Blom, N.; Gammeltoft, S.; Brunak, S. Sequence and structure-based prediction of eukaryotic protein phosphorylation sites. *J. Mol. Biol.* **1999**, *294*, 1351–1362. [CrossRef]
40. Szilagyi, A.; Skolnick, J. Efficient prediction of nucleic acid binding function from low-resolution protein structures. *J. Mol. Biol.* **2006**, *358*, 922–933. [CrossRef]

41. Neuhoff, V.; Arold, N.; Taube, D.; Ehrhardt, W. Improved staining of proteins in polyacrylamide gels including isoelectric focusing gels with clear background at nanogram sensitivity using Coomassie Brilliant Blue G-250 and R-250. *Electrophoresis* **1988**, *9*, 255–262. [CrossRef]
42. Laemmli, U.K. Cleavage of structural proteins during the assembly of the head of bacteriophage T4. *Nature* **1970**, *227*, 680–685. [CrossRef]
43. Wisniewski, J.R.; Zougman, A.; Nagaraj, N.; Mann, M. Universal sample preparation method for proteome analysis. *Nat. Methods* **2009**, *6*, 359–362. [CrossRef]
44. Boersema, P.J.; Aye, T.T.; van Veen, T.A.; Heck, A.J.; Mohammed, S. Triplex protein quantification based on stable isotope labeling by peptide dimethylation applied to cell and tissue lysates. *Proteomics* **2008**, *8*, 4624–4632. [CrossRef]
45. Pauker, V.I.; Bertzbach, L.D.; Hohmann, A.; Kheimar, A.; Teifke, J.P.; Mettenleiter, T.C.; Karger, A.; Kaufer, B.B. Imaging Mass Spectrometry and Proteome Analysis of Marek's Disease Virus-Induced Tumors. *mSphere* **2019**, *4*. [CrossRef]
46. Henning, A.K.; Groschup, M.H.; Mettenleiter, T.C.; Karger, A. Analysis of the bovine plasma proteome by matrix-assisted laser desorption/ionisation time-of-flight tandem mass spectrometry. *Vet. J.* **2014**, *199*, 175–180. [CrossRef]
47. Aken, B.L.; Achuthan, P.; Akanni, W.; Amode, M.R.; Bernsdorff, F.; Bhai, J.; Billis, K.; Carvalho-Silva, D.; Cummins, C.; Clapham, P.; et al. Ensembl 2017. *Nucl. Acids Res.* **2017**, *45*, D635–D642. [CrossRef]
48. Ishihama, Y.; Oda, Y.; Tabata, T.; Sato, T.; Nagasu, T.; Rappsilber, J.; Mann, M. Exponentially modified protein abundance index (emPAI) for estimation of absolute protein amount in proteomics by the number of sequenced peptides per protein. *Mol. Cell. Proteom.* **2005**, *4*, 1265–1272. [CrossRef]
49. Chakraborty, P.; Kuo, R.; Vervelde, L.; Dutia, B.M.; Kaiser, P.; Smith, J. Macrophages from Susceptible and Resistant Chicken Lines have Different Transcriptomes following Marek's Disease Virus Infection. *Genes* **2019**, *10*, 74. [CrossRef]
50. Tahiri-Alaoui, A.; Smith, L.P.; Kgosana, L.; Petherbridge, L.J.; Nair, V. Identification of a neurovirulence factor from Marek's disease virus. *Avian Dis.* **2013**, *57*, 387–394. [CrossRef] [PubMed]
51. Hong, Y.; Coussens, P.M. Identification of an immediate-early gene in the Marek's disease virus long internal repeat region which encodes a unique 14-kilodalton polypeptide. *J. Virol.* **1994**, *68*, 3593–3603. [PubMed]
52. Tahiri-Alaoui, A.; Smith, L.P.; Baigent, S.; Kgosana, L.; Petherbridge, L.J.; Lambeth, L.S.; James, W.; Nair, V. Identification of an intercistronic internal ribosome entry site in a Marek's disease virus immediate-early gene. *J. Virol.* **2009**, *83*, 5846–5853. [CrossRef]
53. Hearn, C.; Preeyanon, L.; Hunt, H.D.; York, I.A. An MHC class I immune evasion gene of Mareks disease virus. *Virology* **2015**, *475*, 88–95. [CrossRef]
54. Becker, Y.; Asher, Y.; Tabor, E.; Davidson, I.; Malkinson, M. Open reading frames in a 4556 nucleotide sequence within MDV-1 BamHI-D DNA fragment: Evidence for splicing of mRNA from a new viral glycoprotein gene. *Virus Genes* **1994**, *8*, 55–69. [CrossRef]
55. Schippers, T.; Jarosinski, K.; Osterrieder, N. The ORF012 gene of Marek's disease virus type 1 produces a spliced transcript and encodes a novel nuclear phosphoprotein essential for virus growth. *J. Virol.* **2015**, *89*, 1348–1363. [CrossRef] [PubMed]
56. Jarosinski, K.W.; Osterrieder, N. Marek's disease virus expresses multiple UL44 (gC) variants through mRNA splicing that are all required for efficient horizontal transmission. *J. Virol.* **2012**, *86*, 7896–7906. [CrossRef] [PubMed]
57. Liu, J.L.; Lin, S.F.; Xia, L.; Brunovskis, P.; Li, D.; Davidson, I.; Lee, L.F.; Kung, H.J. MEQ and V-IL8: Cellular genes in disguise? *Acta Virol.* **1999**, *43*, 94–101. [PubMed]
58. Tombacz, D.; Csabai, Z.; Szucs, A.; Balazs, Z.; Moldovan, N.; Sharon, D.; Snyder, M.; Boldogkoi, Z. Long-Read Isoform Sequencing Reveals a Hidden Complexity of the Transcriptional Landscape of Herpes Simplex Virus Type 1. *Front. Microbiol.* **2017**, *8*, 1079. [CrossRef] [PubMed]
59. Gimeno, I.; Silva, R.F. Deletion of the Marek's disease virus UL41 gene (vhs) has no measurable effect on latency or pathogenesis. *Virus Genes* **2008**, *36*, 499–507. [CrossRef] [PubMed]
60. Sadek, J.; Read, G.S. The Splicing History of an mRNA Affects Its Level of Translation and Sensitivity to Cleavage by the Virion Host Shutoff Endonuclease during Herpes Simplex Virus Infections. *J. Virol.* **2016**, *90*, 10844–10856. [CrossRef] [PubMed]
61. Mandel, C.R.; Bai, Y.; Tong, L. Protein factors in pre-mRNA 3′-end processing. *Cell. Mol. Life Sci.* **2008**, *65*, 1099–1122. [CrossRef]

62. Proudfoot, N.J. Ending the message: Poly(A) signals then and now. *Genes Dev.* **2011**, *25*, 1770–1782. [CrossRef]
63. Kamil, J.P.; Tischer, B.K.; Trapp, S.; Nair, V.K.; Osterrieder, N.; Kung, H.J. vLIP, a viral lipase homologue, is a virulence factor of Marek's disease virus. *J. Virol.* **2005**, *79*, 6984–6996. [CrossRef]
64. Hardy, W.R.; Sandri-Goldin, R.M. Herpes-Simplex Virus Inhibits Host-Cell Splicing, and Regulatory Protein Icp27 Is Required for This Effect. *J. Virol.* **1994**, *68*, 7790–7799. [PubMed]
65. Zhu, H.; Li, H.; Han, Z.; Shao, Y.; Wang, Y.; Kong, X. Identification of a spliced gene from duck enteritis virus encoding a protein homologous to UL15 of herpes simplex virus 1. *Virol. J.* **2011**, *8*, 156. [CrossRef] [PubMed]
66. Zhu, G.S.; Iwata, A.; Gong, M.; Ueda, S.; Hirai, K. Marek's disease virus type 1-specific phosphorylated proteins pp38 and pp24 with common amino acid termini are encoded from the opposite junction regions between the long unique and inverted repeat sequences of viral genome. *Virology* **1994**, *200*, 816–820. [CrossRef]
67. Lv, H.; Zhang, Y.; Sun, G.; Bao, K.; Gao, Y.; Qi, X.; Cui, H.; Wang, Y.; Li, K.; Gao, L.; et al. Genetic evolution of Gallid herpesvirus 2 isolated in China. *Infect. Genet. Evol.* **2017**, *51*, 263–274. [CrossRef] [PubMed]
68. Morgan, R.W.; Xie, Q.; Cantello, J.L.; Miles, A.M.; Bernberg, E.L.; Kent, J.; Anderson, A. Marek's disease virus latency. *Curr. Top. Microbiol. Immunol.* **2001**, *255*, 223–243.
69. Cantello, J.L.; Parcells, M.S.; Anderson, A.S.; Morgan, R.W. Marek's disease virus latency-associated transcripts belong to a family of spliced RNAs that are antisense to the ICP4 homolog gene. *J. Virol.* **1997**, *71*, 1353–1361. [PubMed]
70. Burnside, J.; Bernberg, E.; Anderson, A.; Lu, C.; Meyers, B.C.; Green, P.J.; Jain, N.; Isaacs, G.; Morgan, R.W. Marek's disease virus encodes MicroRNAs that map to meq and the latency-associated transcript. *J. Virol.* **2006**, *80*, 8778–8786. [CrossRef] [PubMed]
71. Yao, Y.; Zhao, Y.; Xu, H.; Smith, L.P.; Lawrie, C.H.; Watson, M.; Nair, V. MicroRNA profile of Marek's disease virus-transformed T-cell line MSB-1: Predominance of virus-encoded microRNAs. *J. Virol.* **2008**, *82*, 4007–4015. [CrossRef]
72. Jarosinski, K.W.; Schat, K.A. Multiple alternative splicing to exons II and III of viral interleukin-8 (vIL-8) in the Marek's disease virus genome: The importance of vIL-8 exon I. *Virus Genes* **2007**, *34*, 9–22. [CrossRef]
73. Anobile, J.M.; Arumugaswami, V.; Downs, D.; Czymmek, K.; Parcells, M.; Schmidt, C.J. Nuclear localization and dynamic properties of the Marek's disease virus oncogene products Meq and Meq/vIL8. *J. Virol.* **2006**, *80*, 1160–1166. [CrossRef]
74. Okada, T.; Takagi, M.; Murata, S.; Onuma, M.; Ohashi, K. Identification and characterization of a novel spliced form of the meq transcript in lymphoblastoid cell lines derived from Marek's disease tumours. *J. Gen. Virol.* **2007**, *88*, 2111–2120. [CrossRef]
75. Trapp, S.; Parcells, M.S.; Kamil, J.P.; Schumacher, D.; Tischer, B.K.; Kumar, P.M.; Nair, V.K.; Osterrieder, N. A virus-encoded telomerase RNA promotes malignant T cell lymphomagenesis. *J. Exp. Med.* **2006**, *203*, 1307–1317. [CrossRef] [PubMed]
76. Tycowski, K.T.; Guo, Y.E.; Lee, N.; Moss, W.N.; Vallery, T.K.; Xie, M.; Steitz, J.A. Viral noncoding RNAs: More surprises. *Genes Dev.* **2015**, *29*, 567–584. [CrossRef] [PubMed]
77. Zhuang, G.; Sun, A.; Teng, M.; Luo, J. A Tiny RNA that Packs a Big Punch: The Critical Role of a Viral miR-155 Ortholog in Lymphomagenesis in Marek's Disease. *Front. Microbiol.* **2017**, *8*, 1169. [CrossRef] [PubMed]
78. Figueroa, T.; Boumart, I.; Coupeau, D.; Rasschaert, D. Hyperediting by ADAR1 of a new herpesvirus lncRNA during the lytic phase of the oncogenic Marek's disease virus. *J. Gen. Virol.* **2016**, *97*, 2973–2988. [CrossRef]
79. Wyler, E.; Menegatti, J.; Franke, V.; Kocks, C.; Boltengagen, A.; Hennig, T.; Theil, K.; Rutkowski, A.; Ferrai, C.; Baer, L.; et al. Widespread activation of antisense transcription of the host genome during herpes simplex virus 1 infection. *Genome Biol.* **2017**, *18*, 209. [CrossRef] [PubMed]
80. Greninger, A.L.; Knudsen, G.M.; Roychoudhury, P.; Hanson, D.J.; Sedlak, R.H.; Xie, H.; Guan, J.; Nguyen, T.; Peddu, V.; Boeckh, M.; et al. Comparative genomic, transcriptomic, and proteomic reannotation of human herpesvirus 6. *BMC Genom.* **2018**, *19*, 204. [CrossRef]
81. Olah, P.; Tombacz, D.; Poka, N.; Csabai, Z.; Prazsak, I.; Boldogkoi, Z. Characterization of pseudorabies virus transcriptome by Illumina sequencing. *BMC Microbiol.* **2015**, *15*, 130. [CrossRef] [PubMed]
82. Shabalina, S.A.; Ogurtsov, A.Y.; Spiridonov, N.A.; Koonin, E.V. Evolution at protein ends: Major contribution of alternative transcription initiation and termination to the transcriptome and proteome diversity in mammals. *Nucl. Acids Res.* **2014**, *42*, 7132–7144. [CrossRef] [PubMed]

5

Recent Advances in Organic Piezoelectric Biomaterials for Energy and Biomedical Applications

Dong-Myeong Shin [1,*], Suck Won Hong [2] and Yoon-Hwae Hwang [3]

1 Department of Mechanical Engineering, The University of Hong Kong, Hong Kong 999077, China
2 Department of Cogno-Mechatronics Engineering, Department of Optics and Mechatronics Engineering, Pusan National University (PNU), Busan 46241, Korea; swhong@pusan.ac.kr
3 Department of Nanoenergy Engineering & BK21 PLUS Nanoconvergence Technology Division, Pusan National University (PNU), Busan 46241, Korea; yhwang@pusan.ac.kr
* Correspondence: dmshin@hku.hk

Abstract: The past decade has witnessed significant advances in medically implantable and wearable devices technologies as a promising personal healthcare platform. Organic piezoelectric biomaterials have attracted widespread attention as the functional materials in the biomedical devices due to their advantages of excellent biocompatibility and environmental friendliness. Biomedical devices featuring the biocompatible piezoelectric materials involve energy harvesting devices, sensors, and scaffolds for cell and tissue engineering. This paper offers a comprehensive review of the principles, properties, and applications of organic piezoelectric biomaterials. How to tackle issues relating to the better integration of the organic piezoelectric biomaterials into the biomedical devices is discussed. Further developments in biocompatible piezoelectric materials can spark a new age in the field of biomedical technologies.

Keywords: piezoelectric materials; organic materials; biomaterials; energy applications; biomedical applications

---

1. Introduction

Piezoelectric materials are a class of solid materials that can accumulate an electric charge in response to applied mechanical agitation, facilitating the conversion from mechanical energy to electrical energy and vice versa. Piezoelectricity has been found in both organic and inorganic materials, where the physical principles of piezoelectricity are varied upon material classification. In inorganic piezoelectric materials, the piezoelectric effect arises from the rearrangement of ions in the dielectric materials that possess a lack of inversion symmetry in crystalline structure [1]. In contrast, the reorientation of molecular dipole mainly induces polarization in organic piezoelectric materials under applied mechanical stress [2,3]. These materials have taken over the entire market of electromechanical devices, such as sensors [4–6], actuators [7], energy harvesting [8–10] and storage [11,12]. Recently, medically implantable and mountable devices have attracted considerable attention [13,14], and are the newly emerging applications for piezoelectric materials.

Organic piezoelectric biomaterials offer several benefits over inorganic piezoelectric materials, which include a high biocompatibility, excellent flexibility, environmental friendliness, and a high level of processability. Ever since the discovery of polarization in asymmetric biological tissue in 1941 [15], many researchers have looked not only to unveil the primary principle underlying the piezoelectricity of those materials, but also to enhance its physical and chemical properties by designing a molecular structure, nanostructuring, and adding dopants [2,16]. Although organic piezoelectric biomaterials exhibit weak piezoelectricity compared to inorganic counterparts, recent research suggests that biocompatible piezoelectric materials, which are interfaced with the biological system of human beings, can serve as the functional materials in the field of medically implantable and mountable applications when they are well-processed. Organic piezoelectric materials are applicable in broad range of devices, including nano- to millimeter-scaled devices, so there might be some challenges in the device fabrication due to local damage and nonlocal elasticity [17,18]. However, our manuscript will only detail views on organic piezoelectric materials.

The rapid development in organic piezoelectric biomaterials calls for a comprehensive review that can provide a useful reference for researchers in relevant fields. Herein, we provide a thorough review of organic piezoelectric biomaterials that are used in energy and biomedical applications. We review the working principle and properties of the different types of organic piezoelectric biomaterials. Efforts to improve the piezoelectric performance of each materials are discussed. The applications of these materials are introduced in terms of energy harvesting, sensor, and cell and tissue regeneration. Meanwhile, the challenges that need to be addressed for practical application are also presented.

2. Mechanism of Piezoelectricity in Biomaterials

Piezoelectricity in organic biomaterials mainly originated from the reorientation of the molecular dipole [2,3] while the breaking of structural symmetry in crystal lattices results in piezoelectricity in traditional inorganic materials [1]. As a piezoelectric biomaterial is deformed under stress, the molecular chains with a permanent dipole in the material are aligned along one direction, yielding or changing in net polarization so that it is able to represent the piezoelectric behavior regardless of the absence of non-centrosymmetry. Therefore, the piezoelectric biomaterials are required to possess the presence of permanent molecular dipoles, the ability to orient the molecular dipoles, and the ability to maintain the dipole alignment [19]. The following sections provide detail on piezoelectric mechanisms and the properties of these materials, including the proteins, peptides, and biopolymers, and the piezoelectric constants summarized in Table 1.

2.1. *Piezoelectric Proteins*

It is well known that the collagen, being a main structural protein in the extracellular matrix in the tissues, mainly causes the piezoelectric effect in bone, but its clear fundamental principle has not yet been discovered. Several hypotheses have been established to elucidate the origin of piezoelectricity in the collagen fibril [20], which involve the noncentrosymmetric structure, the existence of polar bonding at the molecular level, reorientation of the C=O–NH bond in the α-helix structure, and the polarization of hydrogen bonds in collagen [21–24]. A recent study reveals that the piezoelectric effect in collagen comes from the reorientation of, and a magnitude change, in the permanent dipoles of individual charged and polar residues towards the long axis of the collagen fibril [20], as shown in Figure 1a. The shear piezoelectric constant reported is varied from d_{14} = 0.2 pC/N to d_{14} = 2.0 pC/N for collagens impregnated by bone and tendon [25], respectively.

The M13 bacteriophage is recently emerging as the functional material for multiple applications in energy harvesting [26–28], chemical sensor [29–31], and tissue regeneration [32,33]. It is a filamentous bacterial virus with the well-defined dimension of 880 nm in length and 6.6 nm in width, consisting of single-stranded DNA which is wrapped up with 2700 copies of major proteins (pVIII) and lidded with five copies of minor proteins (pIII/pVI or pVIII/pIX) on the top or bottom ends, respectively. The major protein has an ~20° tilt angle with respect to the DNA axial direction, and is arranged with a combined five-fold rotational and two-fold screw symmetry. Each major protein holds a dipole moment directed toward the DNA axis, leading to permanent polarization in both the axial and radial direction of the phage (Figure 1b). The piezoelectric constant for radial direction is d_{33} = ~7.8 pm/V [26], and the constant for axial direction is improved by up to around three times as high as that of radial direction [27]. Recently, Lee et al. improved the value for axial direction up to d_{33} = ~26.4 pm/V by unidirectionally sticking the M13 bacteriophages [28].

2.2. Piezoelectric Peptides

Glycine (G) is a zwitterionic amino acid and a good model building block for investigating the process of polymorphic crystallization [34–36]. At ambient conditions, the crystalline glycines have three distinct structures, α-, β-, and γ-structures, where the β- and γ-structures have shear piezoelectricity due to their acentric structures [37–40]. Herein, the piezoelectricity arises from the displacement of ion in the crystal, as it does in the inorganic materials, and such displacement creates a dipole in local, and a net polarization in bulk, material, as shown in Figure 1c. The β-structured glycine has a high shear piezoelectric constant, d_{16} = ~190 pm/V [41], which is comparable to the normal piezoelectric constant of barium titanate ($BaTiO_3$) [42].

Diphenylalanine (FF) is composed of two phenylalanine (F) amino acids and can be self-assembled into semi-crystalline peptide nanotubes and microrods, exhibiting multiple advantages including morphological diversity, functional diversity, high biocompatibility, and a high Young's modulus [16,43]. The nanostructured diphenylalanines are widely studied piezoelectric materials that have a non-centrosymmetric hexagonal space group ($P6_1$) [44]. This crystalline class serves to demonstrate their diverse physical effects including piezoelectricity, second harmonic generation, optical activity, pyroelectricity, ferroelectricity, and enantiomorphism [45] (Figure 1d). The peptide nanotubes have achieved a high shear piezoelectric constant up to d_{15} = ~60 pm/V [46]. In order to improve the scalability as well as the uniformity of semi-crystalline film, the unidirectionally polarized and aligned diphenylalanine nanotubes films were fabricated using the meniscus-driven self-assembly process [47], representing a similar value of d_{15} = ~45 pm/V to that of the highly crystalline structure. In addition, Nguyen et al. have devoted to obtaining the normal piezoelectricity (reached up to d_{33} = ~17.9 pm/V) of the peptide nanostructures by the vertical alignment of individual microrods [48,49].

2.3. Other Piezoelectric Biopolymers

The poly(vinylidene fluoride) (PVDF) films have shown one of the highest piezoelectric performances among all piezoelectric polymers found to date [50]. The PVDF has five different crystalline structures, α-, β-, γ-, δ-, and ε-structures where the β-structured PVDF has a normal piezoelectricity of $d_{33} = -33$ pC/N [51]. A dipole moment is induced perpendicular to the polymer chain in each unit of PVDF due to the presence of a branched fluorine atom with a large van der Waals radius together with the electronegativity [52–54]. In the β-phase, the orthorhombic crystal structure, with aligned fluorine and hydrogen branches parallel to each other, contributes to the net dipole moment and piezoelectricity [55], whereas the net dipole moment is cancelled out by the anti-parallel alignment of the dipole in the α-phase hexagonal structure [56], as depicted in Figure 1e. The nature of the negative piezoelectric effect is related to the redistribution of the electron molecular orbitals and total charges under an electrical field applied [57]. The copolymer approaches have been applied to enhance the piezoelectric constant by conjugating with trifuoroethylene (TrFE) [58,59], hexafluoropropylene (HPF) [60,61], or chlorotrifluoroethylene (CTFE) [62].

The poly(*L*-lactic acid) (PLLA) is a polymorphic polymer with excellent biodegradability and biocompatibility. The thermodynamically stable conformation is the α-crystalline structure where the dipoles introduced by carbonyl groups (C=O) are not aligned along the main polymer chain. The external stimuli, such as electrospinning (Figure 1f), allows for the dipoles to be unidirectionally oriented along the stretched direction [63], which is termed the β-crystalline structure, resulting in the shear piezoelectricity of d_{14} = 12 pC/N [64,65]. It is worth to note that the β-crystalline PLLA, along with decent piezoelectricity, requires no polling process due to its helical structure [3], widening the application area in the biocompatible mobile devices [66].

Natural polymers are gaining more importance owing to their biocompatibility, as recent research is aiming at the investigation of the utility of piezoelectric materials for body implantable or mountable devices. The piezoelectricity of silk arises from the combined effects of a high degree of silk II, β-sheet crystallinity, and a crystalline orientation [67]. The reported piezoelectric constant is

$d_{14} = -1.5$ pC/N [67]. Cellulose, which is the most abundant natural polymer on earth, is also known to have a shear piezoelectricity of $d_{14} = 0.2$ pC/N [68].

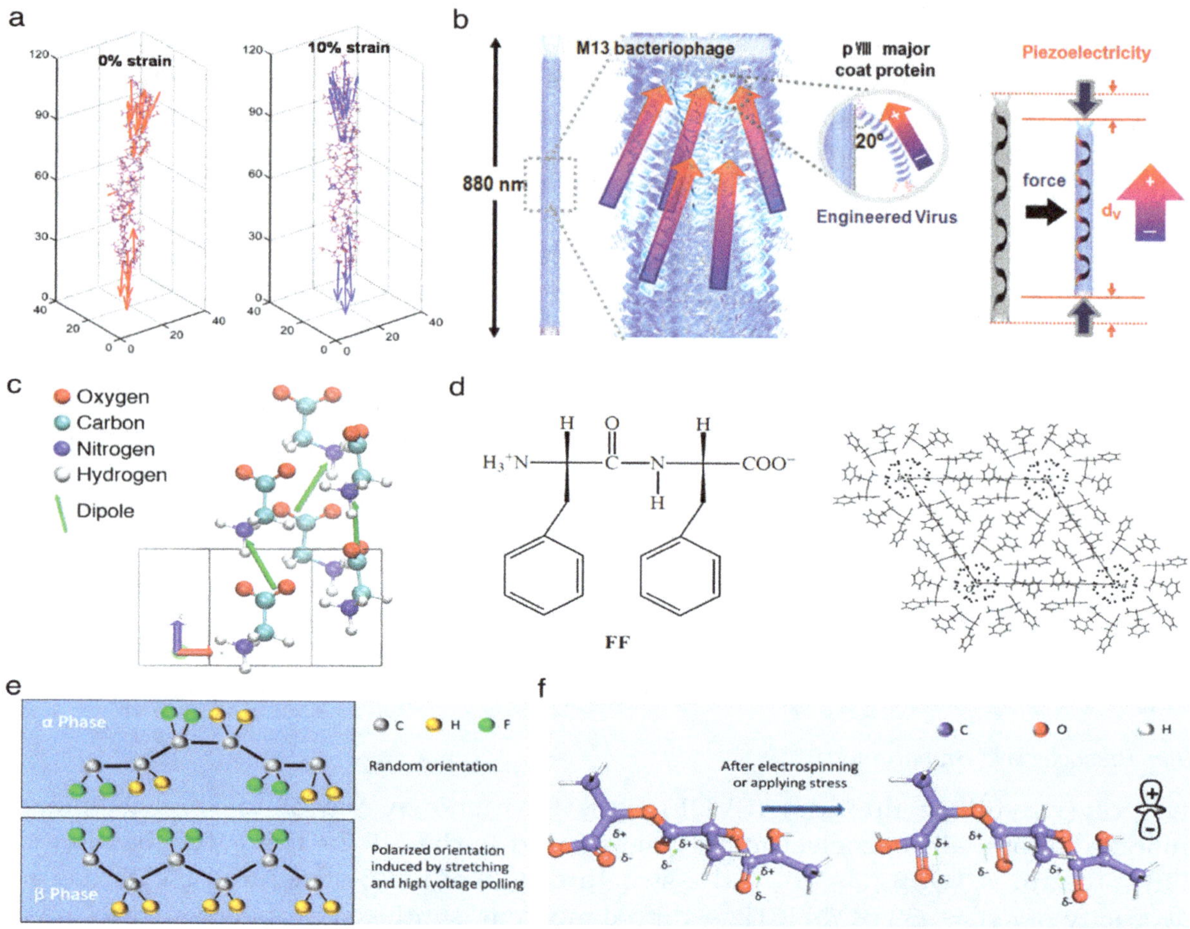

Figure 1. Piezoelectricity in the organic piezoelectric biomaterials. (**a**) Molecular origin of the piezoelectric effect in collagen. Reproduced with permission from [20]. Copyright American Chemical Society, 2016. (**b**) Schematic illustration of piezoelectric M13 bacteriophage. Reproduced with permission from [27]. Copyright The Royal Society of Chemistry, 2015. (**c**) Unit cell of β-glycine crystal has two molecules, where two molecular dipole moments form the net dipole moment along the *z*-axis. Reproduced with permission from [40]. Copyright Nature Publishing Group, 2019. (**d**) Unit cell and molecular packing of diphenylalanine. Reproduced with permission from [44]. Copyright Wiley–VCH, 2001. (**e**) Structures of non-piezoelectric (α-phase) and piezoelectric (β-phase) poly(vinylidene fluoride) (PVDF). Reproduced with permission from [2]. Copyright Wiley–VCH, 2018. (**f**) Molecular structure of poly(L-lactic acid) (PLLA) chain. Reproduced with permission from [2]. Copyright Wiley–VCH, 2018.

Table 1. Comparison of piezoelectric constants for various organic piezoelectric biomaterials.

Piezoelectric Organic Biomaterials	Piezoelectric Constant		References
	Normal Piezoelectric	Shear Piezoelectric	
Collagen	-	d_{14} = 0.2–2.0 pC/N	[25]
M13 bacteriophage	d_{33} = 7.8–26.4 pm/V		[26–28]
Glycine	-	d_{16} = ~190 pm/V	[41]
Diphenylalanine	d_{33} = ~17.9 pm/V	d_{15} = 45–60 pm/V	[46–49]
PVDF	d_{33} = −3 pC/N d_{31} = 23 pC/N	-	[51]
PVDF–TrFE	d_{33} = −25 to −40 pC/N d_{31} = 12–25 pC/N	-	[52,58,59]
PVDF–HPF	d_{33} = −24 pC/N d_{31} = 30 pC/N	-	[60,61]
PVDF–CTFE	d_{33} = −140 pC/N	-	[62]
PLLA	-	d_{14} = 12 pC/N	[64,65]
Silk	-	d_{14} = −1.5 pC/N	[67]
Cellulose	-	d_{14} = 0.2 pC/N	[68]

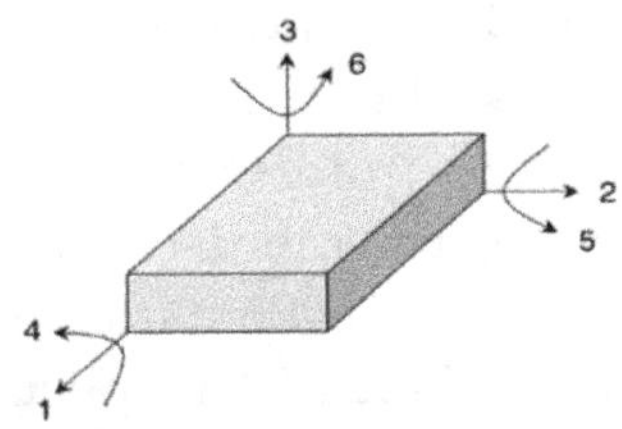

Note: 1. The piezoelectric constant d_{ij} is the ratio of the strain in the j-axis to the electric field applied along the i-axis. In other words, the i and j correspond to the response and excitation of materials, respectively. The inset image below depicts the coordination system. 2. The TrFE, HPF, and CTFE stand for trifuoroethylene, hexafluoropropylene, and chlorotrifluoroethylene, respectively.

3. Device Applications

3.1. Energy Harvesting

Self-powered electronics (SPEs) have gained an attraction as an alternative powering technology due to their independence, sustainability, and maintenance-free nature. SPEs are defined as electronics that can be operated themselves without feeding from external electrical power. In these electronics, electrical energy is provided from renewable resources, such as solar, thermal and mechanical energy. Nanogenerators that are converting mechanical to electrical energy have led the SPE field as the nanogenerators possess multiple advantages, including easy fabrication, portability, and high conversion efficiency, over conventional renewable energy technologies. Ever since the discovery of the piezoelectric and triboelectric nanogenerators (PENG and TENG, respectively) [69,70], their rapid development has shifted the paradigm for mechanical energy harvesting from fast and periodic energy resources to slow and random energy resources. A number of publications have tried over the last decade to develop the various nanogenerators possessing novel device structures and advanced materials [27,71–76]. Organic piezoelectric biomaterials have taken on an important role as the functional materials for applications in humans, being implantable and mountable nanogenerators due to their remarkable biodegradability and biocompatibility.

Vivekananthan et al. [77] reported a piezoelectric collagen nanofibril film that is capable of both converting mechanical energy to electrical energy and functioning as a humidity sensor. A schematic of the device is illustrated in Figure 2a. Collagen-based PENG produced electrical outputs of 250 nA and 45 V. In addition, it served as a humidity sensor that showed a linear response with a good sensitivity (0.1287 μA/% RH) in the range of 50–90% room humidity. These results demonstrated a field of eco-friendly multifunctional biomaterials, towards the development of noninvasive, implantable, smart bio-medical systems.

Lee et al. [26] firstly used a self-assembled M13 bacteriophage film for a piezoelectric energy harvester in 2012. The nanogenerator yields a current of up to 6 nA and a voltage of up to 400 mV. The M13 bacteriophages are expected to have a high piezoelectric response compared to laterally assembled phages, due to their high elasticity properties along the axial direction of the DNA [78]. Shin et al. [27] reported vertically aligned M13 bacteriophage nanopillars using enforced infiltration. The vertically aligned M13 bacteriophage-based nanogenerator exhibits electrical outputs that are up to about 2.6-fold greater than those of the laterally assembled, bacteriophages-based nanogenerator. There was still a limitation on the ability to control the directionality of individual M13 bacteriophage, but Lee et al. [28] addressed this issue using genetic engineering techniques (Figure 2b–d). The resulting structure-based PENG produced up to 2.8 V of potential, 120 nA of current, and 236 nW of power from 17 N of force. The apparent versatility of the M13 bacteriophage suggests that the piezoelectric M13 bacteriophages can serve as functional nanomaterials for numerous electronic and optoelectronic applications.

Piezoelectric peptide nanostructures have also been implemented into the PENGs. Nguyen et al. [49] fabricated the vertical FF microrod arrays by applying an electric field, and then the arrays were integrated into the PENGs, representing an open-circuit voltage of 1.4 V and a power density of 3.3 nW. The performance voltage of the FF-based PENG was improved up to 2.2 V in tandem with a TENG comprising the polyethylene terephthalate and Kapton films as triboelectrically active materials [79]. Recently, Lee et al. [47] developed large-scale, unidirectionally polarized, aligned FF nanotubes and fabricated peptide-based PENGs. They used the meniscus-driven self-assembly process to fabricate horizontally aligned FF nanotubes. The fabricated FF nanotubes-based PENGs can generate voltage, current, and power of up to 2.8 V, 37.4 nA, and 8.2 nW, respectively. Hence, the FF nanostructures will act as a compatible energy source for biomedical applications in the future.

The PVDF and its copolymers have been adopted for flexible PENGs due to their inherent flexibility, high processability, and mechanical rigidity [80–82]. Chang et al. [83] developed a method to directly fabricate the PVDF nanofibers with a β-crystalline structure using the near-field electrospinning process, which provides a peak current of 3 nA and a peak voltage of 30 mV after integration into the PENG. A hybrid nanogenerator, demonstrated by Hansen et al. [84], is made of a piezoelectric PVDF nanofibers-based PENG and a flexible biofuel cell, and was used for powering a single nanowire-based ultraviolet sensor to build an SPEs.

As shown in Figure 2e,f, Ishida et al. [85] demonstrated a self-powered pedometer, which consisted of a PVDF sheet, a 2 V organic circuit, and a flexible printed circuit board. This work suggested that the PVDF sheet can not only harvest the mechanical energy from footsteps but also serve as a footstep sensor. Sun et al. [86] and Xue et al. [87] envisioned a PVDF nanostructures-based PENG to harvest energy from human respiration. Persano et al. [88] demonstrated the textile-based PENG, featuring the highly aligned electrospun fibers of the PVDF–TrFE, exhibiting superior flexibility and mechanical robustness.

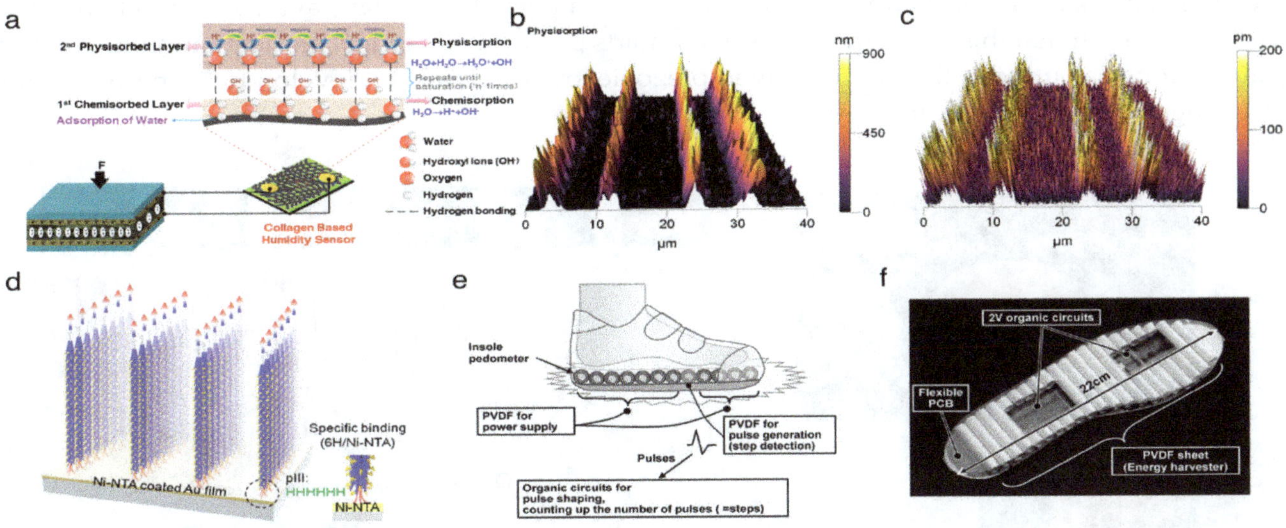

Figure 2. Applications of organic piezoelectric biomaterials in energy harvesting. (**a**) Schematic of sustainable energy harvesting and battery-free humidity sensor using biocompatible collagen nanofibrils. The collagen nanofibrils deposited on cotton cloth serve as a humidity sensor by measuring current signal at a fixed bias voltage. In order to demonstrate the self-powered sensing system, the energy harvester comprising collagen nanofibrils film sandwiched between Al electrodes is parallelly connected to the humidity sensor. Reproduced with permission from [77]. Copyright American Chemical Society, 2018. (**b–d**) Vertical self-assembly of polarized M13 bacteriophage nanostructure for energy harvesting. Reproduced with permission from [28]. Copyright American Chemical Society, 2019. (**b**) 3D-atomic force microscope (AFM) topography image of vertically aligned M13 bacteriophages. (**c**) Piezoresponse force microscope amplitude image corresponding to the 3D–AFM topography image. (**d**) The direction of polarization of the vertically aligned M13 bacteriophage with specific binding between the 6H tag on phage tail and the Ni-nitrilotriacetic acid (NTA) substrate. (**e**,**f**) Insole pedometer with piezoelectric energy harvester. Reproduced with permission from [85]. Copyright IEEE, 2012. (**e**) Schematics of the proposed insole pedometer. The pieces of PVDF sheet are used for the piezoelectric energy harvester as well as the pulse generator to detect steps in which each PVDF piece was rolled to increase the total area. The organic circuits are integrated with PVDF pieces to count the number of steps. (**f**) Photograph of the prototype insole pedometer.

3.2. Sensors

The organic biomaterials have been studied as the platform materials for biomedical pressure-sensing applications due to their high flexibility and high sensitivity to small force. The aligned PVDF–TrFE nanofibers on polyimide substrate were employed to build a flexible and lightweight pressure sensor [88]. This pressure sensor can measure small pressures down to ~0.1 Pa over the course of cyclic bending. The wearable piezoelectric PVDF sensor can serve as a healthcare monitoring device to monitor respiration signals, human gestures, and vocal cord vibrations, as demonstrated by Liu et al. [89] (Figure 3a–c). Bodkhe et al. [90] developed a pressure sensor comprised of 10% of barium titanate nanoparticle and β-crystalline phase PVDF ball mill nanocomposites using 3D printing techniques, which generated a voltage of 4 V upon gentle finger taps. The composite films of PVDF and graphene oxide developed by Park et al. [91] were used as multifunctional electronic skins to monitor multiple stimuli, including static/dynamic pressure and temperature, exhibiting a high sensitivity for monitoring simultaneous artery pulse pressures and temperature (Figure 3d,e). Recently, biodegradable and implantable sensors have gained great interest in medical applications where there is a demand for short-term functionality because biodegradable sensors are not required for the medical surgery of removal. Curry et al. [92] fabricated the piezoelectric pressure sensor featuring all the biodegradable materials of piezoelectric PLLA, molybdenum electrodes, and polylactic acid encapsulators. This device was capable of measuring a wide range of pressure, from 0 to 18 kPa. More interestingly, the sensor was completely degraded over a period of 56 days at an elevated temperature of 74 °C, indicating that this biodegradable sensor holds promise for clinical implementation. A number of publications has utilized the nature-driven piezoelectric materials to develop human physiological monitoring and electronic skins [93–96].

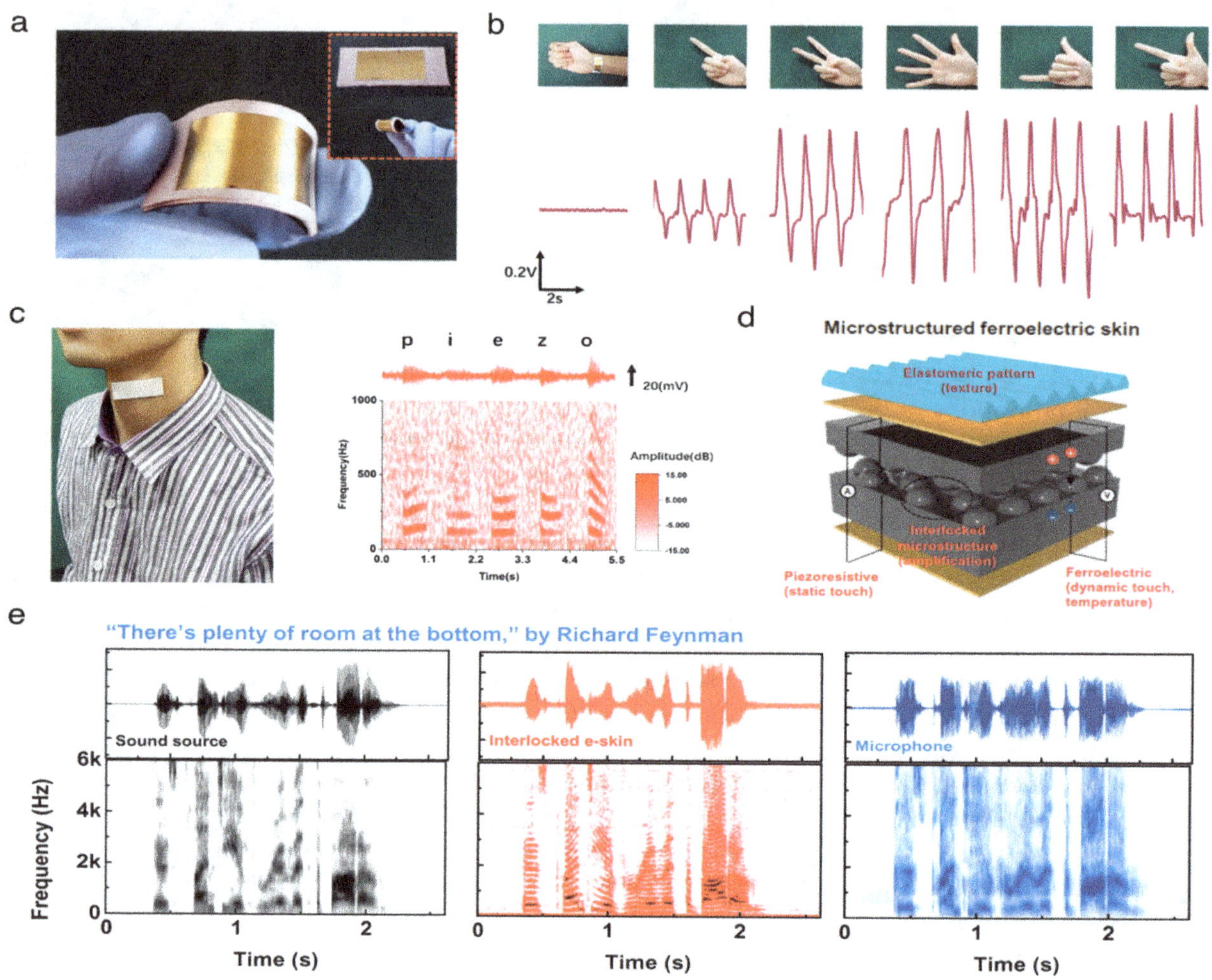

Figure 3. Applications of organic piezoelectric biomaterials in sensors. (**a–c**) Flexible piezoelectric nanogenerator in a wearable, self-powered active sensor for healthcare monitoring. Reproduced with permission from [89]. Copyright IOP Publishing, 2017. (**a**) Photographs of the flexible piezoelectric nanogenerator. (**b**) hand gesture sensing. (**c**) human voice recording. (**d,e**) Electronic skins for discriminating static/dynamic pressure stimuli. Reproduced with permission from [91]. Copyright The American Association for the Advancement of Science, 2015. (**d**) Schematic illustration of flexible and multimodal ferroelectric e-skin. (**e**) The waveform and short-time Fourier transform (STFT) signals of the sound source, readout signals from the interlocked e-skin, and microphone.

3.3. Cell and Tissue Regeneration

Organic piezoelectric biomaterials have been chosen as the functional materials for fabricating a scaffold to grow and differentiate cells in the field of tissue engineering [16]. Several studies have shown that the biocompatible piezoelectric materials can serve as tissue stimulators and scaffolds to promote tissue regeneration. Damaraju et al. [97] found that the cell growth on the β–phase PVDF nanofibers film exhibited higher alkaline phosphatase activity and earlier mineralization compared to the growth on random-phase PVDF film. Then, Damaraju et al. [98] showed that the 3D fibrous scaffolds decorated with electrospun PVDF–TrFE fibers stimulated the differentiation of human mesenchymal stem cells. The electromechanical actuation under high voltage helped osteogenic differentiation, whereas the actuation under low voltage aided chondrogenic differentiation (Figure 4a). Muscle cell adhesion and proliferation were improved due to the fact that the negatively charged β-phase PVDF fibers helped to elongate the muscle cells along the aligned fibers [99]. Hoop et al. [100] demonstrated that wireless stimulations helped to induce the potential in piezoelectric β-phase PVDF, improving the neurite generation in PC12 cells using the ultrasonic technique (Figure 4b–d). Similarly, the PVDF–TrFE fibril scaffolds promoted the differentiation of neural cells, neurite extension and neuronal differentiation due to the piezoelectric effect of the scaffolds [101]. The mechanical stimulation facilitates the enhanced bone cell culture on the piezoelectric PVDF substrate by applying a voltage of 5 V.

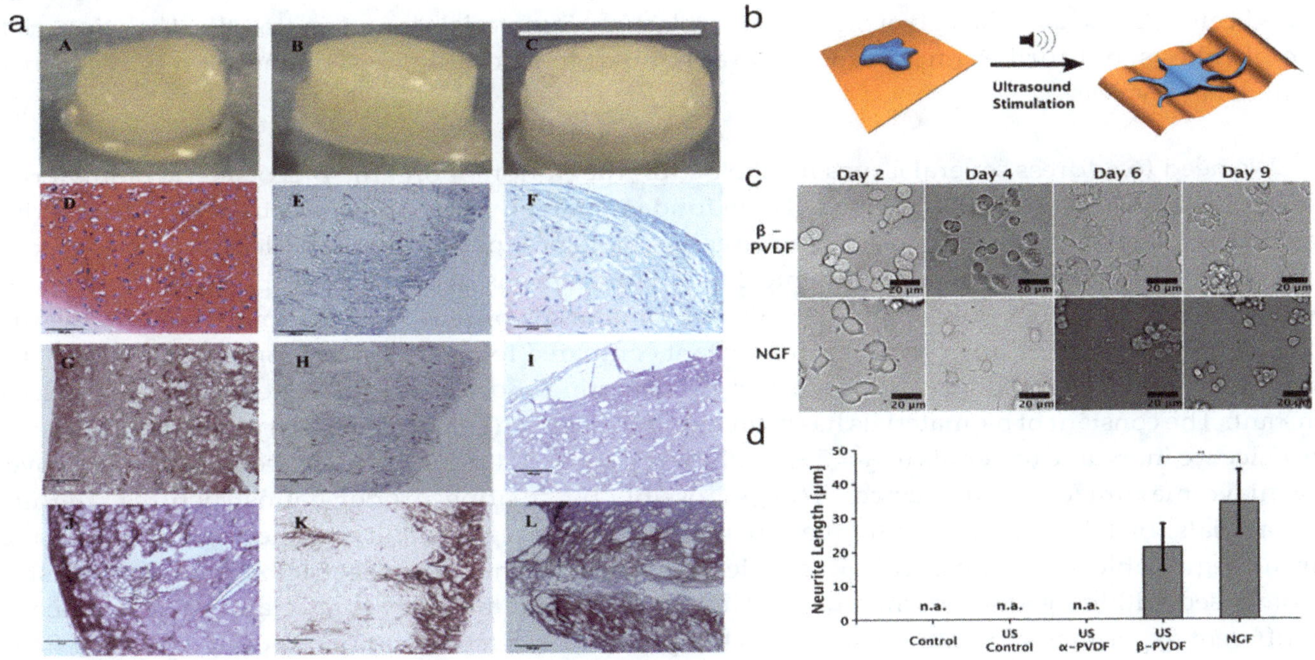

Figure 4. Applications of organic piezoelectric biomaterials in cell and tissue regenerations. (**a**) Representative gross images and histological images of scaffolds after 28 days undergoing chondrogenesis in dynamic conditions. As-spun PVDF–TrFE (left), annealed PVDF–TrFE (middle) and polycaprolactone (right) scaffolds. Reproduced with permission from [98]. Copyright Elsevier, 2017. (**b–d**) Ultrasound-mediated piezoelectric differentiation of neuron-like PC12 cells on PVDF membranes. Reproduced with permission from [100]. Copyright Nature Publishing Group, 2017. (**b**) Schematic of ultrasound stimulation of the piezoelectric β-PVDF membrane. (**c**) Comparison images of PC12 cells cultured under mechanical stimuli on PVDF substrate and neuronal growth factor stimuli. (**d**) Comparison of average neurite length of PC12 cells.

There have also been several attempts to utilize the biodegradable piezoelectric PLLA polymer as the tissue stimulator. Ikada et al. [102] intramedullary implanted PLLA rods in the cut tibiae of cats for internal fixation for up to eight weeks. The high aspect ratio of the PLLA rod enabled the enhanced fracture healing, indicated by improved callus formation, whereas the isotropic PLLA and a polyethylene control rod exhibited no effect on callus formation. Barroca et al. [103,104] observed that surface charges can change the orientation of the adsorbed proteins, resulting in the modulation of cell-binding domains. Indeed, the negatively charged PLLA improves the protein adsorption and cellular adhesion as well as proliferation.

4. Conclusions and Future Perspective

The present review has sought to offer insight into the importance of organic piezoelectric biomaterials in biomedical applications. We have reviewed the origin of piezoelectricity in organic piezoelectric biomaterials, including proteins, peptides, and biopolymers. The intrinsic piezoelectric property of those materials has been presented, and engineering and scientific endeavors to enhance these properties have also been reported. In summary, from current research, organic piezoelectric biomaterials have been likely to impact three major fields across many disciplines. First, they can serve as the functional materials for the power supply of implantable and mountable self-powered electronics because of their sensitivity to mechanical agitation and remarkable biocompatibility. Second, they have been utilized as platform materials for pressure sensing in biomedical applications, which is likely largely due to their high flexibility and high sensitivity to small forces in tandem with their biodegradability. Lastly, in cases where piezoelectric materials were integrated as the scaffolds for cell and tissue regeneration, those materials act as a tissue stimulator to promote the differentiation of the desired cells. An industry based on organic piezoelectric biomaterials is anticipated due to their variety of applications, but further improvements are required to smoothly implement them into practical biomedical devices.

We need to address several issues for the better integration of organic piezoelectric biomaterials into biomedical devices. Here are a few: (1) the fundamental physics of piezoelectricity in biomaterials. Even though researchers have focused on uncovering biological piezoelectricity, plenty of work remains to be done to exploit their electromechanical behavior in terms of unit cell properties. Such studies are in progress using not only single crystals of biomaterials but also calculations based on the first principle. (2) A relatively low piezoelectric constant compared to piezoelectric inorganic materials. The output performances in the applications of energy harvesting and sensors are related to the piezoelectric constant. The constant of biomaterials has been found to be much smaller than that of the state-of-the-art piezoelectric inorganic materials (d_{33} = 593 pC/N and d_{31} = −274 pC/N [21]), which has to be improved to achieve maximized performance. This is possible by creating proper nanostructures, aligning biomaterials, or fabricating a multilayer structure. (3) Biodegradability of the controlled manner. For biodegradable sensor and scaffold applications, the organic piezoelectric biomaterials must be decomposed within the desired time frame. The degradation rate of these materials can be engineered by different experimental treatments, such as temperature, stretching ratio, or poling electrical fields. Although researchers are still facing challenging issues, the promising physical properties of organic piezoelectric biomaterials have suggested feasible biomedical applications in energy harvesting, sensor, and tissue regeneration. We truly believe that organic piezoelectric biomaterials will continue their rapid growth in the next decade.

Author Contributions: D.-M.S., S.W.H. and Y.-H.H. formulated the project. D.-M.S. wrote the manuscript, and all authors contributed to revising the manuscript. All authors have read and agreed to the published version of the manuscript.

References

1. Li, C.; Weng, G. Antiplane crack problem in functionally graded piezoelectric materials. *J. Appl. Mech.* **2002**, *69*, 481–488. [CrossRef]
2. Chorsi, M.T.; Curry, E.J.; Chorsi, H.T.; Das, R.; Baroody, J.; Purohit, P.K.; Ilies, H.; Nguyen, T.D. Piezoelectric biomaterials for sensors and actuators. *Adv. Mater.* **2019**, *31*, 1802084. [CrossRef] [PubMed]
3. Jacob, J.; More, N.; Kalia, K.; Kapusetti, G. Piezoelectric smart biomaterials for bone and cartilage tissue engineering. *Inflamm. Regen.* **2018**, *38*, 2. [CrossRef] [PubMed]
4. Sirohi, J.; Chopra, I. Fundamental understanding of piezoelectric strain sensors. *J. Intell. Mater. Syst. Struct.* **1999**, *11*, 246–257. [CrossRef]
5. Cannata, D.; Benetti, M.; Verona, E.; Varriale, A.; Staiano, M.; D'Auria, S.; Di Pietrantonio, F. Odorant detection via Solidly Mounted Resonator biosensor. In Proceedings of the 2012 IEEE International Ultrasonics Symposium, Dresden, Germany, 7–10 October 2012; IEEE: Piscataway, NJ, USA, 2012; p. 1537.
6. Chen, D.; Wang, J.; Xu, Y. Highly sensitive lateral field excited piezoelectric film acoustic enzyme biosensor. *IEEE Sens. J.* **2013**, *13*, 2217–2222. [CrossRef]
7. Gan, J.; Zhang, X. A review of nonlinear hysteresis modeling and control of piezoelectric actuators. *AIP Adv.* **2019**, *9*, 040702. [CrossRef]

8. Dagdeviren, C.; Yang, B.D.; Su, Y.; Tran, P.L.; Joe, P.; Anderson, E.; Xia, J.; Doraiswamy, V.; Dehdashti, B.; Feng, X.; et al. Conformal piezoelectric energy harvesting and storage from motions of the heart, lung, and diaphragm. *Proc. Natl. Acad. Sci. USA* **2014**, *111*, 1927–1932. [CrossRef]
9. Hwang, G.T.; Park, H.; Lee, J.H.; Oh, S.; Park, K.I.; Byun, M.; Park, H.; Ahn, G.; Jeong, C.K.; No, K.; et al. Self-powered cardiac pacemaker enabled by flexible single crystalline PMN-PT piezoelectric energy harvester. *Adv. Mater.* **2014**, *26*, 4880–4887. [CrossRef]
10. Zhu, G.; Wang, A.C.; Liu, Y.; Zhou, Y.; Wang, Z.L. Functional electrical stimulation by nanogenerator with 58 V output voltage. *Nano Lett.* **2012**, *12*, 3086–3090. [CrossRef] [PubMed]
11. He, Y.-B.; Li, G.-R.; Wang, Z.-L.; Su, C.-Y.; Tong, Y.-X. Single-crystal zno nanorod/amorphous and nanoporous metal oxide shell composites: Controllable electrochemical synthesis and enhanced supercapacitor performances. *Energy Environ. Sci.* **2011**, *4*, 1288–1292. [CrossRef]
12. Ryu, J.; Kim, S.W.; Kang, K.; Park, C.B. Synthesis of diphenylalanine/cobalt oxide hybrid nanowires and their application to energy storage. *ACS Nano* **2011**, *4*, 159–164. [CrossRef]
13. Kim, D.H.; Lu, N.; Ma, R.; Kim, Y.S.; Kim, R.H.; Wang, S.; Wu, J.; Won, S.M.; Tao, H.; Islam, A.; et al. Epidermal electronics. *Science* **2011**, *333*, 838–843. [CrossRef]
14. Choi, C.; Lee, Y.; Cho, K.W.; Koo, J.H.; Kim, D.H. Wearable and implantable soft bioelectronics using two-dimensional materials. *Acc. Chem. Res.* **2019**, *52*, 73–81. [CrossRef]
15. Martin, A.J.P. Tribo-electricity in wool and hair. *Proc. Phys. Soc.* **1941**, *53*, 186. [CrossRef]
16. Yuan, H.; Lei, T.; Qin, Y.; He, J.H.; Yang, R. Design and application of piezoelectric biomaterials. *J. Phys. D Appl. Phys.* **2019**, *52*, 194002. [CrossRef]
17. de Sciarra, F.M. A nonlocal model with strain-based damage. *Int. J. Solids Struct.* **2009**, *46*, 4107–4122. [CrossRef]
18. Barretta, R.; Fabbrocino, F.; Luciano, R.; de Sciarra, F.M. Closed-form solutions in stress-driven two-phase integral elasticity for bending of functionally graded nano-beams. *Phys. E Low Dimens. Syst. Nanostruct.* **2018**, *97*, 13–30. [CrossRef]
19. Fousek, J.; Cross, L.; Litvin, D. Possible piezoelectric composites based on the flexoelectric effect. *Mater. Lett.* **1999**, *39*, 287. [CrossRef]
20. Zhou, Z.; Qian, D.; Minary-Jolandan, M. Molecular mechanism of polarization and piezoelectric effect in super-twisted collagen. *ACS Biomater. Sci. Eng.* **2016**, *2*, 929–936. [CrossRef]
21. Bystrov, V.; Bdikin, I.; Heredia, A.; Pullar, R.; Mishina, E.; Sigov, A.; Kholkin, A. Piezoelectricity and ferroelectricity in biomaterials: From proteins to self-assembled peptide nanotubes. In *Piezoelectric Nanomaterials for Biomedical Applications, Nanomedicine and Nanotoxicology*; Ciofani, G., Menciassi, A., Eds.; Springer: Berlin, Germany, 2012; pp. 187–211.
22. Wojnar, R. Piezoelectric phenomena in biological tissues. In *Piezoelectric Nanomaterials for Biomedical Applications, Nanomedicine and Nanotoxicology*; Ciofani, G., Menciassi, A., Eds.; Springer: Berlin, Germany, 2012; pp. 173–2011.
23. Lemanov, V.; Popov, S.; Pankova, G. Piezoelectric properties of crystals of some protein aminoacids and their related compounds. *Phys. Solid State* **2002**, *44*, 1929–1935. [CrossRef]
24. Namiki, K.; Hayakawa, R.; Wada, Y. Molecular theory of piezoelectricity of α-helical polypeptide. *J. Polym. Sci. Polym. Phys. Ed.* **1980**, *18*, 993–1004. [CrossRef]
25. Fukada, E. History and recent progress in piezoelectric polymers. *IEEE Trans. Ultrason. Ferroelectr. Freq. Control.* **2000**, *47*, 1277–1290. [CrossRef]
26. Lee, B.Y.; Zhang, J.; Zueger, C.; Chung, W.-J.; Yoo, S.Y.; Wang, E.; Meyer, J.; Ramesh, R.; Lee, S.-W. Virus-based piezoelectric energy generation. *Nat. Nanotechnol.* **2012**, *7*, 351. [CrossRef]
27. Shin, D.-M.; Han, H.J.; Kim, W.-G.; Kim, E.; Kim, C.; Hong, S.W.; Kim, H.K.; Oh, J.-W.; Hwang, Y.-H. Bioinspired piezoelectric nanogenerators based on vertically aligned phage nanopillars. *Energy Environ. Sci.* **2015**, *8*, 3198–3203. [CrossRef]
28. Lee, J.-H.; Lee, J.H.; Xiao, J.; Desai, M.S.; Zhang, X.; Lee, S.-W. Vertical self-assembly of polarized phage nanostructure for energy harvesting. *Nano Lett.* **2019**, *19*, 2661–2667. [CrossRef]
29. Moon, J.-S.; Kim, W.-G.; Shin, D.-M.; Lee, S.-Y.; Kim, C.; Lee, Y.; Han, J.; Kim, K.; Yoo, S.Y.; Oh, J.-W. Bioinspired M-13 bacteriophage-based photonic nose for differential cell recognition. *Chem. Sci.* **2017**, *8*, 921–927. [CrossRef]

30. Moon, J.-S.; Lee, Y.; Shin, D.-M.; Kim, C.; Kim, W.-G.; Park, M.; Han, J.; Song, H.; Kim, K.; Oh, J.-W. Identification of endocrine disrupting chemicals using a virus-based colorimetric sensor. *Chem. Asian J.* **2016**, *11*, 3097–3101. [CrossRef]
31. Kim, W.-G.; Zueger, C.; Kim, C.; Wong, W.; Devaraj, V.; Yoo, H.-W.; Hwang, S.; Oh, J.-W.; Lee, S.-W. Experimental and numerical evaluation of a genetically engineered M13 bacteriophage with high sensitivity and selectivity for 2,4,6-trinitrotoluene. *Org. Biomol. Chem.* **2019**, *17*, 5666–5670. [CrossRef]
32. Shin, Y.C.; Kim, C.; Song, S.-J.; Jun, S.; Kim, C.-S.; Hong, S.W.; Hyon, S.-H.; Han, D.-W.; Oh, J.-W. Ternary aligned nanofibers of RGD peptide-displaying M13 bacteriophage/PLGA/graphene oxide for facilitated myogenesis. *Nanotheranostics* **2018**, *2*, 144. [CrossRef]
33. Raja, I.S.; Kim, C.; Song, S.-J.; Shin, Y.C.; Kang, M.S.; Hyon, S.-H.; Oh, J.-W.; Han, D.-W. Virus-incorporated biomimetic nanocomposites for tissue regeneration. *Nanomaterials* **2019**, *9*, 1014. [CrossRef]
34. Chew, J.W.; Black, S.N.; Chow, P.S.; Tan, R.B.H.; Carpenter, K.J. Stable polymorphs: Difficult to make and difficult to predict. *CrystEngComm* **2007**, *9*, 128–130. [CrossRef]
35. Poornachary, S.K.; Chow, P.S.; Tan, R.B.H. Influence of solution speciation of impurities on polymorphic nucleation in glycine. *Cryst. Growth Des.* **2008**, *8*, 179–185. [CrossRef]
36. Dowling, R.; Davey, R.J.; Curtis, R.A.; Han, G.; Poornachary, S.K.; Chow, P.S.; Tan, R.B.H. Acceleration of crystal growth rates: An unexpected effect of tailor-made additives. *Chem. Commun.* **2010**, *46*, 5924–5926. [CrossRef]
37. Iitaka, Y. A new form of glycine. *Proc. Jpn. Soc.* **1954**, *30*, 109–112. [CrossRef]
38. Kumar, R.A.; Vizhi, R.E.; Vijayan, N.; Babu, D.R. Structural, dielectric and piezoelectric properties of nonlinear optical γ-glycine single crystals. *Phys. Condens. B Matter* **2011**, *406*, 2594–2600. [CrossRef]
39. Iitaka, Y. The crystal structure of γ-glycine. *Acta Crystallogr.* **1958**, *11*, 225–226. [CrossRef]
40. Guerin, S.; Tofail, S.A.M.; Thompson, D. Organic piezoelectric materials: Milestones and potential. *NPG Asia Mater.* **2019**, *11*, 10. [CrossRef]
41. Guerin, S.; Stapleton, A.; Chovan, D.; Mouras, R.; Gleeson, M.; McKeown, C.; Noor, M.R.; Silien, C.; Rhen, F.M.F.; Kholkin, A.L.; et al. Control of piezoelectricity in amino acids by supramolecular packing. *Nat. Mater.* **2018**, *17*, 180. [CrossRef]
42. Newnham, R.E. *Properties of Materials: Anisotropy, Symmetry, Structure*, 1st ed.; Oxford University Press: New York, NY, USA, 2005.
43. Yan, X.; Zhu, P.; Fei, J.; Li, J. Self-assembly of peptide-inorganic hybrid spheres for adaptive encapsulation of guests. *Adv. Mater.* **2010**, *22*, 1283–1287. [CrossRef]
44. Görbitz, C.H. Nanotube formation by hydrophobic dipeptides. *Chem. Eur. J.* **2001**, *7*, 5153–5159. [CrossRef]
45. Amdursky, N.; Beker, P.; Rosenman, G. Physics of peptide nanostructures and their nanotechnology applications. In *Peptide Materials: From Nanostuctures to Applications*, 1st ed.; Aleman, C., Bianco, A., Venanzi, M., Eds.; Wiley-Blackwell: Hoboken, NJ, USA, 2013; pp. 1–37.
46. Kholkin, A.; Amdursky, N.; Bdikin, I.; Gazit, E.; Rosenman, G. Strong piezoelectricity in bioinspired peptide nanotubes. *ACS Nano* **2010**, *4*, 610–614. [CrossRef]
47. Lee, J.-H.; Heo, K.; Schulz-Schönhagen, K.; Lee, J.H.; Desai, M.S.; Jin, H.-E.; Lee, S.-W. Diphenylalanine peptide nanotube energy harvesters. *ACS Nano* **2018**, *12*, 8138–8144. [CrossRef]
48. Nguyen, V.; Jenkins, K.; Yang, R. Epitaxial growth of vertically aligned piezoelectric diphenylalanine peptide microrods with uniform polarization. *Nano Energy* **2015**, *17*, 323–329. [CrossRef]
49. Nguyen, V.; Zhu, R.; Jenkins, K.; Yang, R. Self-assembly of diphenylalanine peptide with controlled polarization for power generation. *Nat. Commun.* **2016**, *7*, 13566. [CrossRef]
50. Nalwa, H.S. *Ferroelectric Polymers: Chemistry: Physics, and Applications*, 1st ed.; CRC Press: New York, NY, USA, 1995.
51. Someya, T. *Stretchable Electronics*, 1st ed.; Wiley-VCH: Weinheim, Germany, 2013.
52. Martins, P.; Lopes, A.C.; Lanceros-Mendez, S. Electroactive phases of poly (vinylidene fluoride): Determination, processing and applications. *Prog. Polym. Sci.* **2014**, *39*, 683–706. [CrossRef]
53. Cui, Z.; Hassankiadeh, N.T.; Zhuang, Y.; Drioli, E.; Lee, Y.M. Crystalline polymorphism in poly(vinylidenefluoride) membranes. *Prog. Polym. Sci.* **2015**, *51*, 94–126. [CrossRef]
54. Wan, C.; Bowen, C.R. Multiscale-structuring of polyvinylidene fluoride for energy harvesting: The impact of molecular-, micro-and macro-structure. *J. Mater. Chem. A* **2017**, *5*, 3091–3128. [CrossRef]

55. Lovinger, A.J. Conformational defects and associated molecular motions in crystalline poly (vinylidene fluoride). *J. Appl. Phys.* **1981**, *52*, 5934–5938. [CrossRef]
56. Ameduri, B. From vinylidene fluoride (VDF) to the applications of VDF-containing polymers and copolymers: Recent developments and future trends. *Chem. Rev.* **2009**, *109*, 6632–6686. [CrossRef]
57. Bystrov, V.S.; Paramonova, E.V.; Bdikin, I.K.; Bystrova, A.V.; Pullar, R.C.; Kholkin, A.L. Molecular modeling of the piezoelectric effect in the ferroelectric polymer poly(vinylidene fluoride) (PVDF). *J. Mol. Model.* **2013**, *19*, 3591–3602. [CrossRef]
58. Lovinger, A.J. Ferroelectric polymers. *Science* **1983**, *220*, 1115–1121. [CrossRef]
59. Lutkenhaus, J.L.; McEnnis, K.; Serghei, A.; Russell, T.P. Confinement effects on crystallization and Curie transitions of poly(vinylidene fluoride-co-trifluoroethylene). *Macromolecules* **2010**, *43*, 3844–3850. [CrossRef]
60. Kunstler, W.; Wegener, M.; Seiss, M.; Gerhard-Multhaupt, R. Preparation and assessment of piezo- and pyroelectric poly(vinylidene fluoride-hexafluoropropylene) copolymer films. *Appl. Phys. A* **2002**, *73*, 641–645. [CrossRef]
61. Huan, Y.; Liu, Y.; Yang, Y. Simultaneous stretching and static electric field poling of poly(vinylidene fluoride-hexafluoropropylene) copolymer films. *Polym. Eng. Sci.* **2007**, *47*, 1630–1633. [CrossRef]
62. Li, Z.; Wang, Y.; Cheng, Z.-Y. Electromechanical properties of poly(vinylidene-fluoride-chlorotrifluoroethylene) copolymer. *Appl. Phys. Lett.* **2006**, *88*, 062904. [CrossRef]
63. Sultana, A.; Ghosh, S.K.; Sencadas, V.; Zheng, T.; Higgins, M.J.; Middya, T.R.; Mandal, D. Human skin interactive self-powered wearable piezoelectric bio-e-skin by electrospun poly-L-lactic acid nanofibers for non-invasive physiological signal monitoring. *J. Mater. Chem. B* **2017**, *5*, 7352–7359. [CrossRef]
64. Tajitsu, Y. Piezoelectricity of chiral polymeric fiber and its application in biomedical engineering. *IEEE Trans. Ultrason. Ferroelectr. Freq. Control* **2008**, *55*, 1000–1008. [CrossRef]
65. Fukada, E. New piezoelectric polymers. *Jpn. J. Appl. Phys.* **1998**, *37*, 2775. [CrossRef]
66. Yoshida, T.; Imoto, K.; Tahara, K.; Naka, K.; Uehara, Y.; Kataoka, S.; Date, M.; Fukada, E.; Tajitsu, Y. Piezoelectricity of poly(L-lactic acid) composite film with stereocomplex of poly(L-lactide) and poly(D-lactide). *Jpn. J. Appl. Phys.* **2010**, *49*, 09MC11. [CrossRef]
67. Yucel, T.; Cebe, P.; Kaplan, D.L. Structural origins of silk piezoelectricity. *Adv. Funct. Mater.* **2011**, *21*, 779–785. [CrossRef]
68. Kim, J.; Yun, S.; Ounaies, Z. Discovery of cellulose as a smart material. *Macromolecules* **2006**, *39*, 4202–4206. [CrossRef]
69. Wang, Z.L.; Song, J. Piezoelectric nanogenerators based on zinc oxide nanowire arrays. *Science* **2006**, *312*, 242–246. [CrossRef]
70. Fan, F.-R.; Tian, Z.-Q.; Wang, Z.L. Flexible triboelectric generator. *Nano Energy* **2012**, *1*, 328–334. [CrossRef]
71. Kim, G.H.; Shin, D.-M.; Kim, H.-K.; Hwang, Y.-H. Effect of the dielectric layer on the electrical output of a ZnO nanosheet-based nanogenerator. *J. Korean Phys. Soc.* **2015**, *67*, 1920–1924. [CrossRef]
72. Kim, T.; Jeon, S.; Lone, S.; Doh, S.J.; Shin, D.-M.; Kim, H.K.; Hwang, Y.-H.; Hong, S.W. Versatile nanodot-patterned Gore-Tex fabric for multiple energy harvesting in wearable and aerodynamic nanogenerators. *Nano Energy* **2018**, *54*, 209–217. [CrossRef]
73. Park, H.-Y.; Kim, H.K.; Hwang, Y.-H.; Shin, D.-M. Water-through triboelectric nanogenerator based on Ti-mesh for harvesting liquid flow. *J. Korean Phys. Soc.* **2018**, *72*, 449–503. [CrossRef]
74. Phan, H.; Shin, D.-M.; Jeon, S.H.; Kang, T.Y.; Han, P.; Kim, G.H.; Kim, H.K.; Kim, K.; Hwang, Y.-H.; Hong, S.W. Aerodynamic and aeroelastic flutters driven triboelectric nanogenerators for harvesting broadband airflow energy. *Nano Energy* **2017**, *33*, 476–484. [CrossRef]
75. Shin, D.-M.; Tsege, E.L.; Kang, S.H.; Seung, W.; Kim, S.-W.; Kim, H.-K.; Hong, S.W.; Hwang, Y.-H. Freestanding ZnO nanorod/graphene/ZnO nanorod epitaxial double heterostructure for improved piezoelectric nanogenerators. *Nano Energy* **2015**, *12*, 268–277. [CrossRef]
76. Tsege, E.L.; Shin, D.-M.; Lee, S.; Kim, H.-K.; Hwang, Y.-H. Highly Durable Ti-mesh based triboelectric nanogenerator for self-powered device applications. *J. Nanosci. Nanotechnol.* **2016**, *16*, 4864–4869. [CrossRef]
77. Vivekananthan, V.; Alluri, N.R.; Purusothaman, Y.; Chandrasekhar, A.; Selvarajan, S.; Kim, S.-J. Biocompatible collagen nanofibrils: An approach for sustainable energy harvesting and battery-free humidity sensor applications. *ACS Appl. Mater. Interfaces* **2018**, *10*, 18650–18656. [CrossRef]
78. Shi, Y.; He, S.; Hearst, J.E. Statistical mechanics of the extensible and shearable elastic rod and of DNA. *J. Chem. Phys.* **1996**, *105*, 714–731. [CrossRef]

79. Vu, N.; Kelly, S.; Yang, R. Piezoelectric peptide-based nanogenerator enhanced by single-electrode triboelectric nanogenerator. *APL Mater.* **2017**, *5*, 074108.
80. Fang, J.; Wang, X.; Lin, T. Electrical power generator from randomly oriented electrospun poly(vinylidene fluoride) nanofibre membranes. *J. Mater. Chem.* **2011**, *21*, 11088–11091. [CrossRef]
81. Pan, C.-T.; Yen, C.-K.; Wang, S.-Y.; Lai, Y.-C.; Lin, L.; Huang, J.C.; Kuo, S.-W. Near-field electrospinning enhances the energy harvesting of hollow PVDF piezoelectric fibers. *RSC Adv.* **2015**, *5*, 85073–85081. [CrossRef]
82. Liu, Z.H.; Pan, C.T.; Lin, L.W.; Huang, J.C.; Ou, Z.Y. Direct-write PVDF nonwoven fiber fabric energy harvesters via the hollow cylindrical near-field electrospinning process. *Smart Mater. Struct.* **2014**, *23*, 25003. [CrossRef]
83. Chang, C.; Tran, V.H.; Wang, J.; Fuh, Y.-K.; Lin, L. Direct-write piezoelectric polymeric nanogenerator with high energy conversion efficiency. *Nano Lett.* **2010**, *10*, 726–731. [CrossRef]
84. Hansen, B.J.; Liu, Y.; Yang, R.; Wang, Z.L. Hybrid nanogenerator for concurrently harvesting biomechanical and biochemical energy. *ACS Nano* **2010**, *4*, 3647–3652. [CrossRef]
85. Ishida, K.; Huang, T.-C.; Honda, K.; Shinozuka, Y.; Fuketa, H.; Yokota, T.; Zschieschang, U.; Klauk, H.; Tortissier, G.; Sekitani, T.; et al. Insole pedometer with piezoelectric energy harvester and 2 V organic circuits. *IEEE J. Solid State Circuits* **2013**, *48*, 255–264. [CrossRef]
86. Sun, C.; Shi, J.; Bayerl, D.J.; Wang, X. PVDF microbelts for harvesting energy from respiration. *Energy Environ. Sci.* **2011**, *4*, 4508–4512. [CrossRef]
87. Xue, H.; Yang, Q.; Wang, D.; Luo, W.; Wang, W.; Lin, M.; Liang, D.; Luo, Q. A wearable pyroelectric nanogenerator and self-powered breathing sensor. *Nano Energy* **2017**, *38*, 147–154. [CrossRef]
88. Persano, L.; Dagdeviren, C.; Su, Y.; Zhang, Y.; Girardo, S.; Pisignano, D.; Huang, Y.; Rogers, J.A. High performance piezoelectric devices based on aligned arrays of nanofibers of poly (vinylidenefluoride-co-trifluoroethylene). *Nat. Commun.* **2013**, *4*, 1633. [CrossRef]
89. Liu, Z.; Zhang, S.; Jin, Y.M.; Ouyang, H.; Zou, Y.; Wang, X.X.; Xie, L.X.; Li, Z. Flexible piezoelectric nanogenerator in wearable self-powered active sensor for respiration and healthcare monitoring. *Semicond. Sci. Technol.* **2017**, *32*, 064004. [CrossRef]
90. Bodkhe, S.; Turcot, G.; Gosselin, F.P.; Therriault, D. One-step solvent evaporation-assisted 3D printing of piezoelectric PVDF nanocomposite structures. *ACS Appl. Mater. Interfaces* **2017**, *9*, 20833–20842. [CrossRef]
91. Park, J.; Kim, M.; Lee, Y.; Lee, H.S.; Ko, H. Fingertip skin-inspired microstructured ferroelectric skins discriminate static/dynamic pressure and temperature stimuli. *Sci. Adv.* **2015**, *1*, e1500661. [CrossRef]
92. Curry, E.J.; Ke, K.; Chorsi, M.T.; Wrobel, K.S.; Miller, A.N.; Patel, A.; Kim, I.; Feng, J.; Yue, L.; Wu, Q.; et al. Biodegradable piezoelectric force sensor. *Proc. Natl. Acad. Sci. USA* **2018**, *115*, 909–914. [CrossRef]
93. Joseph, J.; Singh, S.G.; Vanjari, S.R.K. Leveraging innate piezoelectricity of ultra-smooth silk thin films for flexible and wearable sensor applications. *IEEE Sens. J.* **2017**, *17*, 8306–8313. [CrossRef]
94. Wang, X.; Gu, Y.; Xiong, Z.; Cui, Z.; Zhang, T. Silk-molded flexible, ultrasensitive, and highly stable electronic skin for monitoring human physiological signals. *Adv. Mater.* **2014**, *26*, 1336–1342. [CrossRef]
95. Ghosh, S.K.; Mandal, D. Sustainable energy generation from piezoelectric biomaterial for noninvasive physiological signal monitoring. *ACS Sustain. Chem. Eng.* **2017**, *5*, 8836–8843. [CrossRef]
96. Moreno, S.; Baniasadi, M.; Mohammed, S.; Mejia, I.; Chen, Y.; Quevedo-Lopez, M.A.; Kumar, N.; Dimitrijevich, S.; Minary-Jolandan, M. Flexible electronics: Biocompatible collagen films as substrates for flexible implantable electronics. *Adv. Electron. Mater.* **2015**, *1*, 1500154. [CrossRef]
97. Damaraju, S.M.; Wu, S.; Jaffe, M.; Arinzeh, T.L. Structural changes in pvdf fibers due to electrospinning and its effect on biological function. *Biomed. Mater.* **2013**, *8*, 045007. [CrossRef]
98. Damaraju, S.M.; Shen, Y.; Elele, E.; Khusid, B.; Eshghinejad, A.; Li, J.; Jaffe, M.; Arinzeh, T.L. Three-dimensional piezoelectric fibrous scaffolds selectively promote mesenchymal stem cell differentiation. *Biomaterials* **2017**, *149*, 51–62. [CrossRef] [PubMed]
99. Martins, P.M.; Ribeiro, S.; Ribeiro, C.; Sencadas, V.; Gomes, A.C.; Gama, F.M.; Lancerosmendez, S. Effect of poling state and morphology of piezoelectric poly(vinylidene fluoride) membranes for skeletal muscle tissue engineering. *RCS Adv.* **2013**, *3*, 17938–17944. [CrossRef]
100. Hoop, M.; Chen, X.Z.; Ferrari, A.; Mushtaq, F.; Ghazaryan, G.; Tervoort, T.; Poulikakos, D.; Nelson, B.; Pane, S. Ultrasound-mediated piezoelectric differentiation of neuron-like PC12 cells on PVDF membranes. *Sci. Rep.* **2017**, *7*, 4028. [CrossRef]

101. Lee, Y.S.; Arinzeh, T.L. The influence of piezoelectric scaffolds on neural differentiation of human neural stem/progenitor cells. *Tissue Eng. Part A* **2012**, *18*, 2063–2072. [CrossRef]
102. Ikada, Y.; Shikinami, Y.; Hara, Y.; Tagawa, M.; Fukada, E. Enhancement of bone formation by drawn poly(L-lactide). *J. Biomed. Mater. Res. Part A* **1996**, *30*, 553–558. [CrossRef]
103. Barroca, N.; Daniel-da-Silva, A.; Gomes, P.; Fernandes, M.; Lanceros-Mendez, S.; Sharma, P.; Gruverman, A.; Fernandes, M.; Vilarinho, P. Suitability of PLLA as piezoelectric substrates for tissue engineering evidenced by microscopy techniques. *Microsc. Microanal.* **2012**, *18*, 63–64. [CrossRef]
104. Barroca, N.; Vilarinho, P.M.; Daniel-da-Silva, A.L.; Wu, A.; Fernandes, M.H.; Gruverman, A. Stability of electrically induced-polarization in poly (L-lactic) acid for bone regeneration. *Appl. Phys. Lett.* **2011**, *98*, 133705. [CrossRef]

6

QuantIF: An Image J Macro to Automatically Determine the Percentage of Infected Cells after Immunofluorescence

Lynda Handala [1,†], Tony Fiore [1,†], Yves Rouillé [2,*] and Francois Helle [1,*]

1 EA4294, Agents Infectieux, Résistance et Chimiothérapie, Centre Universitaire de Recherche en Santé, Centre Hospitalier Universitaire et Université de Picardie Jules Verne, 80054 Amiens, France; l.handala@gmail.com (L.H.); toni.fiore37@gmail.com (T.F.)

2 University of Lille, CNRS, INSERM, CHU Lille, Pasteur Institute of Lille, U1019-UMR8204-CIIL-Center for Infection and Immunity of Lille, 59019 Lille, France

* Correspondence: yves.rouille@ibl.cnrs.fr (Y.R.); francois.helle@u-picardie.fr (F.H.)

† These authors contributed equally to this work.

Abstract: Counting labeled cells, after immunofluorescence or expression of a genetically fluorescent reporter protein, is frequently used to quantify viral infection. However, this can be very tedious without a high content screening apparatus. For this reason, we have developed QuantIF, an ImageJ macro that automatically determines the total number of cells and the number of labeled cells from two images of the same field, using DAPI- and specific-stainings, respectively. QuantIF can automatically analyze hundreds of images, taking approximately one second for each field. It is freely available as supplementary data online at MDPI.com and has been developed using ImageJ, a free image processing program that can run on any computer with a Java virtual machine, which is distributed for Windows, Mac, and Linux. It is routinely used in our labs to quantify viral infections in vitro, but can easily be used for other applications that require quantification of labeled cells.

Keywords: virus; infection; fluorescent reporter protein; image quantification; Hepatitis C virus; Yellow Fever Virus; polyomavirus; Coxsackievirus B4

---

1. Introduction

When evaluating viral infections in vitro, fluorescence microscopy is commonly used to monitor the expression of a viral protein following immunostaining. However, this method requires a high content screening apparatus to count large numbers of fluorescent cells. Manual evaluation is feasible when analyzing few images, but it can result in subjective evaluation by the researcher. Furthermore, it is very time-consuming when working with hundreds of images, containing thousands of cells per image.

ImageJ is a free image-processing program that was developed 20 years ago by Wayne S. Rasband at the National Institute of Health, and has become a valuable tool for researchers [1,2]. It is a Java-based software that can run on any computer using a Java virtual machine. It is thus available for Windows, Mac, and Linux. ImageJ can convert images into numerical values that can be exported and further processed with other software for statistical analysis. Furthermore, a major strength of ImageJ is the possibility to record macros that enable the automatization of image analysis.

In this technical note, we present QuantIF, an ImageJ macro for determining the percentage of fluorescent cells following immunofluorescence staining. QuantIF can be used when the specific staining in the cytoplasm and/or nucleus of a cell is diffuse. The macro automatically determines the total number of cells and fluorescently labeled cells for a series of images corresponding to different conditions. For each condition, two pictures of the same field must be taken, the first one corresponding to the specific staining and the second one corresponding to the DAPI staining. In this way, the series of images to be analyzed are placed in the same folder, with images corresponding to the specific staining in odd rank and images of DAPI staining in even rank. When the macro is run, it automatically processes all images in the folder, taking around one second to analyze both images of each field. Ultimately, all results are saved as a ".xls" file that can be processed for statistical analysis.

2. Macro Description

QuantIF was developed using ImageJ version 1.52e and Java version 8. It is freely available as supplementary data online at MDPI.com. In order to use the QuantiIF macro, it is necessary to save the QuantIF.ijm file in the "Plugins" folder of ImageJ. The macro will then appear in the "Plugins" menu. When QuantIF is launched, the folder containing the images for analysis must be selected. Then, parameter values should be entered in a dialog box (Figure 1a), (i) the type/name of the specific staining, (ii) the staining threshold, and (iii) the size limits of nuclei. Once the parameters have been entered, the macro starts analyzing the images. They are first converted to 8-bit images, displaying 256 gray levels. Indeed, we recommend directly exporting images as 8-bit TIFF files, from the microscope software. The background of the images is then removed by running the efficient "Subtract Background" ImageJ command.

QuantIF relies on the "Analyze Particles" tool of ImageJ, which requires binary, black and white, images. For this reason, images are converted to binary masks by implementing the Huang's fuzzy thresholding method [3]. An automatic threshold is set for DAPI staining images since strong and contrasted signals are expected for all these images (Figure 1b,c). In contrast, for immunostaining images, the automatic threshold is generally not applicable since some images may show no signal (in negative controls for instance). For this reason, a manual thresholding is implemented with the staining threshold value entered in the parameter's dialog box (Figure 1a). The threshold value must range between 0 and 255; pixels with values under and above the threshold are converted to white and black, respectively (Figure 1d,e). The "Watershed" command is also applied to the DAPI staining mask in order to separate nearby nuclei [4]. However, it is important to avoid cell overconfluence to obtain interpretable results (see below). Furthermore, the commands "Dilate", "Close", and "Fill Holes" are applied to the mask of the specific staining in order to completely include the area corresponding to the nuclei. To analyze similar particles in the DAPI- and specific-staining masks and avoid counting autofluorescent debris, a new mask corresponding to the nuclei of immunostained cells is created. This is performed by executing the "Image Calculator" command and the "AND" operator using the DAPI- and specific-staining masks (Figure 1f). Finally, the total number of DAPI-stained and immunostained cells' nuclei are counted by implementing the "Analyze Particles" tool to the DAPI staining mask (Figure 1g) and the immunostained cells' nuclei mask (Figure 1h), respectively. The size limits for the nuclei entered in the parameters dialog box correspond to the minimum and maximum pixel area sizes that are taken into account to exclude anything that is not an object of interest. Additionally, to help exclude unwanted objects, roundness values have been set between 0.7 and 1.0.

After processing, the numbers of DAPI-stained nuclei and immunostained cells' nuclei for each condition are saved as a ".xls" file in the folder that has been analyzed (Figure 1i). In addition, "Total Area", "Average Size", and "%Area" values are saved in the file. While "Total Area" values are not useful, the "Average Size" values can help in choosing the size limits for the nuclei that must be entered in the parameters dialog box. Furthermore, "%Area" values of the DAPI staining masks give an idea on cell confluence, which should not typically exceed 30% for optimal results. To help researchers find the best parameters for their analyses, the different masks can be saved in the folder that is being analyzed. To do so, the "//" symbols preceding the "saveAs" line commands must be deleted in the QuantIF.ijm file.

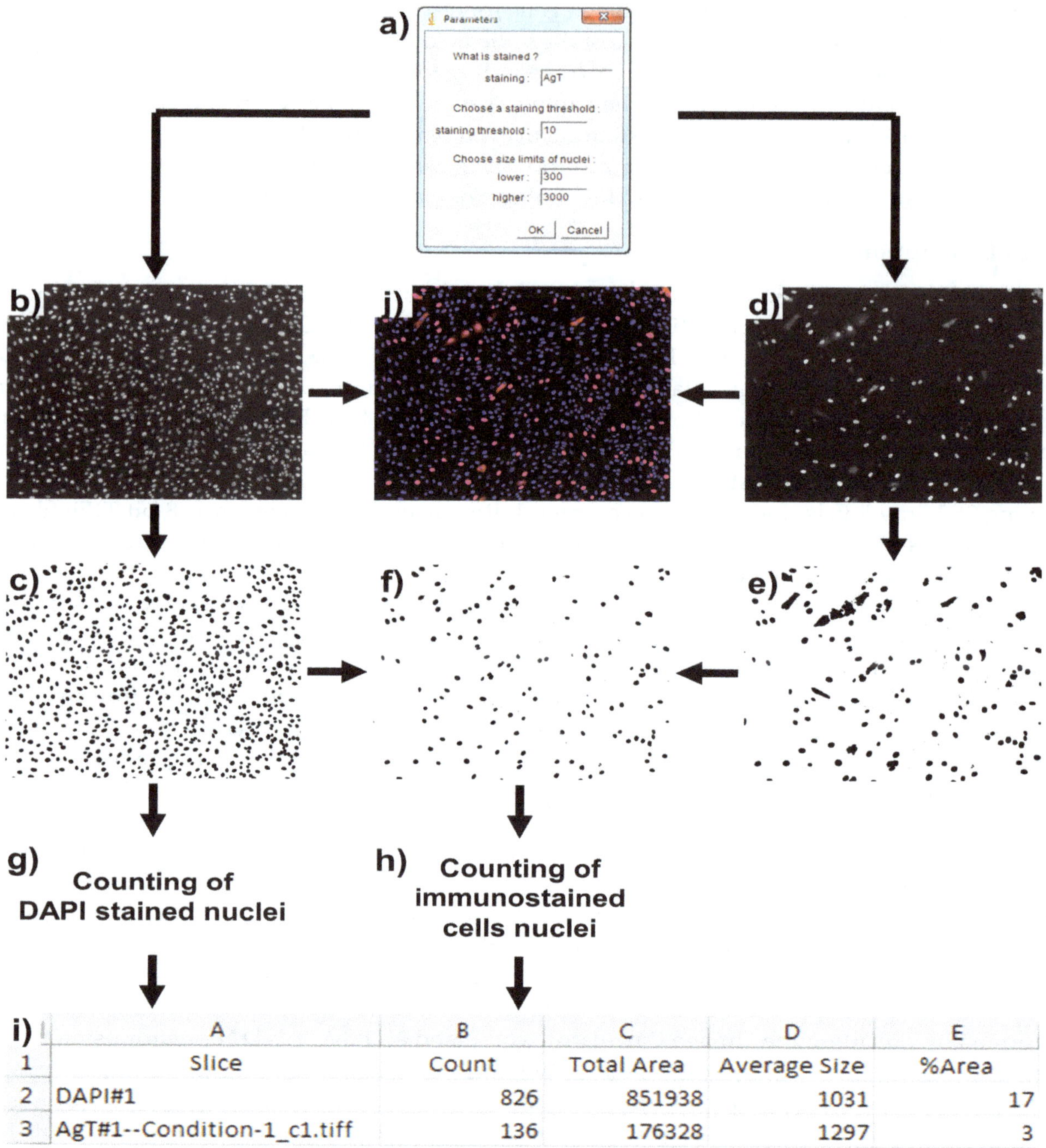

Figure 1. Description of the QuantIF macro. After entering the parameters into the dialog box (**a**), two images of each field are analyzed. The DAPI staining image (**b**) is converted to a DAPI staining mask (**c**), and the specific staining image (**d**) is converted to a specific staining mask (**e**), by implementing the Huang's fuzzy thresholding method. A third mask corresponding to the nuclei of the immunostained cells is created using the "Image Calculator" command and the "AND" operator (**f**). Finally, DAPI stained nuclei and immunostained cell nuclei are counted using the "Analyze Particles" tool (**g**,**h**). After processing, the numbers of DAPI-stained nuclei and immunostained cell nuclei for each condition are saved as a ".xls" file in the folder that has been analyzed (**i**). A merge of the DAPI and specific staining images is shown for informational purposes (**j**).

3. Discussion and Conclusions

QuantIF is a free, simple, and robust automated tool to estimate the proportion of virally infected cells after immunofluorescence. It is routinely used in our labs to quantify Hepatitis C Viral infections following detection of the E1 envelope glycoprotein that localizes predominantly to the endoplasmic reticulum in HCV-infected cells [5,6]. Similarly, we use it to evaluate Yellow Fever Virus infections using anti-E staining [5]. QuantIF is also used to quantify BKPyV and SV40 infections after detection of the VP1 or AgT proteins that show cytoplasmic and/or nuclear staining patterns (Figure 1d), as well as Coxsackievirus B4 infections using anti-VP1 staining [5]. QuantIF can also be used to quantify infection when using recombinant viruses expressing fluorescent reporter proteins [5]. Furthermore, it can serve many researchers for other applications that require counting labeled cells.

Author Contributions: Conceptualization, Y.R. and F.H.; validation, L.H., T.F., Y.R., and F.H.; writing, Y.R. and F.H.

Acknowledgments: We thank Zuzana Saidak and Thomas Belcher for proofreading the manuscript.

References

1. Schindelin, J.; Rueden, C.T.; Hiner, M.C.; Eliceiri, K.W. The ImageJ ecosystem: An open platform for biomedical image analysis. *Mol. Reprod. Dev.* **2015**, *82*, 518–529. [CrossRef] [PubMed]
2. Schneider, C.A.; Rasband, W.S.; Eliceiri, K.W. NIH Image to ImageJ: 25 years of image analysis. *Nat. Methods* **2012**, *9*, 671–675. [CrossRef] [PubMed]
3. Huang, L.-K.; Wang, M.-J.J. Image thresholding by minimizing the measures of fuzziness. *Pattern Recognit.* **1995**, *28*, 41–51. [CrossRef]
4. Vincent, L.; Soille, P. Watersheds in digital spaces: An efficient algorithm based on immersion simulations. *IEEE Trans. Pattern Anal. Mach. Intell.* **1991**, *13*, 583–598. [CrossRef]
5. Ferlin, J.; Farhat, R.; Belouzard, S.; Cocquerel, L.; Bertin, A.; Hober, D.; Dubuisson, J.; Rouille, Y. Investigation of the role of GBF1 in the replication of positive-sense single-stranded RNA viruses. *J. Gen. Virol.* **2018**, *99*, 1086–1096. [CrossRef] [PubMed]
6. Rouille, Y.; Helle, F.; Delgrange, D.; Roingeard, P.; Voisset, C.; Blanchard, E.; Belouzard, S.; McKeating, J.; Patel, A.H.; Maertens, G.; et al. Subcellular localization of hepatitis C virus structural proteins in a cell culture system that efficiently replicates the virus. *J. Virol.* **2006**, *80*, 2832–2841. [CrossRef] [PubMed]

7

Hydrophobization of Tobacco Mosaic Virus to Control the Mineralization of Organic Templates

Petia Atanasova [1,*], Vladimir Atanasov [2], Lisa Wittum [3], Alexander Southan [4], Eunjin Choi [1], Christina Wege [3], Jochen Kerres [2], Sabine Eiben [3] and Joachim Bill [1]

[1] Institute for Materials Science, University of Stuttgart, Heisenbergstr. 3, 70569 Stuttgart, Germany; choi@is.mpg.de (E.C.); bill@imw.uni-stuttgart.de (J.B.)

[2] Institute of Chemical Process Engineering, University of Stuttgart, Böblinger Straße 78, 70199 Stuttgart, Germany; vladimir.atanasov@icvt.uni-stuttgart.de (V.A.); jochen.kerres@icvt.uni-stuttgart.de (J.K.)

[3] Institute of Biomaterials and Biological Systems, University of Stuttgart, Pfaffenwaldring 57, 70569 Stuttgart, Germany; lisawittum1@yahoo.de (L.W.); christina.wege@bio.uni-stuttgart.de (C.W.); sabine.eiben@gmx.de (S.E.)

[4] Institute of Interfacial Process Engineering and Plasma Technology, University of Stuttgart, Pfaffenwaldring 31, 70569 Stuttgart, Germany; alexander.southan@igvp.uni-stuttgart.de

* Correspondence: atanasova@imw.uni-stuttgart.de

Abstract: The robust, anisotropic tobacco mosaic virus (TMV) provides a monodisperse particle size and defined surface chemistry. Owing to these properties, it became an excellent bio-template for the synthesis of diverse nanostructured organic/inorganic functional materials. For selective mineralization of the bio-template, specific functional groups were introduced by means of different genetically encoded amino acids or peptide sequences into the polar virus surface. An alternative approach for TMV surface functionalization is chemical coupling of organic molecules. To achieve mineralization control in this work, we developed a synthetic strategy to manipulate the surface hydrophilicity of the virus through covalent coupling of polymer molecules. Three different types of polymers, namely the perfluorinated (poly(pentafluorostyrene) (PFS)), the thermo-responsive poly(propylene glycol) acrylate (PPGA), and the block-copolymer polyethylene-block-poly(ethylene glycol) were examined. We have demonstrated that covalent attachment of hydrophobic polymer molecules with proper features retains the integrity of the virus structure. In addition, it was found that the degree of the virus hydrophobicity, examined via a ZnS mineralization test, could be tuned by the polymer properties.

Keywords: tobacco mosaic virus; ZnS; bio/inorganic hybrid materials; hydrophobization; polymer coupling

---

1. Introduction

In the process of fast-growing development of new nanostructured functional materials, a huge variety of organic materials, in particular biological objects, have been used to synthesize organic/inorganic hybrids with desired properties for nanotechnological applications [1–4]. Among them, tobacco mosaic virus (TMV) is a robust tube-like plant virus, harmless for humans and animals and can be produced in scalable amounts in a green house. It has been utilized intensively as a bio-template due to its anisotropic structure, high uniformity in size and shape and defined surface chemistry. Applying 'bottom-up' approaches such as evaporative self-assembly of the capsids by convective assembly and controlled evaporation methods, homogeneous virus monolayers, aligned TMV stripes and nanowires have been produced [5–8]. TMV has been used as a template to deposit various inorganic materials (metals and semiconductors) on the exterior or the interior surface of the capsid [9–11]. The virus itself and TMV-based hybrids have found application in construction of functional devices like field-effect transistors (FETs) [12] and sensors [13–16], in Li-ion battery [17], etc. Electroconductive materials have been generated making use of the electrostatic interactions between TMV and the polymers polyaniline and polypyrrole, which enabled in-situ polymerization of the polymers on the TMV surface [18]. Most of these studies have been performed using either wild type (wt) TMV, cysteine mutants within the first 3 amino acids of the coat protein (CP) or a C-terminal lysine mutant. An additional number of TMV CP mutants exist [19–21], however especially those changing the surface charge interfere with the natural assembly process [22], forming nanotubes without incorporation of stabilizing RNA. These nanotubes vary greatly in length and are often liable to disassembly upon small changes in buffer conditions making them unavailable for mineralization studies. Although there has been a quite successful attempt to stabilize these RNA-free virus-like particles by introducing an inter CP disulfide bridge in the inner channel of the nanotubes and thus opening up the possibility to use bacterially expressed CP mutants for templating [23,24].

Another strategy to increase the possible surface properties of TMV is to use the existing stable mutants for chemical modification by coupling, which has been done to add mineralization inducing peptides, chemicals for magnetic resonance imaging, whole enzymes and polymers [25]. Recently, a DNA-controlled "stop-and-go" strategy was established to assemble two distinct, selectively addressable CPs variants with RNA into artificial TMV nanotubes with highly defined longitudinal subdomains [26]. The presence of two domains consisting of CPs with different functional groups on one artificial TMV nanoparticle [27] gives the opportunity of using, e.g., "click reactions" to specifically couple organic molecules to only one part of the particle and thereby to synthesize Janus-type TMV particles with a hydrophilic and a hydrophobic portion. The amphiphilicity of classical Janus particles provides unique chemical and physical properties, not accessible for their homogeneous counterparts. Owing to their amphiphilic, magnetic, catalytic or optical properties, they have found numerous applications in fields like drug delivery and catalysis, as surfactants and building blocks for complex 3D nanostructures, for water-repellent coatings, etc [28]. Furthermore, such particles provide additional perspectives towards exploring the influence of the genetic modification of TMV CPs on the nucleation and growth of inorganic materials within one artificial TMV particle, i.e., on the guidance of mineralization reactions. While TMV CPs can be engineered genetically or modified chemically to introduce different hydrophilic functionalities on the virus surface, TMV hydrophobization is a challenging task. An important requirement by choosing viable reaction conditions is the prevention of the hydrophilic complex protein structure from disassembly in the presence of a hydrophobic surrounding. In this regard, A. J. Patil et al. have modified the surface of cowpea mosaic virus (CPMV) with the anionic polymer-surfactant poly(ethylene glycol) 4-nonylphenyl 3-sulfopropyl ether through electrostatically directed assembly [29]. However, electrostatic interactions do not allow control over a local specific functionalization of the virus surface, which in contrast can be achieved via covalent coupling of organic molecules. Another important point concerns the degree of virus hydrophobization required to suppress subsequent mineralization, if partially coated particles are sought after. Therefore, in this work we used wt-TMV and a thiol-displaying TMV-Cys mutant [27] to examine different synthesis strategies and to establish an approach for covalent coupling of polymer molecules to the virus surface, with the aim to produce stably functionalized viruses with sufficient hydrophobicity to suppress mineral deposition on the viral protein coat. For the bioconjugation reactions, the highly hydrophobic perfluorinated (poly(pentafluorostyrene) (PFS)), the thermo-responsive poly(propylene glycol) acrylate (PPGA), which becomes hydrophobic by increasing the temperature and the block-copolymer polyethylene-block-poly(ethylene glycol) ((PE)-b-(PEG)) with a well distinguished hydrophobic part were used. We demonstrate that the integrity of the virus particles can be preserved after covalent attachment of hydrophobic polymer molecules, compatible with the structure of the virus. The achieved hydrophobicity was verified via mineralization reaction applied to the modified virus surface with ZnS, in order to examine if the degree of virus hydrophobicity can be controlled by the choice of the conjugated polymer.

2. Materials and Methods

Materials Disodium hydrogen phosphate, acetone, sodium dodecyl sulfate, tris(hydroxymethyl)aminomethane, acetic acid, sodium carbonate, ammonium persulfate, tetramethylenediamine, sodium chloride and sodium thiosulfate were obtained from Carl Roth (Karlsruhe, Germany). Triphenylphosphine, hydrazine hydrate, PE-b-PEG-OH, N-hydroxyphthalimide (97%), sodium nitrite (97%, ACS), p-toluenesulfonic acid monohydrate (ACS), p-aminoacetophenone (0.15 M in acetonitrile), diisopropyl azodicarboxylate (95%), zinc chloride (≥98%) and sodium sulfide nonahydrate (≥98%) were obtained from Sigma-Aldrich. Poly(propylene glycol) acrylate (M_n = 475 g mol^{-1}), silver nitrate, formaldehyde, Coomassie-Brilliant-Blue R250 and dihydrogen phosphate were obtained from Merck (Darmstadt, Germany). Acrylamide/Bis Solution, 37.5:1 was purchased from Serva (Heidelberg, Germany). Pentaflourostyrene 98% was purchased from ABCR GmbH. All solvents used in this study were in HPLC grade and used as received.

Substrate cleaning: Silicon wafers (100, p-doped polished wafers, Silchem, Germany) were used as substrates. To get a hydrophilic surface, they were thoroughly cleaned prior to use applying the following procedure: 10 min sonication in ultrapure water, 10 min sonication in ethanol/acetone (1:1, v/v), 10 min O_2 plasma treatment (30 W) and 10 min sonication in ultrapure water. After each sonication step, the substrates were washed 10 times with the corresponding solvent and dried under Ar stream. Silicon substrates with reduced hydrophilicity (not plasma treated) were cleaned only successively with ultrapure water, ethanol and acetone.

Preparation and purification of TMV: wt-TMV of the type strain U1 as well as the TMV-Cys mutant [27] were propagated in Nicotiana tabacum 'Samsun' nn plants and purified by PEG precipitation as described before [30]. The virus was stored in 10 mM sodium potassium phosphate (SPP) buffer at pH 7.4 at 4 °C. Buffer-free virus solutions were obtained by resuspension of TMV in ultrapure water after pelleting the virus in a Beckman ultracentrifuge at 35,000 rpm (corresponding to an average g force of 95,800) for 105 min at 4 °C using a 45 Ti rotor and resuspension in ultrapure water.

TMV immobilization: A droplet (3 μL, 0.2 mg mL^{-1}) buffer-free virus solution was spotted onto a substrate surface and incubated for 10 min in a closed chamber. Then, the droplet was removed and the substrate with the immobilized viruses was dried under argon stream.

Synthesis of poly(pentafluorostyrene)(PFS): The synthesis of PFS has already been reported elsewhere [31]. Briefly, emulsion polymerization reaction of pentafluorostyrene in conditions similar to styrene polymerization was used. The molecular weight was optimized to obtain high values (M_n = 52 kDa, M_w = 124 kDa and MWD = 2.4), but allowing good solubility of the polymer in solvents like tetrahydrofuran (THF) and $CHCl_3$.

Click reaction of TMV-Cys with PFS: Due to the limited solubility of PFS in N,N-dimethylacetamide (DMAc), first 1 wt% PFS in DMAc was prepared at 90 °C for 24 h. The solution was filtered through 0.2 μm filter, and silicon substrates with immobilized viruses were let to react with the filtered polymer solution for 30 min at room temperature. After the treatment, the substrates were washed with DMAc and dried. For reaction in tetrahydrofuran (THF), silicon substrates with immobilized TMV-Cys particles were placed in a vessel with 1 wt% PFS in THF/trimethylamine (TEA) solution at room temperature for 2 h. Then, the substrates were thoroughly washed with THF and dried.

Synthesis of TMV-Cys-poly(propylene glycol) acrylate (PPGA): All solutions were stored in an ice-bath for at least 30 min to prevent warming up of the PPGA. For coupling, 1–4 mg/mL of the virus solutions in 10 mM SPP buffer were incubated with the same volume of PPGA (M_n = 475) diluted in 10 mM SPP buffer to give a molecular ratio from 1:1000 to 1:10,000 of CP to PPGA. The solution was incubated in a cooled shaker at 4 °C and 300 rpm for 72 h. Before loading of the functionalized TMV-Cys on a 16/60 Sephacryl S-500 column, bulk PPGA was removed by precipitation of the virus by ultracentrifugation at 34,000 rpm for 2 h at 4°C in a pre-cooled 45 Ti rotor. As sample and running buffer 50 mM SPP containing 150 mM NaCl was used. Purification was performed on an ÄKTApurifier system in a cooling cabinet at a flowrate of 0.5 mL/min, and 1 mL fractions were collected after 75 min until the end of the purification at 250 min. The fractions containing TMV were determined by UV-Vis analysis. Fractions showing an absorbance ratio of 260 nm and 280 nm of 1.19 were pooled and concentrated using an Amicon® Ultra 4 mL centrifugal filter unit with 50,000 nominal molecular weight limit (NMWL) cut-off (Merck, Darmstadt, Germany).

Synthesis of PE-b-PEG phthalimide **(2)**: A three-neck round flask was flame-dried under nitrogen stream. Triphenylphosphine (PPh_3, 1.377 g, 5.25 mmol), N-hydroxyphthalimide (0.895 g, 5.5 mmol), PE-b-PEG-OH (**1**) (11.25 g, 5 mmol, Mw = 2250 g/mol) and anhydrous CH_2Cl_2 (50 mL) were placed in the flask and stirred under inert atmosphere. After 15 min of stirring at room temperature, diisopropyl azodicarboxylate (DIAD, 1.08 mL, 5.5 mmol) was added in small portions (0.15–0.2 mL) into the reaction mixture. The obtained orange color of the solution was allowed to fade prior to adding the next portion DIAD. The solution was left to react under stirring in inert atmosphere at room temperature for 12 h. Then, the reaction mixture was precipitated in diethyl ether (Et_2O) (1 L) under vigorous stirring in an ice bath for 45 min. The white precipitate was separated from the solvent via suction filtration. The collected product was washed several times with Et_2O. To achieve high conversion, the reaction was repeated by using the product as a starting reagent. The obtained PE-b-PEG phthalimide (**2**) was precipitated. To improve its purity, it was dissolved in small amount of CH_2Cl_2, precipitated in Et_2O twice and dried under vacuum for 3.5 h. Yield = 91.6 %. FT-IR (cm^{-1}) = 3586, 2917, 2849, 1789, 1734, 1640, 1462, 1348, 1324, 1295, 1248, 1187, 1095, 1036, 995, 948, 877, 848, 808, 706, 669, 518. ^1H NMR (500 MHz, $CHCl_3$-d, δ ppm) = 0.88 (t, J = 6.94 Hz, 3 H), 1.25 (s, 54 H), 1.57 (quin, J = 6.94 Hz, 2 H), 3.44 (t, J = 6.94 Hz, 2 H), 3.47–3.52 (m, 1H), 3.53–3.71 (m, 180 H), 3.74–3.81 (m, 1 H), 3.82–3.90 (m, 2 H), 4.34–4.42 (m, 2 H), 7.72–7.87 (m, 4 H).

Synthesis of aminooxy PE-b-PEG **(3)**: A 100 mL flame-dried two neck round flask equipped with stirrer bar was charged with anhydrous CH_2Cl_2 (10 mL) under nitrogen. Then, PE-b-PEG phthalimide (**2**) (1.0675 g, 0.446 mmoL) was added and dissolved at room temperature. After 20 min, hydrazine hydrate (58 μL, 0.468 mmoL) was added to the reaction mixture and stirred in inert atmosphere at room temperature for 1 h. The color changed from orange to yellow, and a white precipitate was monitored. Et_2O (0.5 L) was added to the reaction mixture, placed in an ice bath, under vigorous stirring and left to stir for 2 h. The filtered solid product was rinsed with Et_2O and dried under vacuum. Yield = 48.2%. FT-IR (cm^{-1}) = 2917, 2849, 1462, 1348, 1324, 1297, 1248, 1095, 1040, 995, 948, 846, 720, 532. ^1H NMR (500 MHz, $CHCl_3$-d, δ ppm) = 0.87 (t, J = 6.94 Hz, 3 H), 1.25 (s, 55 H), 1.57 (quin, J = 7.09 Hz, 2 H), 3.44 (t, J = 6.62 Hz, 2 H), 3.47–3.53 (m, 1H), 3.53–3.76 (m, 190 H), 3.76–3.82 (m, 2 H), 3.84–3.89 (m, 1 H), 4.00–4.04 (m, 1 H), 4.06–4.41 (m, 1 H), 4.06–4.41 (m, 1 H), 4.41–4.50 (m, 1 H)

Functionalization of wt-TMV: To introduce a ketone residue to the virus surface, diazonium salt was prepared from aqueous p-toluenesulfonic acid monohydrate (800 μL, 0.84 M), aqueous sodium nitrite (800 μL, 0.46 M) and p-aminoacetophenone (1.6 mL, 0.15 M in acetonitrile), mixed in a reaction tube and placed in an ice bath. The color of the solution was converted to yellow while stirring for 1 h. Then, virus solution (1033 μL, 4.84 mg/mL) was prepared separately by mixing wt-TMV (500 μL, 10 mg/mL) in 10 mM SPP buffer (pH 7.4) with borate buffer (533 μL, 68 mM) containing 100 mM NaCl (pH 9). 160 μL (11.97 μmoL) diazonium salt solution were mixed with the obtained virus solution and kept in a water bath in the fridge at 4°C for 2 h. The color of the solution turned from yellow to dark brown. The reaction solution was purified using a PD MiniTrap G-25 column with an exclusion limit of approximately M_r = 5000 applying a gravity protocol. The column was eluted with 100 mM aqueous potassium phosphate buffer (pH 6). After the purification step, 2 mL containing 2.1 mg/mL TMV-ketone (**4**) were obtained. The synthesis of the product (**4**) was followed by SDS-PAGE analysis.

Synthesis of h-TMV **(4)**: Aminooxy PE-b-PEG (**3**) (0.1614 g, 71.25 μmoL), the TMV-ketone (**4**) (2 mL, 2.1 mg/mL) and 2 mL 100 mM potassium phosphate buffer (pH 6) were mixed in a reaction tube. The mixture was gently rotated overnight at room temperature and purified with a PD MiniTrap G-25 column. Further purification was obtained by size exclusion chromatography at a flow rate of 0.5 mL

min^{-1} using a 16/60 Sephacryl S-500 column and 50 mM SPP containing 150 mM NaCl as running buffer. A Pharmacia Biotech LCC-501 PLUS FPLC system with automatic fraction collector was used at room temperature. Fractions exhibiting 1.19 absorbance ratio of 260 nm and 280 nm, indicative for virus particles composed of 95 w/w % proteins and 5 w/w % RNA, were pooled and precipitated by ultracentrifugation at 34,000 rpm at 4°C in a 45 Ti rotor for 2 h. The functionalized virus particles were then resuspended in ultrapure water. The success of the reaction was proved by SDS-PAGE. The height and integrity of the hydrophobized viruses (**5**) was checked by AFM. In a control experiment, wt-TMV (10 µL, 10 mg/mL), borate buffer (10.7 µL, 68 mM) containing 100 mM NaCl (pH 9), potassium phosphate buffer (160 µL, 100 mM, pH 6) and aminooxy PE-b-PEG **(3)** (3.3 mg, 1.46 µmoL) were mixed and rotated overnight at room temperature. The reaction mixture was purified and characterized according to the procedure used for product (**5**).

Sodiumdodecylsulfate-polyacrylamide-gel electrophoresis (SDS-PAGE): Standard procedures according to Green et al. were applied [32]. Samples containing between 0.2–1 µg protein were heated for 5 min at 95 °C in sample buffer and resolved in 15% discontinuous SDS-PA gels. Fixed gels were stained using a silver staining procedure.

ZnS mineralization: Silicon substrates immobilized with the corresponding virus types were fixed in a holder and mounted perpendicular in the center of a vessel. There, the viruses were incubated in aqueous $ZnCl_2$ precursor solution (30 mL, 100 mM, pH 6.2) for 20 min. Then, equimolar aqueous Na_2S solution (pH 12.5) was added dropwise to the reaction solution applying a peristaltic pump with a constant speed (1 mL min^{-1}) under continuous stirring at room temperature. The addition of Na_2S to the $ZnCl_2$ precursor solution caused a drop in the pH to 3.5, and the reaction pH was maintained until the dropwise addition was stopped. Within the addition of the first Na_2S droplets, a white precipitate was formed in the deposition solution. After finishing the reaction (controlled according to the required reaction time (5 or 30 min)), the substrates were washed thoroughly with ultrapure water and dried under argon stream.

Sample characterization: Atomic force microscopy on a Digital Instruments MultiMode 8 from Bruker with a NanoScope 5 controller operated in tapping mode was used to image the immobilized and the mineralized viruses. Silicon cantilevers and PPP-NCHR-W (Nanosensors) n^+ doped tips with resistivity 0.01–0.02 Ω cm were used. The virus height of wt-TMV and h-TMV was evaluated with the manufacturer's software Nanoscope. The height of a single virus was averaged over the whole virus length with the software. The average height of 10 virus particles from each virus type was used to obtain the virus height distribution of the bare virus. The virus height of wt-TMV and h-TMV after treatment with ZnS deposition solution for 5 min, 20 min and 30 min was determined applying the same procedure, taking 5 to 10 virus particles of each virus type for the evaluation and comparing only wt-TMV and h-TMV particles mineralized simultaneously in the same deposition solution. ^{1}H-NMR measurements were conducted on Bruker Avance 500 in 500 MHz field and analyzed by ACD/Spectrus Processor. Fourier transform infrared spectroscopy (FT-IR) analysis of the samples was performed at a Bruker FT-IR spectrometer. The samples were investigated in the wavelength range: 500–4000/cm^{-1} and the spectra were analyzed by ACD/Spectrus Processor.

3. Results and Discussion

The wt-TMV particle consists of a genomic RNA helix, embedded into about 2130 identical CP subunits. It has a tube-like structure, which is 300 nm long. The inner diameter of the longitudinal central channel is 4 nm, while the outer nanotube diameter is 18 nm when the capsid is dispersed in solution. It may be flattened to a height of around 14 nm upon immobilization on a hydrophilic silicon substrate surface [12,33]. The viral CPs can be engineered genetically to introduce amino acid residues on the virus surface with certain terminal groups appropriate for bioconjugation reactions. In this study, wt-TMV and a TMV-Cys mutant, presenting a thiol group for coupling on each CP were used to find appropriate synthetic pathways for covalent attachment of organic molecules making the virus surface hydrophobic without destroying the integrity of the capsid, in order to suppress the virus mineralization by inorganic deposits as shown in Scheme 1.

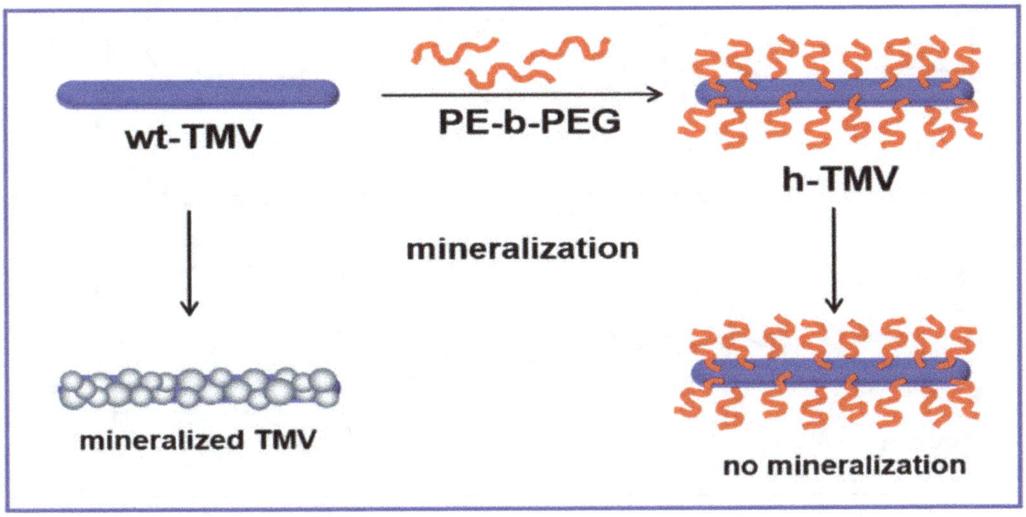

Scheme 1. Schematic representation of tobacco mosaic virus (TMV) functionalization making the virus surface hydrophobic, which restricts subsequent mineralization.

Recently, we reported on selective mineralization of wt-TMV with ZnS at room temperature in aqueous solution and at various reaction pH [10]. $ZnCl_2$ and Na_2S were used as ZnS precursors without the need of any further additives as structure directing or complexing agents. The growth mechanism of the inorganic material on the virus surface was studied comparing the optical properties, band gap (Eg) and particle size of solution-grown ZnS nanoparticles and ZnS nanoparticles mineralized onto the virus template. It was shown that the virus particles played templating function, and heterogeneous nucleation triggered by the virus surface was confirmed. The XRD analysis confirmed formation of ZnS with cubic lattice structure in solution and on the virus surface. Since well-studied, the mineralization of wt-TMV with ZnS applying this reaction was used as a reference in this work and compared to the mineralization behavior of the functionalized virus particles in order to identify, if the change of virus surface hydrophilicity influenced the mineralization process.

3.1. Reaction of TMV with Highly Hydrophobic Polymers

A covalent attachment of a highly hydrophobic polymer such as poly(pentafluorostyrene) (PFS) to the virus surface could provide both specific coupling to the virus surface and high surface hydrophobicity. Therefore, first a click reaction of PFS to a TMV-Cys was examined. Our previous study revealed that the reactivity of the para-fluorine function in the PFS with nucleophiles such as sulfides is enhanced by the reduced electron density of the perfluorinated phenyl ring, allowing the use of relatively mild reaction conditions [31]. The reaction is conducted in *N,N*-dimethylacetamide (DMAc) or tetrahydrofuran/ triethylamine (THF/TEA) at room temperature. For the click reaction, TMV-Cys particles were selected, because each of their CPs provides a single surface-accessible thiol group, introduced by genetic modification of the third CP amino acid from serine to cysteine (S3C) [27]. The bioconjugation was conducted according to the reaction pathway presented in Scheme 2A.

Before starting with the coupling reaction, the stability of TMV-Cys in DMAc and in THF with or without TEA was studied at room temperature. To do this, viruses immobilized on a plasma cleaned silicon substrate (Figure 1a) were incubated in DMAc for 30 min. This led to partial loss of the virus integrity especially visible in Figure 1b, where the viruses became roughened and the virus sharp ends are lost. The mean virus height is reduced from 14.5 nm to around 13 nm. Finally, reaction with PFS in DMAc for the same reaction time led to even further disassembly of the viruses, where the virus coat appeared irregular, with an uneven coverage of the RNA with CPs (Figure 1c).

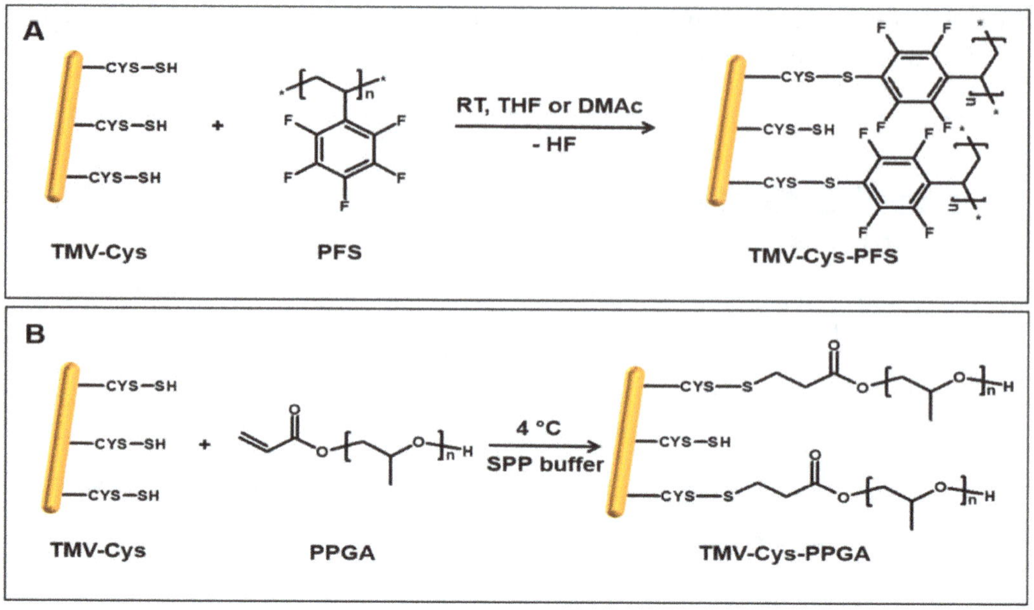

Scheme 2. Click reaction of TMV-Cys mutant with poly(pentafluorostyrene) (PFS) and poly(propylene glycol) acrylate (PPGA).

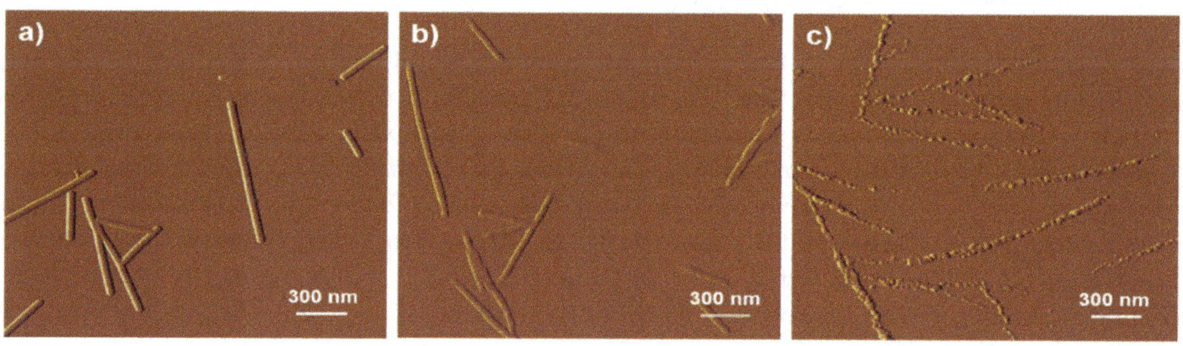

Figure 1. AFM amplitude images of: (**a**) TMV-Cys virus particles, (**b**) TMV-Cys incubated in N,N-dimethylacetamide (DMAc) for 30 min and (**c**) TMV-Cys reacted with poly(pentafluorostyrene) (PFS) in DMAc for 30 min.

In comparison, the virus structure seemed stable in both THF and THF/TEA for a few hours (Figure 2a,b). This is in contradiction to our previous experiments on the stability of TMV dispersed in THF/H_2O mixtures, where TMVs appeared to be unstable. The reason for the increased stability in this case might be explained by the immobilization of the viruses on the substrate. Based on the enhanced overall structural stability and the need of an extended reaction time for better educts conversion, the TMV-Cys particles were let to react with PFS in THF/TEA for 2 h at room temperature. The AFM analysis indicated that the capsids stayed intact and retained their integrity after this treatment without change in virus height (Figure 2c). However, subsequent mineralization resulted in deposition of ZnS on the virus surface, and the virus height increased to 16.5 nm (Figure 2d,e). The latter suggests that although stable in this reaction solution, the viruses could not be functionalized with PFS. This may be attributed to the lower reactivity of PFS in THF/TEA compared to that in DMAc, which not only requires longer reaction time but usually also higher temperatures not suitable for TMV.

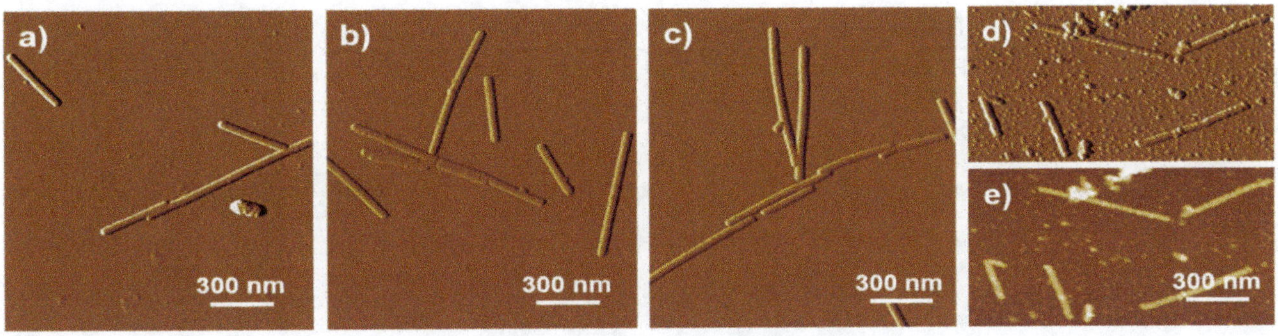

Figure 2. AFM amplitude images of TMV-Cys virus particles treated: (**a**) 1 h in tetrahydrofuran (THF), (**b**) 1 h in THF/TEA, (**c**) 2 h with PFS in THF/TEA at room temperature. (**d**) AFM amplitude image of viruses from (**c**) mineralized 5 min with ZnS and (**e**) AFM height image of (**d**).

3.2. TMV Hydrophobization with Aliphatic Polyethers

As established in the previous section, the highly hydrophobic environment and the aggressive reaction solvent necessary for dissolving the polymer caused strain on the virus structure and subsequent disintegration. As an alternative, so-called thermo-responsive polymers e.g., poly (propylene glycol)methacrylate can be applied [34]. This polymer can change its hydrophobicity with temperature. It is water-soluble at temperatures below ~10 °C and neutral pH, and becomes insoluble at higher temperatures due to conformational changes inducing hydrophobization. This would allow a chemical coupling of the polymer at 4 °C in its hydrophilic form to the virus surface with expectation that the virus will retain its stability at these conditions. Then, a mineralization test at room temperature or at elevated temperatures, when the polymer becomes hydrophobic, will be conducted to examine the obtained hydrophobicity. Due to the higher reactivity of the acrylate group over the methacrylate group as a Michael addition donor, poly(propylene glycol)acrylate (PPGA) (M_n = 475 g mol^{-1}) was chosen for our experiments (Scheme 2B).

To determine the coupling efficiency and specificity, TMV-Cys and wt-TMV were incubated at 4 °C where PPGA is completely soluble (see Supplementary Materials) for 72 h, with PPGA added at different ratios with regard to the number of CPs. SDS-PAGE analysis of the CPs revealed for a ratio of 1000:1 PPGA:CP-Cys only a slight smear above the CP band, which increased substantially at a ratio of 3300:1, but not significantly more at a further excess of PPGA:CP-Cys of 10,000:1 (see Figure 3). The coupling of PPGA to CP-Cys did not result in a defined band due to the polydispersity of PPGA. The absence of any smear above wt-CPs of wt-TMV particles treated in parallel shows on the one hand that the reaction was specific for TMV-Cys, and on the other hand that the TMV structure stayed intact during the reaction, as during TMV particle disassembly an otherwise hidden cysteine residue would become accessible to the reaction with PPGA.

As PPGA was used in high excess, residual unreacted polymer was removed prior to further investigations by a combined ultracentrifugation/fast protein liquid chromatography (UC/FPLC) approach. All purification steps were performed below 5 °C to prevent switching of PPGA into the hydrophobic state. The first ultracentrifugation step was necessary to get rid of the bulk amount of PPGA, which may result in a broadening of the peaks due to overloading size-exclusion chromatography. About 25 % of the TMV initially used for coupling could be obtained in pure form. Figure 4a,b show AFM images of PPGA-functionalized TMV-Cys before and after purification, where residual excess PPGA appears as large spots around the TMV rods in the first image and had disappeared in samples after the cleaning procedure.

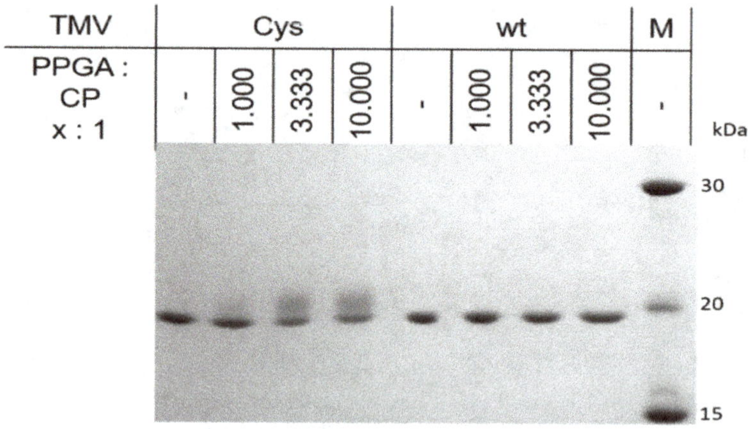

Figure 3. Silver-stained 15%-SDS-PAGE showing the coupling efficiency at different ratios of PPGA to CP using TMV-Cys and no coupling of PPGA to wt-TMV. The product of the conjugation reaction can be seen as a smear above the CP band at about 20 kDa.

After having shown that the thermo-responsive PPGA was successfully coupled to TMV-Cys, the degree of hydrophobicity achieved with PPGA was investigated by subjecting the modified nanotubes to mineralization reaction with ZnS [10]. PPGA-functionalized TMV-Cys viruses were immobilized on a plasma cleaned silicon substrate, which was placed perpendicularly in $ZnCl_2$ aqueous solution in order to avoid attachment of big ZnS agglomerates formed in solution on the substrate surface. The mineralization reaction was allowed to proceed for 5 min, which based on previous experiments is enough to observe differences in the mineralization behavior between untreated TMV-Cys and hydrophobized particles, respectively. The amplitude AFM image in Figure 4c clearly shows that the PPGA-functionalized TMV-Cys surface is covered with small ZnS nanoparticles, and the virus height calculated from the corresponding AFM height images was increased by around 1.6 nm. This leads to the conclusion that, although successfully attached to the virus surface preserving the capsid structure, the thermo-responsive PPGA residues could not provide sufficient hydrophobicity to fully suppress virus mineralization.

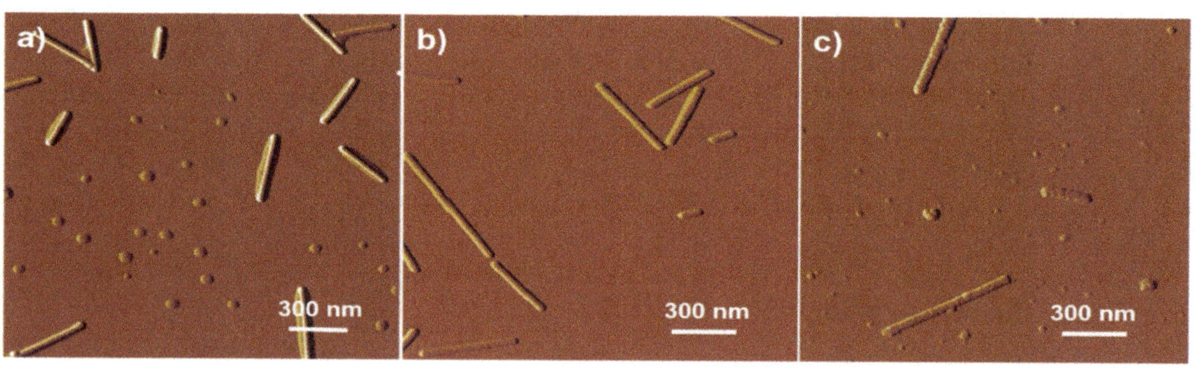

Figure 4. AFM amplitude images of TMV-Cys-PPGA: (**a**) before and (**b**) after purification. (**c**) TMV-Cys-PPGA mineralized with ZnS.

3.3. TMV Hydrophobization with Block-Co-Polymers

Covalent attachment of PPGA to the virus CPs was possible without affecting the overall capsid structure, however, the obtained viruses were not hydrophobic enough to suppress mineralization. A possible reason for this could be the length of the PPGA molecules, and, as obvious from the SDS-PAGE analysis presented in Figure 3, only part of the viral CPs was covalently coupled with PPGA. Another possible reason may be connected to the surface charge of the functionalized viruses. The overall surface charge of organic templates (in our case functionalized TMV) plays an important role for mineralization in solution. The presence of even partial charges like from the free electron pairs of the oxygen in the PPGA polyether can induce mineralization of the PPGA-functionalized viruses. Therefore in the next step, a synthetic pathway developed by Schlick et al. [35]. was used. In their work, the TMV capsid exterior was covered with polyethylene glycol (PEG) to make TMV soluble in various hydrophobic organic solvents and to increase its thermal stability. Although PEG introduced sufficient hydrophobicity on the virus surface to transfer the PEG-TMV conjugate from aqueous solution into organic solvents, the presence of oxygen in the PEG chains would support mineralization of the virus, similar to the case of TMV-Cys-PPGA. Therefore, in order to introduce sufficient hydrophobicity without affecting the integrity of the virus structure, we covalently coupled the block-co-polymer $(PE)_{11}$-$(PEG)_{34}$-OH (M_w 2250 g/moL) to the wt-TMV surface. The non-polar polyethylene (PE) part is expected to ensure sufficient hydrophobicity of the virus surface and to suppress the mineralization of the organic template. The PEG chain, compatible with the virus, should separate the virus from the hydrophobic PE chain. Additionally, the PEG spacer unit is responsible to attach the block-co-polymer to the virus CP [36]. To do this, we used the already reported electrophilic substitution reaction with diazonium salts at the *ortho* position of the phenyl ring of the tyrosine residue (Y139) in TMV CPs [35]. The polymer coupling was done in a three-step procedure as can be seen in Scheme 3.

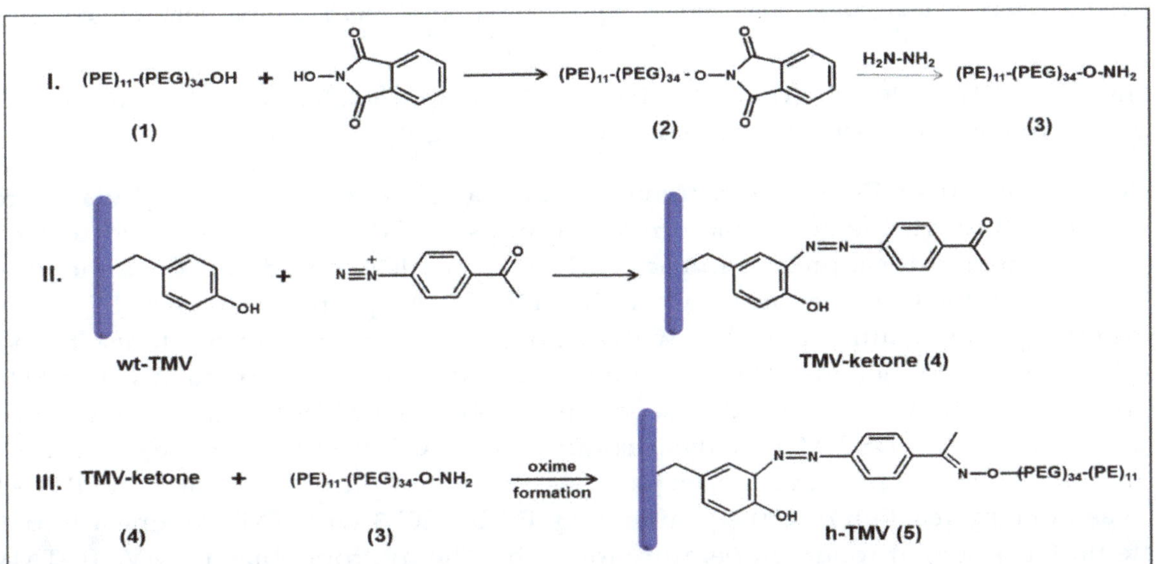

Scheme 3. Synthetic pathway for functionalization of $(PE)_{11}$-$(PEG)_{34}$-OH block-co-polymer **1** and formation of hydrophobized h-TMV.

To this end, the block-co-polymer **1** converted into the alkoxyamine **3** was reacted with the ketone-functionalized wt-TMV **4** to form the ketoxime **5**. In step I, the terminal hydroxyl group of the block-co-polymer **1** was activated with N-hydroxyphtalimide. As a good leaving group, N-hydroxyphtalimide facilitated the subsequent reaction with hydrazine resulting in the formation of alkoxyamine **3**. The success of the reactions was verified by ^{1}H-NMR combined with FT-IR spectroscopy [37]. The appearance of protons with a chemical shift in the low field region around 7.8 ppm and an IR signal in 1730–1790 cm^{-1} corresponding to aromatic and carbonyl residues by product **2** (Figure 5) and their disappearance by product **3** confirmed the attachment and the subsequent replacement of N-hydroxyphtalimide from the polymer.

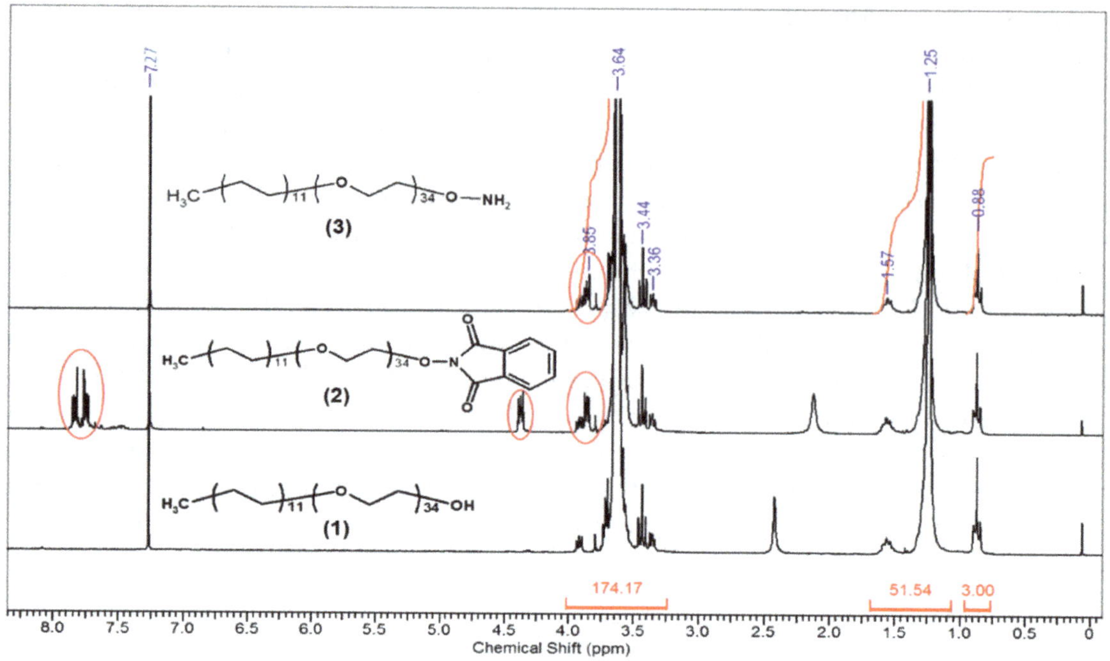

Figure 5. ^{1}H-NMR of **(1)** $(PE)_{11}$-$(PEG)_{34}$-OH block-co-polymer, **(2)** activated with the N-hydroxyphtalimide derivative and **(3)** the corresponding alkoxyamine.

In step II, the virus CPs were functionalized and a terminal ketone group was introduced, which was done through the accessible tyrosine groups of CPs. It was assumed that only one of the four tyrosine residues, present in each TMV CP, should be modified since the other three are hidden between the CPs after assembly to the helical TMV particle structure [35]. The ketone group was included by treatment of TMV with diazonium salt, prepared prior to use from sodium nitrite and p-aminoacetophenone. The virus functionalization was performed at 4 °C, pH 9 for 2 h. PD MiniTrap G-25 columns were used to separate the obtained TMV-ketone **4** from the added in surplus diazonium salt. SDS-PAGE analysis confirmed the CP modification (see Figure S1) while AFM examination showed preserved virus particles. The hydrophobization of wt-TMV (Scheme 3, step III) was done by reaction of 250 eq. aminooxy PE-b-PEG **3** with TMV-ketone **4** in potassium phosphate buffer (pH 6) at room temperature for 15 h. The hydrophobized TMVs (h-TMV) were purified through PD MiniTrap G-25 columns. Then, the h-TMVs were further separated from the not reacted polymer residue via FPLC and concentrated via UC. The shift of the CP band in the SDS-PA gel (Figure S1) to one with higher molecular weight clearly indicated the successful attachment of polymer to the virus CPs compared to the control experiment, where the aminooxy PE-b-PEG **3** was reacted with a not functionalized wt-TMV and no change in the position of wt-TMV CPs was observed. The blurring of the band corresponding to the CPs of h-TMV can be explained by the polydispersity of the block-co-polymer. The AFM height images in Figure 6 confirmed that the h-TMV particles remained intact. The AFM cross-sections have shown that compared to wt-TMV (Figure 6a,c), the polymer attachment on the virus surface led to an increase in virus height of about 3 nm on hydrophilic (O_2-plasma treated) (Figure 6b) but also on hydrophobic (not plasma treated) (Figure 6d) silicon substrates.

As stated in previous reports, very polar and protic solvents lead to precipitation of PEG functionalized TMV [36]. Therefore, the stability of the h-TMV in the precursor solution was examined before starting with the mineralization test with ZnS. A silicon substrate with immobilized h-TMV was placed in $ZnCl_2$ aqueous solution. After 30 min stirring at room temperature, the viruses could still be detected as intact particles via AFM (see Figure S2a). Then, a substrate with h-TMV and one substrate with immobilized wt-TMV, used as a reference, were mineralized simultaneously with ZnS for different reaction times. In Figure 7 are shown the corresponding AFM height, amplitude and phase images of wt-TMV and h-TMV after 5 and 30 min treatment. It can be seen that after 5 min reaction time, the surface of wt-TMV became rough due to attachment of ZnS particles, while the surface of h-TMV remained smooth. This is especially visible comparing their AFM amplitude and phase images. Increase of the reaction time to 30 min maintained the trend. The amount of the ZnS particles on wt-TMV surface was increased, which is well detectible also on the AFM height image. Opposed to this, no specific ZnS deposition on h-TMV was observed.

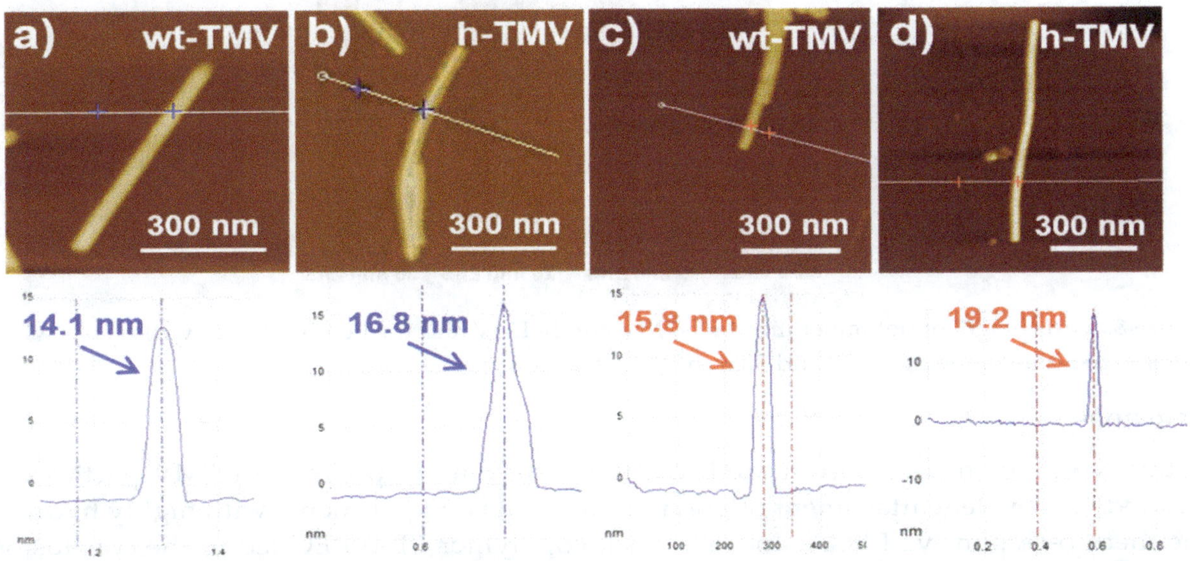

Figure 6. AFM height images with the corresponding cross-sections showing the virus height of: (**a**) wt-TMV and (**b**) h-TMV immobilized on hydrophilic (O_2-plasma treated) silicon substrates. (**c**) wt-TMV and (**d**) h-TMV immobilized on hydrophobic (not plasma treated) silicon substrates.

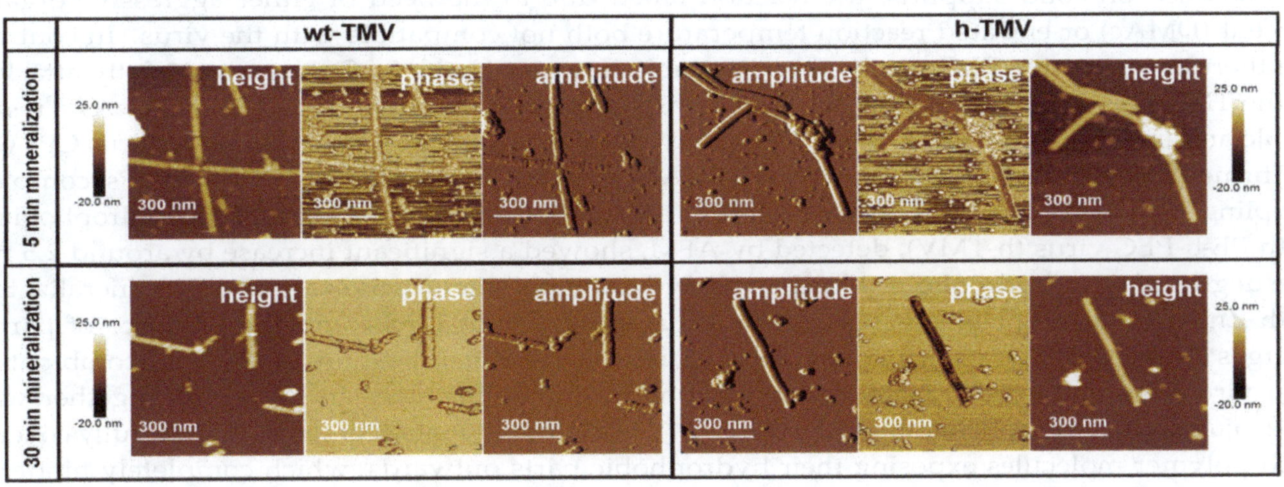

Figure 7. AFM height, amplitude and phase images of wt-TMV and h-TMV mineralized simultaneously with ZnS for 5 and 30 min, respectively.

The observation of a selective deposition of ZnS nanoparticles only on the surface of wt-TMV, but not on the surface of the hydrophobized counterpart was further supported by evaluation and comparison of the virus height of wt-TMV and h-TMV treated with ZnS deposition solution for a certain reaction time (Figure 8). Since small fluctuations in the mineralization conditions might cause deviations in the deposition rate and hence might influence the resulting virus height, for better assessment, the virus heights of wt-TMV and h-TMV mineralized in the same deposition solution and within the same experiment were compared. The chart diagram in Figure 8 shows that in comparison to the hydrophobized viruses, whose surface remain smooth with the reaction time, the wt-TMV particles increase their height linearly with the mineralization time. The latter is a result of selective deposition of ZnS on the virus surface with a constant rate, which is in a good agreement with our previous observations for mineralization of wt-TMV with ZnS applying this reaction [10].

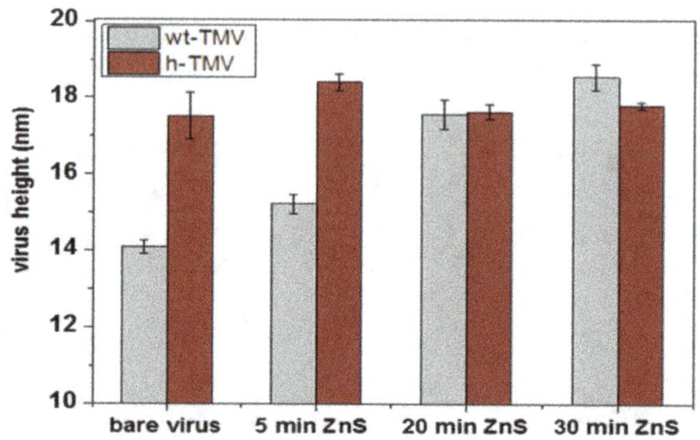

Figure 8. Virus height of not mineralized wt-TMV and h-TMV and the height of both virus types when mineralized with ZnS after 5, 20 and 30 min.

4. Conclusions

In this work, we have demonstrated that the surface hydrophilicity of TMV particles can be manipulated via covalent attachment of polymer molecules. Experiments with highly hydrophobic PFS, the thermo-responsive PPGA, and the block-copolymer PE-b-PEG led to the conclusion that the success of the reaction depends on the polymer features. The use of perfluorinated polymers showed that highly hydrophobic polymers are not appropriate candidates for virus functionalization. Although the virus particles displayed significant stability enhancement in organic solvents when immobilized on solid supports, the reaction failed due to the need of either aggressive organic solvent (DMAc) or elevated reaction temperature both not compatible with the virus. In contrast, the thermo-responsive PPGA polyether, insoluble in water above 8 °C, was successfully attached to the TMV-Cys surface, and also PE-b-PEG was successfully bonded to wt-TMV CPs. As a result, stable and intact functionalized particles were synthesized. The covalent coupling to virus CPs was confirmed via SDS-PAGE analysis. While PPGA only partially coupled to TMV-Cys CPs, complete coupling efficiency was reached with PE-b-PEG and wt-TMV CPs. The height of the hydrophobized with PE-b-PEG virus (h-TMV), detected by AFM, showed a significant increase by around 2.5 nm. The degree of virus hydrophobicity, achieved with both polymers, was examined by mineralization with ZnS. The incomplete coverage of the virus surface with PPGA and the presence of partial charges in the polymer chains could not suppress virus mineralization. However, combination of a virus-compatible PEG block and a pure hydrophobic PE block in PE-b-PEG together with bioconjugation of the polymer to most CPs in TMV led to the formation of intact h-TMV, fully covered with polymer molecules exposing their hydrophobic parts outwards, which completely hindered subsequent mineralization of the virus surface. This synthetic approach could be used for the synthesis of Janus-type TMV particles with a hydrophobic and a hydrophilic part in future experiments, to study the mineralization behavior of TMV mutants within a single artificial TMV particle with an intrinsic hydrophobic reference. Furthermore, the self-organization of pure and mineralized Janus-type TMV particles in complex highly organized structures as well as their potential integration in interfacial catalysis and as drug cargo is envisioned.

Supplementary Materials:
Figure S1: SDS-PAGE of samples containing: (**1**) wt-TMV, (**2**) TMV-ketone, (**3**) TMV-ketone after purification with a PD MiniTrap G-25 column, (**4**) h-TMV, (**5**) h-TMV after purification with a PD MiniTrap G-25 column and (**6**) control experiment including reaction of wt-TMV with PE-b-PEG. Figure S2: Lower magnification AFM height images showing the stability of the h-TMV particles: (**a**) after 30 min stirring in 100 mM $ZnCl_2$ precursor solution and (**b**) after 5 min mineralization with ZnS.

Author Contributions: Conceptualization, P.A., S.E., V.A. and A.S.; Investigation, P.A., S.E., V.A., L.W. and E.C.; Validation, E.C. and L.W.; Writing—Original Draft, P.A., S.E. and V.A.; Writing—Review & Editing, J.B., C.W., P.A., S.E., V.A., E.C. and A.S.; Supervision, J.B., J.K., P.A. and S.E.

References

1. Yan, H.; Park, S.H.; Finkelstein, G.; Reif, J.H.; LaBean, T.H. DNA-templated self-assembly of protein arrays and highly conductive nanowires. *Science* **2003**, *301*, 1882–1884. [CrossRef]
2. Tseng, R.J.; Tsai, C.L.; Ma, L.P.; Ouyang, J.Y. Digital memory device based on tobacco mosaic virus conjugated with nanoparticles. *Nat. Nanotechnol.* **2006**, *1*, 72–77. [CrossRef]
3. Mao, C.B.; Solis, D.J.; Reiss, B.D.; Kottmann, S.T.; Sweeney, R.Y.; Hayhurst, A.; Georgiou, G.; Iverson, B.; Belcher, A.M. Virus-based toolkit for the directed synthesis of magnetic and semiconducting nanowires. *Science* **2004**, *303*, 213–217. [CrossRef]
4. Rothemund, P.W.K. Folding DNA to create nanoscale shapes and patterns. *Nature* **2006**, *440*, 297–302. [CrossRef] [PubMed]
5. Atanasova, P.; Stitz, N.; Sanctis, S.; Maurer, J.H.; Hoffmann, R.C.; Eiben, S.; Jeske, H.; Schneider, J.J.; Bill, J. Genetically improved monolayer-forming tobacco mosaic viruses to generate nanostructured semiconducting bio/inorganic hybrids. *Langmuir* **2015**, *31*, 3897–3903. [CrossRef] [PubMed]
6. Wargacki, S.P.; Pate, B.; Vaia, R.A. Fabrication of 2D ordered films of tobacco mosaic virus (TMV): Processing morphology correlations for convective assembly. *Langmuir* **2008**, *24*, 5439–5444. [CrossRef] [PubMed]
7. Lin, Z. *Evaporative Self-Assembly of Ordered Complex Structures*; World Scientific: Singapore; Hackensack, NJ, USA, 2012; 380p.
8. Lomonossoff, G.P.; Wege, C. Chapter Six—TMV Particles: The Journey From Fundamental Studies to Bionanotechnology Applications. In *Advances in Virus Research*; Palukaitis, P., Roossinck, M.J., Eds.; Academic Press: Cambridge, MA, USA, 2018; pp. 149–176.
9. Fan, X.Z.; Pomerantseva, E.; Gnerlich, M.; Brown, A.; Gerasopoulos, K.; McCarthy, M.; Culver, J.; Ghodssi, R. Tobacco mosaic virus: A biological building block for micro/nano/bio systems. *J. Vac. Sci. Technol. A* **2013**, *31*. [CrossRef]
10. Atanasova, P.; Kim, I.; Chen, B.; Eiben, S.; Bill, J. Controllable virus-directed synthesis of nanostructured hybrids induced by organic/inorganic interactions. *Adv. Biosys.* **2017**, *1*, 1700106. [CrossRef]
11. Bittner, A.M.; Alonso, J.M.; Górzny, M.L.; Wege, C. Nanoscale science and technology with plant viruses and bacteriophages. In *Structure and Physics of Viruses: An Integrated Textbook*; Mateu, M.G., Ed.; Springer Science+Business Media: Dordrecht, The Netherlands, 2013; pp. 667–702.
12. Atanasova, P.; Rothenstein, D.; Schneider, J.J.; Hoffmann, R.C.; Dilfer, S.; Eiben, S.; Wege, C.; Jeske, H.; Bill, J. Virus-templated synthesis of ZnO nanostructures and formation of field-effect transistors. *Adv. Mater.* **2011**, *23*, 4918–4922. [CrossRef]
13. Zang, F.H.; Gerasopoulos, K.; Fan, X.Z.; Brown, A.D.; Culver, J.N.; Ghodssi, R. An electrochemical sensor for selective TNT sensing based on tobacco mosaic virus-like particle binding agents. *Chem. Commun.* **2014**, *50*, 12977–12980. [CrossRef] [PubMed]
14. Bruckman, M.A.; Liu, J.; Koley, G.; Li, Y.; Benicewicz, B.; Niu, Z.W.; Wang, Q.A. Tobacco mosaic virus based thin film sensor for detection of volatile organic compounds. *J. Mater. Chem.* **2010**, *20*, 5715–5719. [CrossRef]
15. Koch, C.; Eber, F.J.; Azucena, C.; Forste, A.; Walheim, S.; Schimmel, T.; Bittner, A.M.; Jeske, H.; Gliemann, H.; Eiben, S.; et al. Novel roles for well-known players: From tobacco mosaic virus pests to enzymatically active assemblies. *Beilstein J. Nanotech.* **2016**, *7*. [CrossRef] [PubMed]
16. Eiben, S.; Koch, C.; Altintoprak, K.; Southan, A.; Tovar, G.; Laschat, S.; Weiss, I.; Wege, C. Plant virus-based materials for biomedical applications: Trends and prospects. *Adv. Drug Delivery Rev.* **2018**. [CrossRef]
17. Chen, X.L.; Gerasopoulos, K.; Guo, J.C.; Brown, A.; Wang, C.S.; Ghodssi, R.; Culver, J.N. Virus-enabled silicon anode for lithium-ion batteries. *ACS Nano* **2010**, *4*, 5366–5372. [CrossRef] [PubMed]
18. Niu, Z.; Liu, J.; Lee, L.A.; Bruckman, M.A.; Zhao, D.; Koley, G.; Wang, Q. Biological templated synthesis of water-soluble conductive polymeric nanowires. *Nano Lett.* **2007**, *7*, 3729–3733. [CrossRef]
19. Turpen, T.H.; Reinl, S.J.; Charoenvit, Y.; Hoffman, S.L.; Fallarme, V.; Grill, L.K. Malarial epitopes expressed on the surface of recombinant tobacco mosaic-virus. *Biotechnology* **1995**, *13*, 53–57. [PubMed]
20. Lee, L.A.; Nguyen, Q.L.; Wu, L.; Horvath, G.; Nelson, R.S.; Wang, Q. Mutant plant viruses with cell binding motifs provide differential adhesion strengths and morphologies. *Biomacromolecules* **2012**, *13*, 422–431. [CrossRef]

21. Lu, B.; Taraporewala, Z.F.; Stubbs, G.; Culver, J.N. Intersubunit interactions allowing a carboxylate mutant coat protein to inhibit tobamovirus disassembly. *Virology* **1998**, *244*, 13–19. [CrossRef] [PubMed]
22. Bendahmane, M.; Koo, M.; Karrer, E.; Beachy, R.N. Display of epitopes on the surface of tobacco mosaic virus: Impact of charge and isoelectric point of the epitope on virus-host interactions. *J. Mol. Biol.* **1999**, *290*, 9–20. [CrossRef]
23. Zhou, K.; Li, F.; Dai, G.; Meng, C.; Wang, Q. Disulfide bond: Dramatically enhanced assembly capability and structural stability of tobacco mosaic virus nanorods. *Biomacromolecules* **2013**, *14*, 2593–2600. [CrossRef]
24. Zhang, J.; Zhou, K.; Wang, Q. Tailoring the self-assembly behaviors of recombinant tobacco mosaic virus by rationally introducing covalent bonding at the protein–protein interface. *Small* **2016**, *12*, 4955–4959. [CrossRef]
25. Chu, S.; Brown, A.D.; Culver, J.N.; Ghodssi, R. Tobacco mosaic virus as a versatile platform for molecular assembly and device fabrication. *Biotechnol. J.* **2018**, *13*, 1800147. [CrossRef]
26. Schneider, A.; Eber, F.J.; Wenz, N.L.; Altintoprak, K.; Jeske, H.; Eiben, S.; Wege, C. Dynamic DNA-controlled "stop-and-go" assembly of well-defined protein domains on RNA-scaffolded TMV-like nanotubes. *Nanoscale* **2016**, *8*, 19853–19866. [CrossRef] [PubMed]
27. Geiger, F.C.; Eber, F.J.; Eiben, S.; Mueller, A.; Jeske, H.; Spatz, J.P.; Wege, C. TMV nanorods with programmed longitudinal domains of differently addressable coat proteins. *Nanoscale* **2013**, *5*, 3808–3816. [CrossRef]
28. Hu, J.; Zhou, S.X.; Sun, Y.Y.; Fang, X.S.; Wu, L.M. Fabrication, properties and applications of Janus particles. *Chem. Soc. Rev.* **2012**, *41*, 4356–4378. [CrossRef]
29. Patil, A.J.; McGrath, N.; Barclay, J.E.; Evans, D.J.; Colfen, H.; Manners, I.; Perriman, A.W.; Mann, S. Liquid viruses by nanoscale engineering of capsid surfaces. *Adv. Mater.* **2012**, *24*, 4557–4563. [CrossRef]
30. Gooding, G.V.; Hebert, T.T. A simple technique for purification of tobacco mosaic virus in large quantities. *Phytopathology* **1967**, *57*, 1285.
31. Atanasov, V.; Burger, M.; Lyonnard, S.; Porcar, L.; Kerres, J. Sulfonated poly(pentafluorostyrene): Synthesis & characterization. *Solid State Ionics* **2013**, *252*, 75–83. [CrossRef]
32. Green, M.R.; Sambrook, J. *Molecular Cloning: A Laboratory Manual*, 4th ed.; Cold Spring Harbor Laboratory Press: Cold Spring Harbor, NY, USA, 2012.
33. Knez, M.; Sumser, M.P.; Bittner, A.M.; Wege, C.; Jeske, H.; Hoffmann, D.M.P.; Kuhnke, K.; Kern, K. Binding the tobacco mosaic virus to inorganic surfaces. *Langmuir* **2004**, *20*, 441–447. [CrossRef]
34. Suljovrujic, E.; Micic, M. Smart poly(oligo(propylene glycol) methacrylate) hydrogel prepared by gamma radiation. *Nucl. Instrum. Meth. B* **2015**, *342*, 206–214. [CrossRef]
35. Schlick, T.L.; Ding, Z.B.; Kovacs, E.W.; Francis, M.B. Dual-surface modification of the tobacco mosaic virus. *J. Am. Chem. Soc.* **2005**, *127*, 3718–3723. [CrossRef] [PubMed]
36. Holder, P.G.; Finley, D.T.; Stephanopoulos, N.; Walton, R.; Clark, D.S.; Francis, M.B. Dramatic Thermal stability of virus-polymer conjugates in hydrophobic solvents. *Langmuir* **2010**, *26*, 17383–17388. [CrossRef] [PubMed]
37. Iha, R.K.; Van Horn, B.A.; Wooley, K.L. Complex, degradable polyester materials via ketoxime ether-based functionalization: Amphiphilic, multifunctional graft copolymers and their resulting solution-state aggregates. *J. Polym. Sci. Pol. Chem.* **2010**, *48*, 3553–3563. [CrossRef]

8

Conserved Secondary Structures in Viral mRNAs

Michael Kiening [1], Roman Ochsenreiter [2], Hans-Jörg Hellinger [4], Thomas Rattei [4], Ivo Hofacker [2,3] and Dmitrij Frishman [1,5,*]

1 Department of Bioinformatics, Wissenschaftszentrum Weihenstephan, Technische Universität München, Maximus-von-Imhof-Forum 3, D-85354 Freising, Germany; m.kiening@wzw.tum.de

2 Research Group Bioinformatics and Computational Biology, Faculty of Computer Science, University of Vienna, Währingerstr. 29, 1090 Vienna, Austria; romanoch@tbi.univie.ac.at (R.O.); ivo@tbi.univie.ac.at (I.H.)

3 Department of Theoretical Chemistry, Faculty of Chemistry, University of Vienna, Währingerstrasse 17, 1090 Vienna, Austria

4 Division of Computational Systems Biology, Department of Microbiology and Ecosystem Science, University of Vienna, Althanstraße 14, 1090 Vienna, Austria; joerg.hellinger@univie.ac.at (H.-J.H.); thomas.rattei@univie.ac.at (T.R.)

5 St. Petersburg State Polytechnic University, St. Petersburg 195251, Russia

* Correspondence: d.frishman@wzw.tum.de

Abstract: RNA secondary structure in untranslated and protein coding regions has been shown to play an important role in regulatory processes and the viral replication cycle. While structures in non-coding regions have been investigated extensively, a thorough overview of the structural repertoire of protein coding mRNAs, especially for viruses, is lacking. Secondary structure prediction of large molecules, such as long mRNAs remains a challenging task, as the contingent of structures a sequence can theoretically fold into grows exponentially with sequence length. We applied a structure prediction pipeline to Viral Orthologous Groups that first identifies the local boundaries of potentially structured regions and subsequently predicts their functional importance. Using this procedure, the orthologous groups were split into structurally homogenous subgroups, which we call subVOGs. This is the first compilation of potentially functional conserved RNA structures in viral coding regions, covering the complete RefSeq viral database. We were able to recover structural elements from previous studies and discovered a variety of novel structured regions. The subVOGs are available through our web resource RNASIV (RNA structure in viruses).

Keywords: mRNA structure; structure database; secondary structure; viral mRNA; subVOG; structurally related; RNA structure; structurally homogenous; structurally related; mRNA families

---

1. Introduction

Secondary structures formed in single-stranded mRNA molecules through complementary self-interactions, both in the untranslated (UTR) and coding (CDS) regions of mRNAs, have been implicated in a variety of regulatory functions [1]. For example, riboswitches modulate gene expression through conformational changes in response to various stimuli [2]. Translation initiation, elongation, and termination as well as translation efficiency depend on higher order mRNA secondary structures in non-coding regions [3,4]. CDS hairpins have also been suggested to play a role in the regulation of translation [5], in particular by causing ribosomal stalling and modulating translational efficiency [6]. The relationship between mRNA structure in the CDS and gene expression has been demonstrated both computationally and experimentally [7–11]. In particular, reduced mRNA stability near the start codon has been observed in a wide range of species, probably as a mechanism to facilitate ribosome binding or start codon recognition by initiator-tRNA [12]. Structured elements within CDS directly influence mRNA abundance [13]. Computational studies show that native mRNAs have lower folding energies and are thus more stable than codon-randomized ones [5]. The three mRNA functional domains—5′UTR, CDS, and 3′UTR—form largely independent folding units, while base pairing across domain borders is rare [14]. The ability of viruses to persist in their host in a genus-specific manner is influenced by the interplay between local structural motifs and genome-scale ordered RNA structures (GORS) [15], which impose additional restraints on the RNA sequence space. Evolutionarily conserved local secondary structures have been identified in CDSs [16] and shown to be functional [17]. An indirect indication of the global importance of RNA structures in the coding regions comes from the recent study of Fricke et al. who identified selection favoring specific pairing patterns between synonymous codons within RNA hairpins [18].

Increasing evidence suggests that secondary structural elements in the CDSs of viral RNAs also constitute a previously underappreciated, evolutionarily conserved level of functional organization of viruses. A large number of conserved secondary structural motifs were computationally identified in the Flavivirus genomes [19–21], predicted to restrain sequence variability [22] and experimentally shown to regulate important biological processes, such as replication and infection [21]. Multiple secondary structures were described in the coding regions of the (+) sense RNA of the Influenza A virus [23]. Another example is a secondary structural element within the coding region of the Dengue virus type 2, which is essential for its replication [24]. More recently, using a comparative genomics approach, Goz and Tuller identified a large number of potentially functionally important regions in the coding regions of Dengue viruses, in which the RNA folding strength is conserved independently of sequence conservation and compositional bias [25]. Specific regions in the HIV structural genes were reported to be under strong selection for stable secondary structures [26]. Recent research shows that mechanisms of translational control by RNA structures can be shared between viruses and cellular organisms [27].

Given the important role played by RNA structures in shaping the evolutionary dynamics of viruses and modulating their interaction with the host, a large-scale investigation of RNA motifs in viruses would be warranted. However, there are two major challenges that need to be addressed before embarking on such an investigation. First, accurate structure prediction for long RNA molecules, such as mRNAs, is generally out of reach for the existing computational methods. Second, conserved stem-loop structures can only be derived from a collection of high-quality alignments of orthologous viral transcripts, which are difficult to obtain, given the rapid pace of viral evolution and the ensuing poor sequence conservation, even between closely related species.

Here, we propose a computational approach to explore the RNA structurome of the viral coding regions, in which local structure predictions are applied to VOG (Viral Orthologous Groups, http://vogdb.org), the first comprehensive collection of orthologous groups derived for all viral proteins contained in the RefSeq [28] database. We utilize RNALalifold [29] to scan long input sequences for locally optimal secondary structures. The identified structural boundaries are more accurate than those derived from using a sliding window of fixed length. Functional importance of structured regions is assessed by RNAz [30]. We present a novel database, RNASIV (RNA structure in viruses; http://rnasiv.bio.wzw.tum.de), which contains the largest currently available collection of predicted RNA structures in viruses. It provides access to 201,708 viral mRNA sequences clustered into 42,293 structurally homogenous groups and is intended to become a useful tool for exploring structure–function relationships in virus families.

2. Materials and Methods

2.1. Viral Orthologous Groups (VOGs)

All genome sequences and their annotations were retrieved from the RefSeq viral database release 79 [31] and grouped into phages and non-phages, based on the available taxonomic information. Assemblies containing inconsistently annotated or completely unannotated polyproteins were identified based on the manually curated information provided by ViralZone [32] and excluded from consideration. Phage and non-phage protein sequences were clustered into phage and non-phage preVOGs, using the NCBI's COG software package with all default settings.

For all phage and non-phage preVOGs, multiple sequence alignments were constructed with Clustal Omega v1.2.4 [33] and used to build HMM-profiles using HMMer 3 [34]. The profiles were subsequently aligned against each other, using HHalign from the HHsuite toolkit [35]. The number of aligned HMM columns was used as an alignment score. All scores for alignments with HHalign probability >85, HHalign *e*-Value < 10^{-5}, and more than 70% of aligned columns between the query and the match HMM were stored as an all-against-all matrix. This matrix was clustered into 21,200 VOGs, using the MCL (Markov Clustering) method [36]. Based on the manual inspection of the homogeneity of the protein function descriptions in the resulting clusters, we selected the inflation value of 2.0 for the MCL clustering. For all VOG member proteins, we determined the closest homolog in the UniProt database [37] from BLAST [38] hits with *E*-values better than 10^{-5} and a minimal query coverage of 90%. Functional descriptions of VOGs were automatically derived based on the most frequent protein description found in the UniProt entries or, if not available, in the RefSeq annotation [31]. The complete VOG dataset, which was used in this study, and supplementary files are available for download at http://vogdb.org.

2.2. Mapping VOG Sequences to Specific Hosts

We used Virus-Host DB [39] to assign host information to VOG proteins. For 20757 VOGs, we were able to map all contained sequences to a specific host, while 428 VOGs contain proteins from at least one viral species for which we could not find host annotation. Most VOGs include viruses infecting hosts from only one domain of life, i.e., bacteria (~72%), eukaryotes (~22%), or archaea (4%), while only 2% of VOGs are taxonomically mixed (Figure 1). Only 12 VOGs contain viruses that infect hosts from all three domains of life. The VOG sizes range from 15 proteins of 12 distinct species, up to 265 proteins belonging to 261 different species (on average, 104 proteins from 95 different species). These VOGs mostly harbor highly conserved core enzymes of double-stranded DNA viruses, such as kinases, ligases, methylases, helicases, hydrolases, and synthases [40]. The other two VOGs additionally contain proteins from viruses belonging to the order of Caudovirales, which belong to the bacteriophages, which are not classified as double-stranded DNA viruses, according to the NCBI taxonomy. We excluded from consideration 15 VOGs containing satellite viruses infecting other viruses.

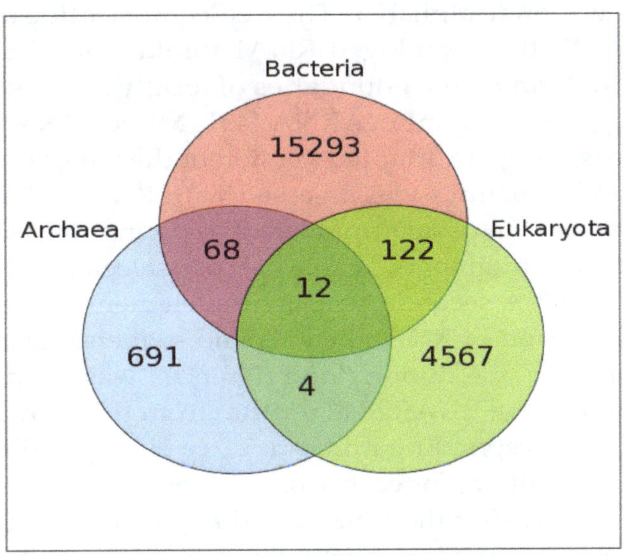

Figure 1. Venn diagram showing the taxonomy of the host organisms within all viral orthologous groups (VOGs). Only those VOGs are included for which host annotation for all viruses is available in the Virus-Host DB.

2.3. Distance Trees of VOG Proteins

Expectedly, we found that RNA structure conservation within VOGs decreases with increasing VOG size. Most VOGs (66%) consist of at least three sequences (size distribution shown in Figure 2) and can therefore potentially be split into smaller groups containing structures that are not conserved across the entire VOG. We therefore utilized distance trees derived by the neighbor-joining algorithm [41] to identify structurally homogeneous subsets of VOGs (subVOGs). All-against-all pairwise alignments of protein sequences were calculated using Clustal Omega and then converted to the nucleotide alphabet. The distance matrices were derived from pairwise sequence identity values, and the trees were created from the matrices using neighbor joining, as implemented in the BioPerl toolkit [42]. The inner nodes of the sequence trees represent possible subVOG candidates, potentially containing structurally homogenous sequences.

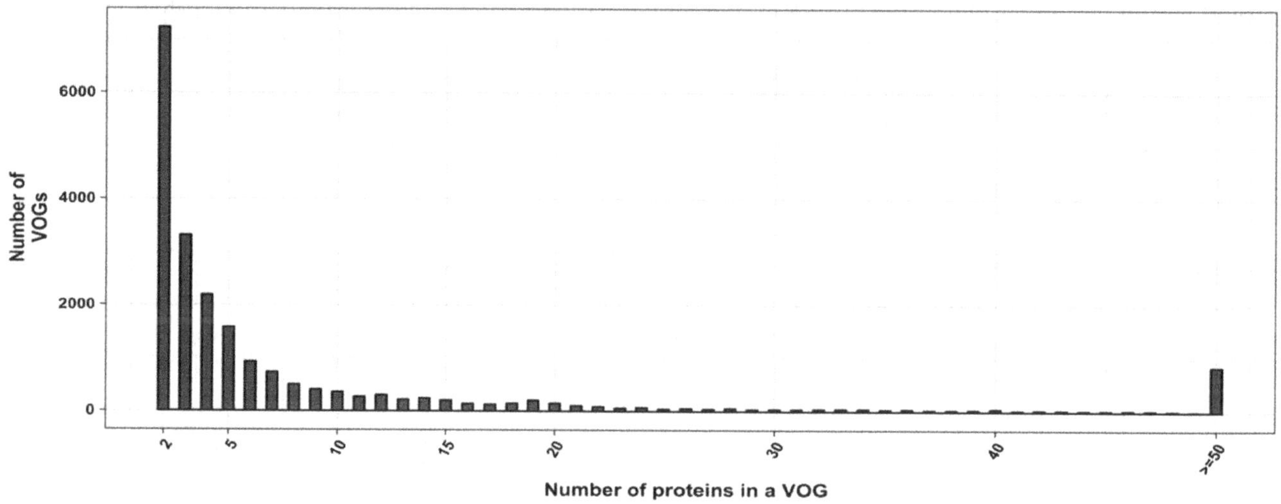

Figure 2. Distribution of VOG sizes.

2.4. Structure Prediction and subVOG Assignment

In order to assess the amount of structural RNA conservation present in subVOG candidates, multiple sequence alignments (MSAs) of proteins were calculated for each inner node of the distance trees and converted to the nucleotide alphabet. The RefSeq nucleotide and protein sequences were obtained from the VOGDB. We then employed RNALalifold from the ViennaRNA package [29], with default parameters, to determine the boundaries of locally stable structures within each MSA, and realigned these local regions using mLocARNA [43]. MLocARNA produces structure-guided multiple sequence alignments, using an adapted version of the Sankoff algorithm. The significance and conservation of the found structures was assessed with RNAz [30]. This procedure is simpler and arguably more accurate than the usual approach of applying RNAz to the entire MSA within a sliding window. RNAz classifies fragments of an MSA pre-selected by RNALalifold as containing or not containing a functional RNA secondary structural element. Realignment with mLocARNA significantly increases the precision of RNAz [30]. As no sequence of a potential subVOG can be regarded as a reference sequence, the option "no reference" was used for the subsequent RNAz analysis. The RNAz method uses the RNAfold algorithm from the ViennaRNA package to calculate secondary structures and the corresponding minimum free energy (MFE) for each individual RNA sequence in the alignment. In addition, for each aligned sequence set, RNAz calculates a consensus secondary structure and its MFE using the RNAalifold algorithm. RNAz assumes that conserved and thermodynamically stable structures are functional, in which case it outputs "RNA". Otherwise, it outputs "OTHER". For this purpose, a class probability value, combining all information on an input alignment is calculated. We used a stringent threshold of 0.9 (default 0.5) for the class probability value, which is recommended for finding high confidence structures [30]. Subsequently, the trees were scanned for subtrees containing at least one conserved structural element, that is, predicted to be functional, and the largest subtrees were designated as structurally homogenous subVOGs. We found that sequences that are only distantly related according to the neighbor-joining tree may still share conserved RNA structures. In order to account for structure-level relationships between sequences, we built covariance models for all conserved structures found within subVOGs, using the tool cmbuild from the infernal package [44], and used them to search against all sequences in the entire VOG database.

2.5. mRNA Stability

Following Tuller et al. [45] and Faure et al. [46], we employed RNAfold to calculate the folding energy of the most and the least stable 30-nucleotide segment of mRNAs (ΔGmin and ΔGmax, respectively), as well as the average folding energy of all possible 30 nucleotide segments (ΔGmean). Faure et al. investigated the effect of mRNA stability on the translation rate and protein folding. During translation, the ribosome sequentially unfolds parts of the mRNA. These parts are typically 30 nucleotides long, which explains the choice of segment length in Faure et al. As this procedure does not take into account the actual boundaries of local structures, but rather limits all structures to the size of 30 nucleotides, we additionally calculated the three energy values for all local optimal structures found with RNALfold.

2.6. mRNA Structures and Protein Function

We investigated the relationship between protein function, described in terms of gene ontology (GO) annotation [47], and mRNA structures. Instead of using the global folding energy for classifying mRNAs as highly or lowly structured [48], we considered structural coverage—the portion of an mRNA covered by functional and conserved structures. GO terms for all VOG proteins were downloaded using QuickGO [49], where available. Based on the Evidence & Conclusion Ontology (ECO) evidence codes [50], two separate datasets were created: (i) Proteins annotated by manually or experimentally derived GO terms (ECO evidence codes: ECO:0000352, ECO:0000269), and (ii) proteins annotated by GO terms with any evidence codes. To find out whether mRNAs of proteins with certain functions tend to harbor more or fewer structures, we pooled together functionally similar GO terms with the average structural coverage of their corresponding mRNAs, using Revigo [51]. Revigo uses a semantic similarity measure to group similar GO terms together, which results in a concise list of distinct functions. To perform this analysis, we calculated the average structural coverage of all subVOG mRNAs with available GO annotation. For the experimental dataset we allowed a coverage value to be associated with a GO term if more than 50% of the sequences in a particular subVOG were annotated with this term. Within the dataset based on all evidence codes, we only allowed GO terms shared by all sequences of a subVOG. We only used mRNAs that were clustered into a subVOG. For sequences that were not part of any subVOG, we did not find conserved structures, although this does not necessarily mean that the mRNA did not contain functional structures. The distributions of standard deviations of the structural coverage values were compared within the actual and randomly generated Revigo clusters. Randomization was performed 1000 times by preserving the size of the clusters and filling them with randomly chosen GO terms.

3. Results

3.1. Overview of the Study

A graphical overview of the study is given in Figure 3. In a first step, we created distance trees for all protein sequences contained in each VOG, using the neighbor joining method, as described in Materials and Methods. All sequences of the inner nodes of each tree, representing potential subVOGs, were multiply aligned, converted to the nucleotide alphabet and processed with RNALalifold to obtain all potentially conserved local optimal structures. Each part of the alignment covering a potential structure was then realigned with the structure-guided alignment method mLocARNA and checked for functionality using RNAz. The use of structure-guided alignments as input for RNAz improves the performance, compared to pure sequence-based alignments [30]. The tree nodes containing the most sequences that yielded conserved structures were taken as final subVOGs. For all obtained subVOG structures, we computed covariance models that could be used to search for similar structures in future research.

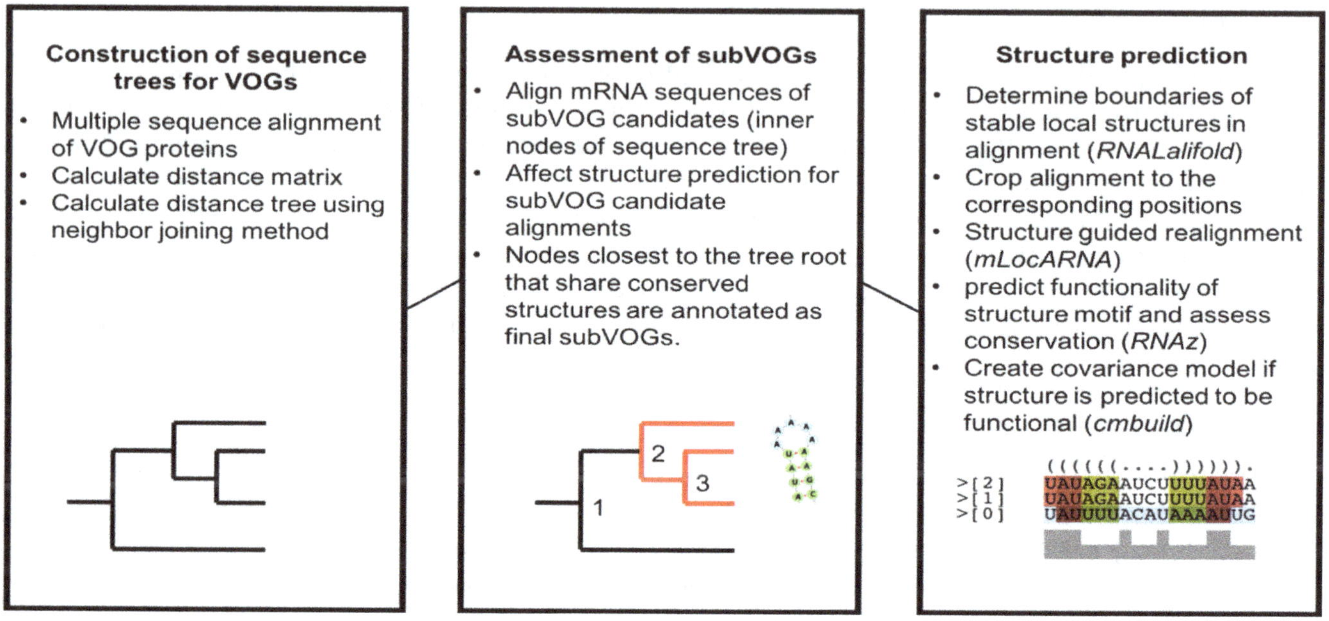

Figure 3. Overview of the analysis of conserved RNA structures in VOGs.

3.2. Structure Conservation in VOGs

The current release of the VOG database, derived from the RefSeq release 77, contains 21,200 VOGs, composed of 251,796 proteins from 6252 phages and eukaryotic viruses (Figure S1). Protein sequences in each VOG were aligned by Clustal Omega, converted to the nucleotide alphabet, and used as input for RNA structure prediction by RNALalifold. As seen in Figure 4, the number of local optimal structures conserved within entire VOGs decreases quickly with the number of aligned sequences, which may in part be the consequence of poor multiple alignment quality in large sets of sequences. Indeed, we found that proteins in smaller VOGs tend to be more closely related (Figure S2). To exclude structures found due to sequence conservation only, the potential functionality of structures was verified with RNAz. However, even those VOGs that only consist of a few sequences do not always contain conserved structures. There are 7232 VOGs with exactly two sequences, and for 1237 of these, we could not find any conserved structures. The remaining 5995 VOGs of size two had an average structural coverage of approximately 25% (Figure 5a). Out of the 13,968 VOGs with more than two sequences, 7238 VOGs were predicted to contain RNA structures conserved across the entire VOG, with an average structural coverage of approximately 18% (Figure 5b). These contain between 3 and 96 sequences, with an average of 6. On average, VOGs contain sequences from three different genera, mostly belonging to the same taxonomic family and thus also to the same order (Figure 6a–c). The 25 most diverse VOGs contain sequences from three different orders and up to 19 taxonomic families. On average, a VOG contains mRNAs from viruses that infect hosts from four different genera, belonging to three different taxonomic families and two orders. The VOG with the highest host diversity corresponds to 209 different host genera from 114 families and 64 orders.

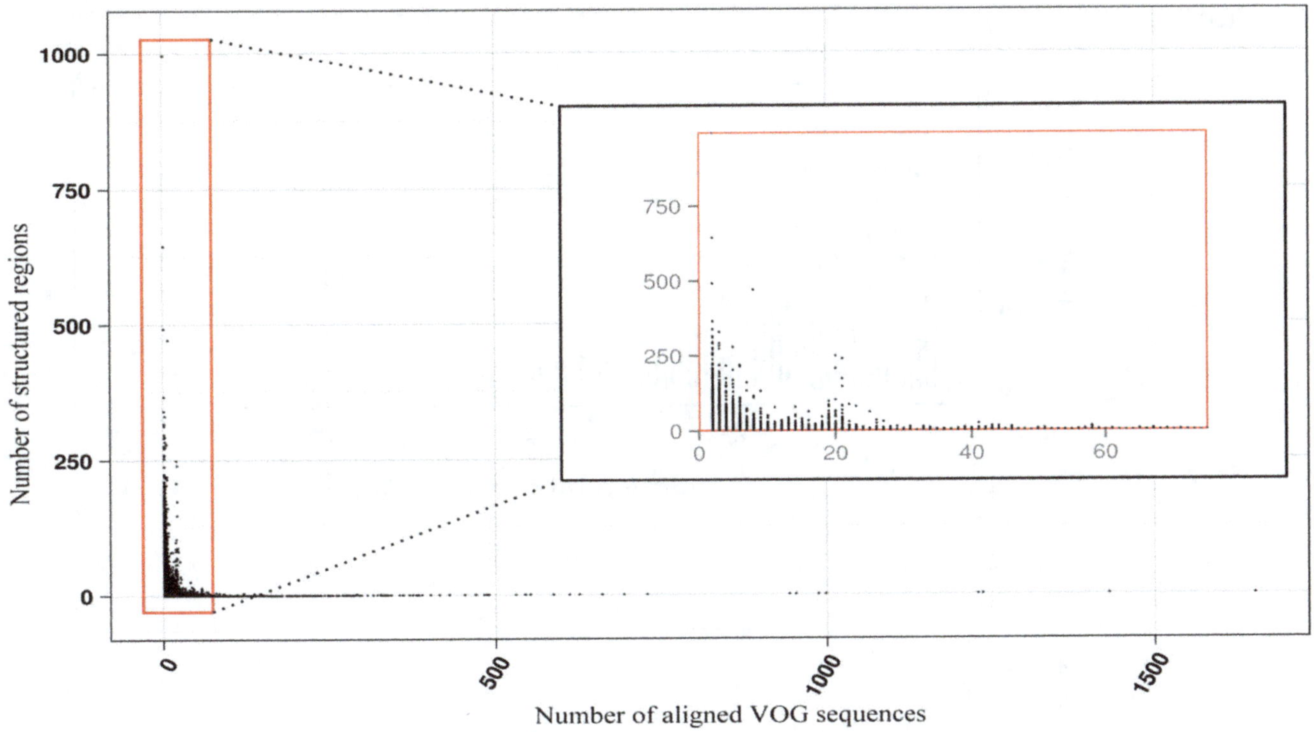

Figure 4. Number of local RNA structures as a function of VOG size.

(a)

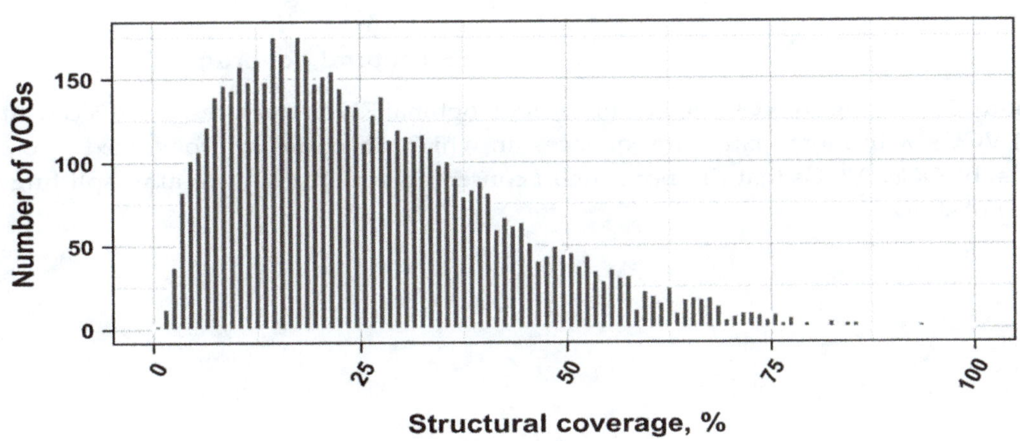

Figure 5. *Cont.*

(b)

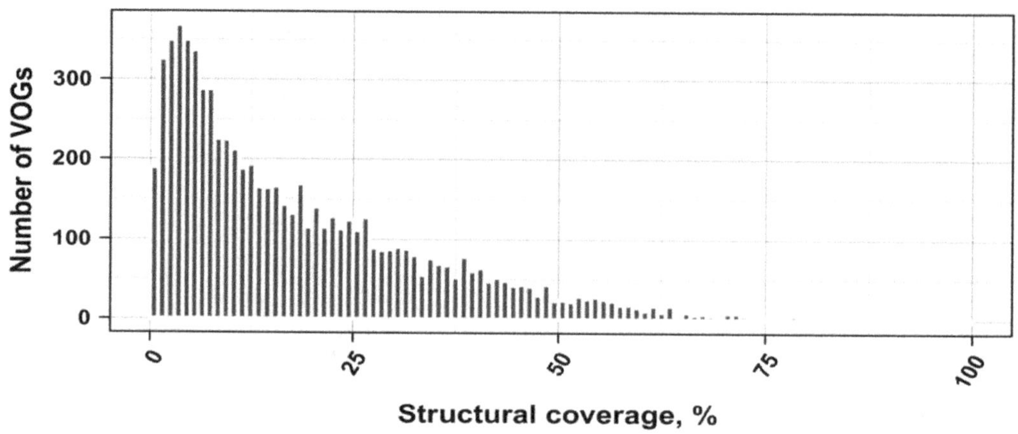

(c)

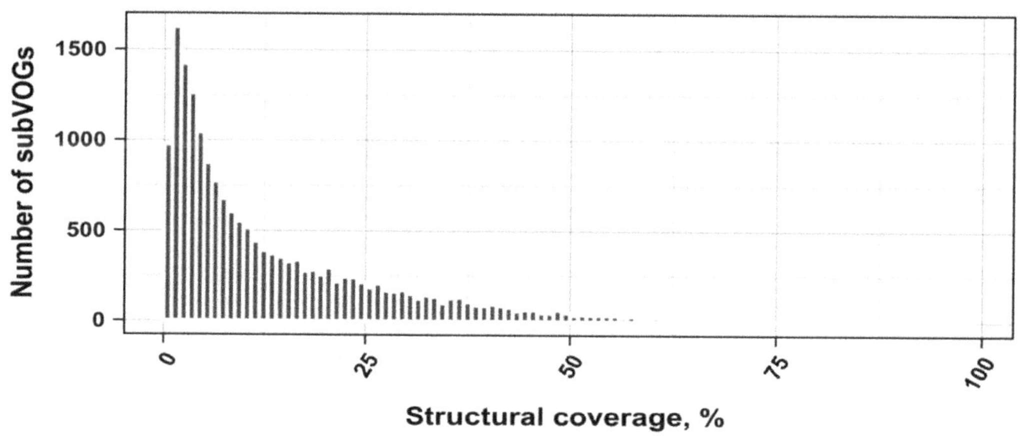

Figure 5. Coverage of VOG alignments by local optimal RNA structures. (**a**) VOGs with two sequences. (**b**) VOGs with more than two sequences, in which structures are conserved across all sequences. (**c**) subVOGs. VOGs that did not contain conserved structures, even after splitting into subVOGs, are not shown.

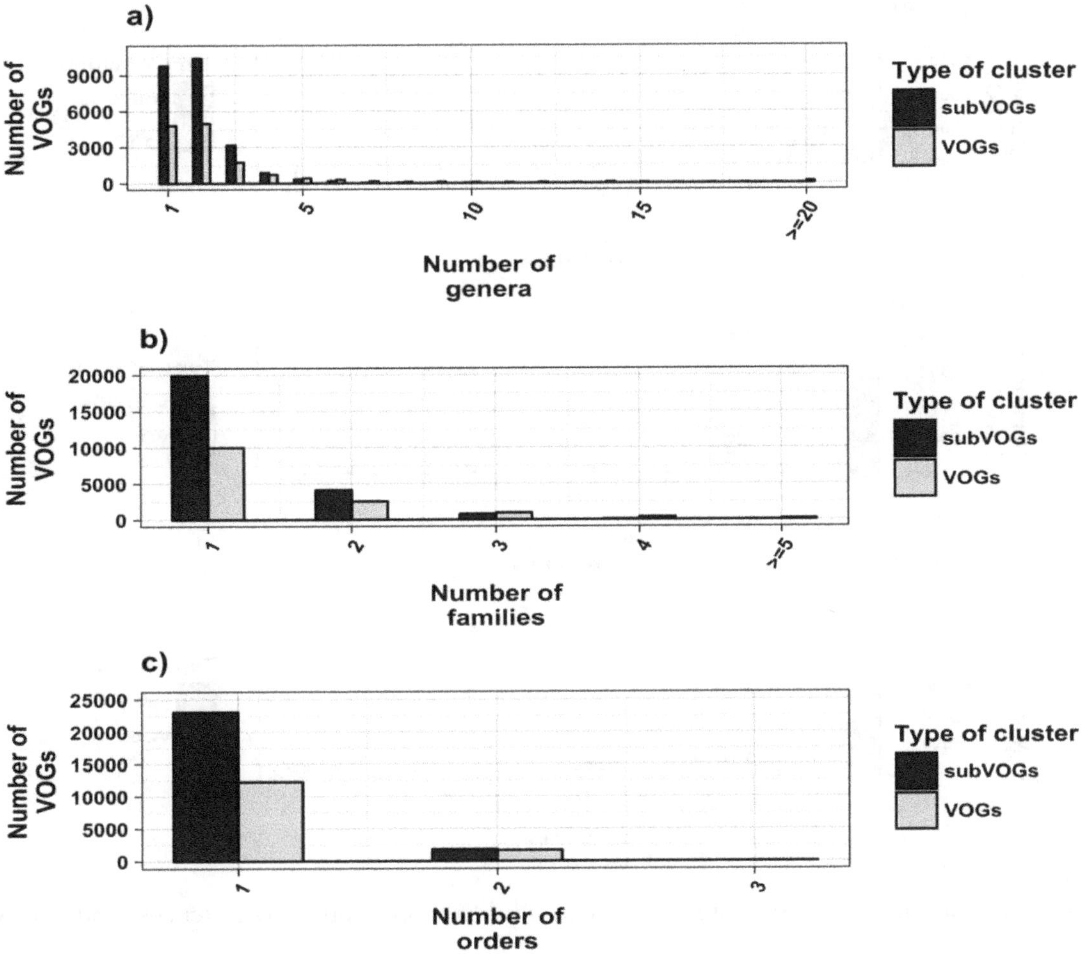

Figure 6. Taxonomic distribution of proteins in VOGs (with more than two sequences) and subVOGs.

3.3. Structure Conservation in subVOGs

We attempted to subdivide 6730 VOGs with more than two sequences and without conserved structures into structurally homogeneous subsets, which we call subVOGs, using phylogenetic trees derived by the neighbor-joining method. This procedure resulted in 17,678 subVOGs with an average structural coverage of approximately 13% (Figure 5c). The average number of genera per subVOG is 2 and the most diverse of them contains sequences from three orders and 14 families. A subVOG contains on average sequences that infect two different host genera, and the most diverse subVOG infects hosts of 42 different genera, belonging to 33 families and 20 different orders (Figure 7a–c). Thus, unsurprisingly, subVOGs, which constitute subsets of full VOGs with increased structural homogeneity, exhibit a reduced taxonomic spread, both of the viruses they contain and their hosts. A large fraction of subVOGs (63%) contains sequences from more than one genus and 21% contain sequences from more than one family. The structural coverage of subVOGs, i.e., the fraction of alignment positions that are located within conserved RNA structures, decreases with increasing taxonomic diversity of the viruses and their hosts (Figure 8). An example that demonstrates the reduction of taxonomic spread between a VOG and its corresponding subVOGs is given in Figure 9. Here, the VOG 00052, which contains 20 proteins from 12 different virus species belonging to 4 different taxonomic families, was split into four structurally homogenous subVOGs. Two of the four subVOGs consist of mRNAs belonging to the genus Avipoxvirus from the family Poxviridae, the third subVOG contains sequences from the family Mimiviridae, and the fourth subVOG consists of two mRNAs belonging to viruses from two different taxonomic families, the Ascoviridae and the Iridoviridae. For two mRNAs, we could not find structures conserved in any of the other VOG members and they are therefore not part of any subVOG.

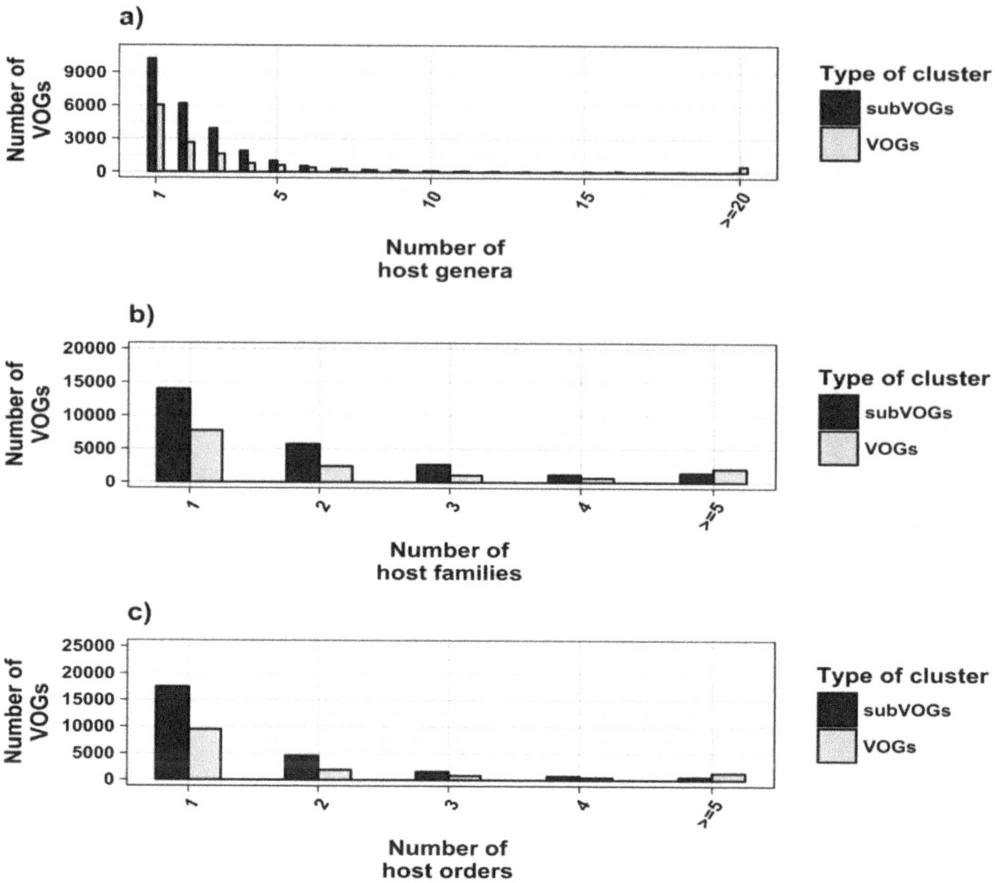

Figure 7. Taxonomic distribution of hosts in VOGs (with more than two sequences) and subVOGs.

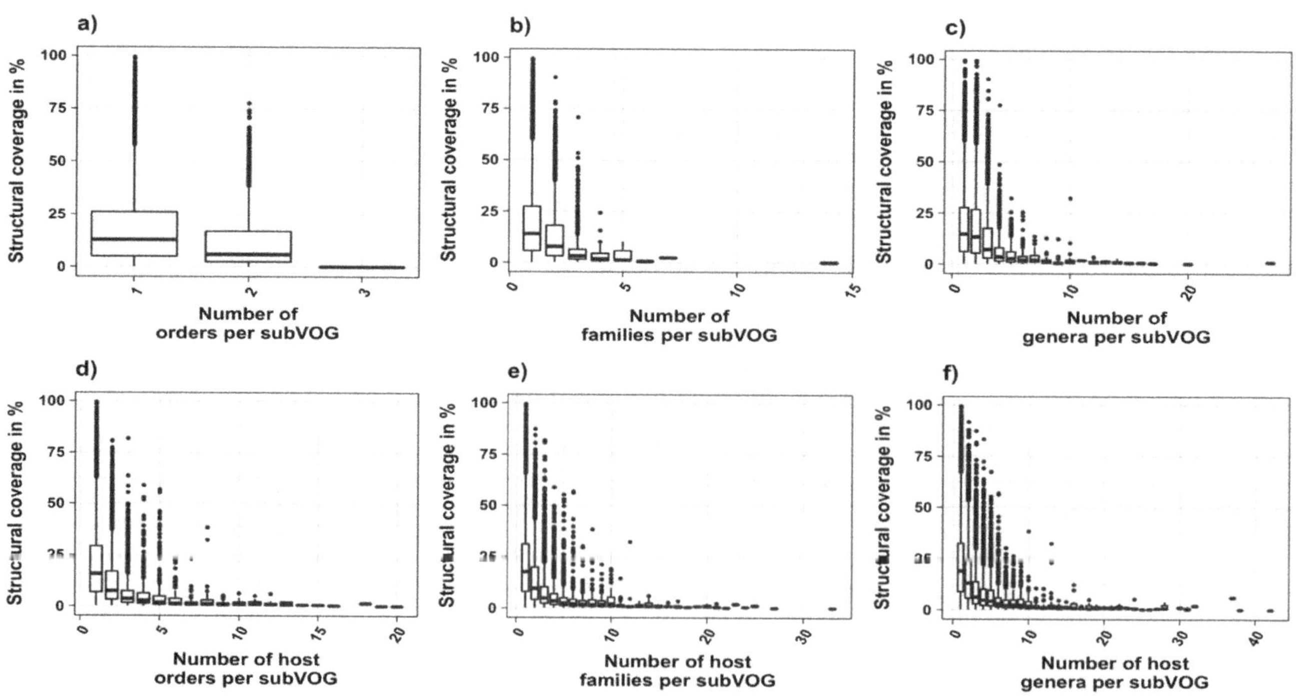

Figure 8. Structural coverage as a function of the taxonomic variety of subVOGs and their host organisms.

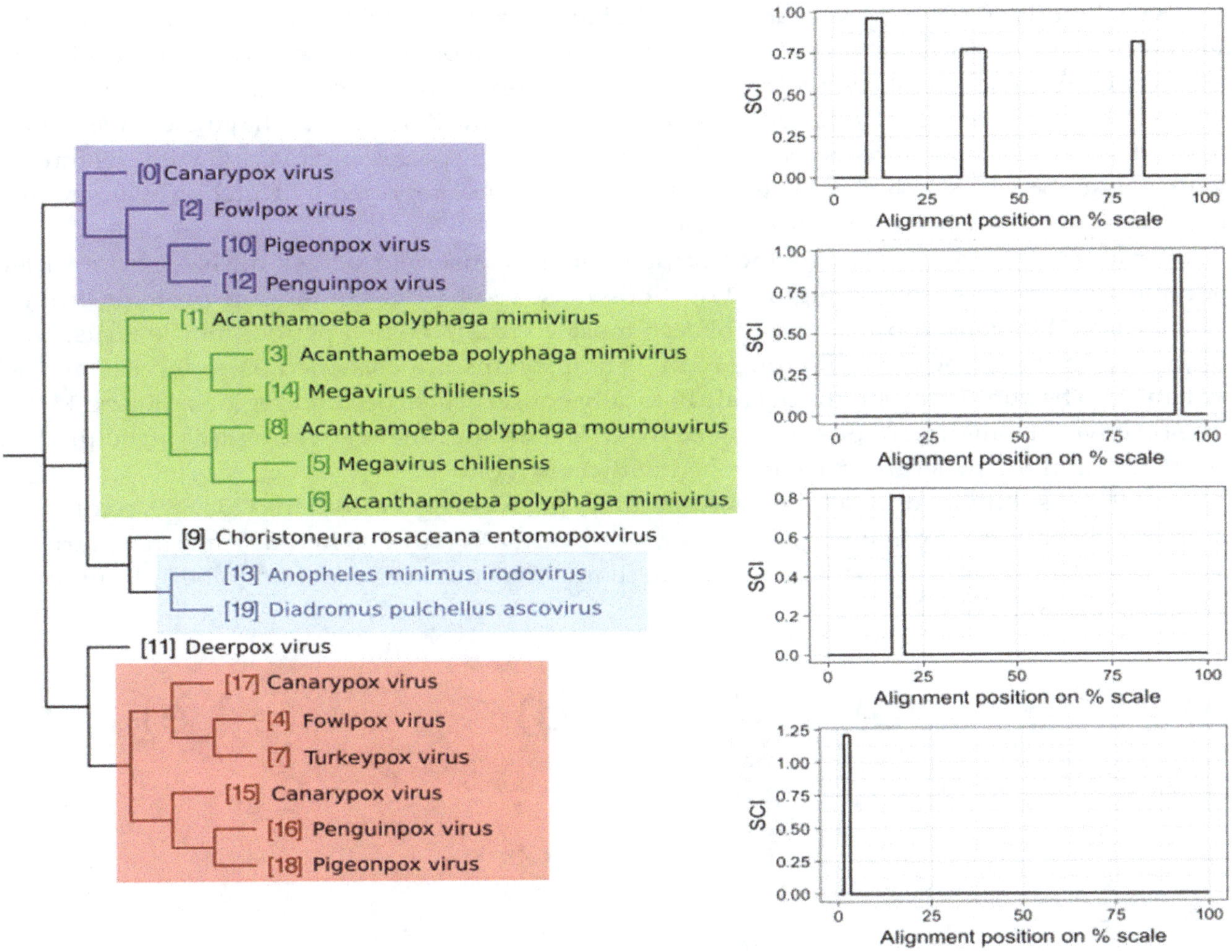

Figure 9. Example of a VOG split into structurally homogenous subVOGs. Shown is the VOG 00052 containing 20 mRNAs, encoding for Kila-N domain proteins, from 12 virus species. On the left, the neighbor-joining tree based on the pairwise sequence identity between the protein sequences is shown. Colored boxes indicate subVOGs, within which conserved structures were predicted. The tree nodes outside colored boxes did not yield any conserved structures. On the right, the structure conservation index (SCI) (black line for each subVOG alignment) is plotted against the alignment position on the percentage scale. Plots are ordered according to the subVOG position in the tree.

As an example, Figure 10 shows the subVOG 1 of VOG11160, which contains two mRNAs encoding the matrix protein 1 from the Influenza A virus (H3N2) and the Influenza B virus. There are three RNA structural motifs described in the literature for the Influenza A mRNA. Nucleotides 105 to 192 form either a multibranch structure, according to Moss et al. [23] and Jiang et al. [52], or a double hairpin structure, proposed by Jiang et al. [52]. Two consecutive stem-loop structures are formed from position 682 to 744, according to Moss et al. [23]. Despite the sequences' dissimilarity between Influenza A and B, both motifs are partly conserved, according to our RNAz analysis of the corresponding subVOG (Figure 10). Our analysis supports the second hairpin loop from the double hairpin structure, described by Jiang et al. (Figure 10a–c). From the second motif, proposed by Moss et al., we also found that the second hairpin structure was partly conserved (Figure 10d–e). The consensus structure of the first motif has a high structure conservation index (SCI) of 0.78, although the part of the alignment covering the structure has a low pairwise identity of 29%. The second motif has an SCI of 0.58 and a pairwise identity of 32%. Our analysis also revealed three further conserved stem-loop structures—position 346 to 369, 438 to 483, and 654 to 674, with SCIs and mPIDs of 0.81 and 29%, 0.66 and 48%, and 0.65 and 33%, respectively.

A recent study of secondary structures in alphaviruses by Kutchko et al. revealed that Sindbis virus mRNAs harbor many functional structures, but they are poorly conserved in the closely related Venezuelan equine encephalitis virus [53]. The corresponding subVOG containing mRNAs coding for the non-structural protein 1 includes orthologous mRNAs from 12 further alphaviruses. We identified three short structures that are conserved in all of the contained species and overlap with the functional structures described by Kutchko et al., while all other structures reported by Kutchko et al. are indeed poorly conserved in further Alphavirus species.

An example of a subVOG in which structures are conserved across mRNAs from different taxonomic families is given in Figure 11. Shown is a subVOG containing proteins from two mosaic viruses (Maracuja mosaic virus, Tobacco mosaic virus), the Bell pepper mottle virus, and the Odontoglossum ringspot virus (Figure 11a,b). The proteins are classified as replicases and RNA polymerases. The subVOG contains overall 15 locally conserved structured regions. Figure 11 shows the region covering alignment positions 4766 to 4815. The alignment covering this structure has an mPID of 72% and the structures are conserved with an SCI of 0.9.

Overall, we subdivided 21,200 VOGs containing, on average, 11 proteins (233,380 in total) into a total of 42,293 subVOGs, containing, on average, five mRNAs (201,708 in total) and three structured regions (147,087 in total). The VOGs with more than two sequences that had to be split up contain, on average, four subVOGs.

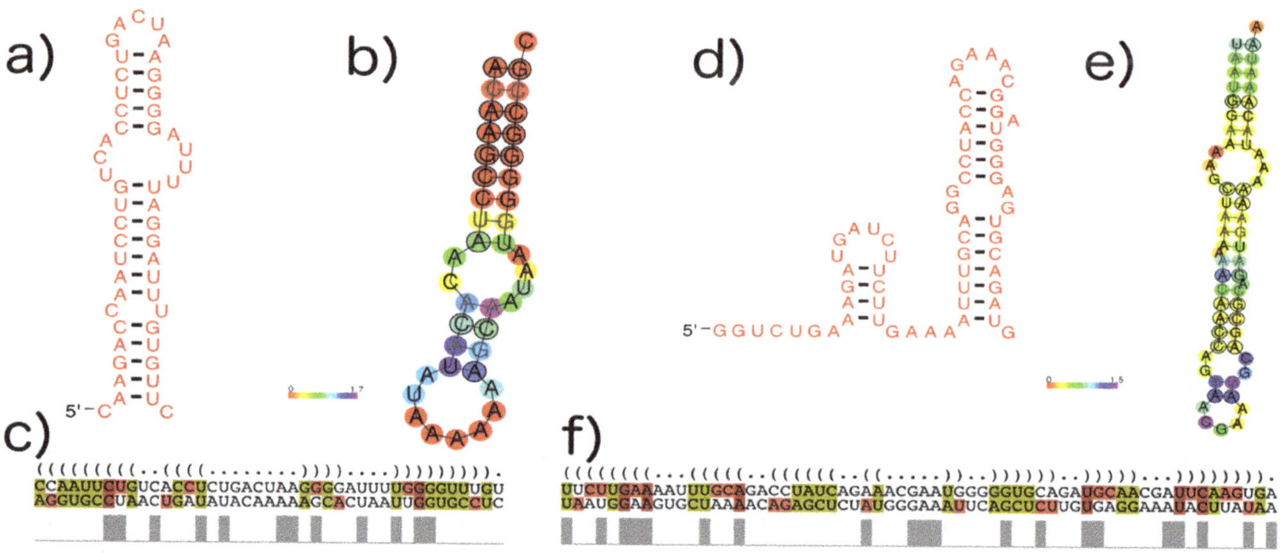

Figure 10. Structures found in Influenza A and B mRNAs encoding the matrix protein (VOG11160). Colors in MSA pictures encode compensatory mutations supporting the consensus structure. Red marks pairs with no sequence variation; ochre, green, turquoise, blue, and violet mark pairs with 2, 3, 4, 5, and 6 different types of pairs, respectively. (**a**) The second of the two consecutive stem loops of the structure proposed by Jiang et al. [52], covering positions 147–192, visualized with R2R [54]; (**b**) The predicted conserved consensus structure for nucleotides 148–188 supports the second hairpin loop of the model of Jiang et al., shown in (**a**). Colors encode the positional entropy; (**c**) Structure-guided alignment and dot bracket structure notation for the consensus structure shown in (**a**). The upper sequence corresponds to Influenza A and the lower sequence to Influenza B; (**d**) Shown are two consecutive hairpin loops for nucleotide positions 682 to 744, proposed by Moss et al. [23], visualized with R2R; (**e**) The predicted conserved structure for nucleotides 697–758 partly supports the model shown in (**e**). Colors encode the positional entropy; (**f**) Structure-guided alignment and dot bracket notation for the consensus structure shown in (**e**). The upper sequence corresponds to Influenza A and the lower sequence to Influenza B.

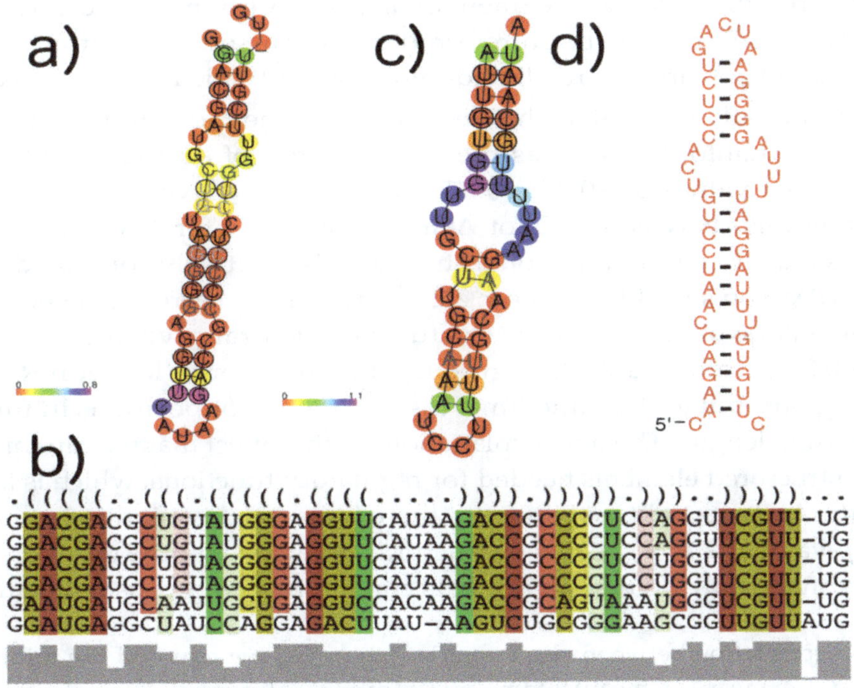

Figure 11. Example structures that were identified within subVOGs. (**a**) Structural annotation of the subVOG 30, belonging to VOG00029, which contains six mRNAs encoding a replicase protein of different Tobamovirus species. Consensus structure visualized by RNAalifold. Colors encode the positional entropy; (**b**) Structure-guided MSA and consensus structure in dot bracket notation corresponding to consensus structure shown in (**a**). Colors encode compensatory mutations supporting the consensus structure. Red marks pairs with no sequence variation; ochre, green, turquoise, blue, and violet mark pairs with 2, 3, 4, 5, and 6 different types of pairs, respectively; (**c**) Consensus structure of subVOG 64 from VOG00003, which contains four mRNAs coding for a p28-like protein of different alphabaculoviruses; (**d**) Structure found in a Heliothis virescens ascovirus 3e, by covariance model search of the structure shown in (**c**), using cmsearch in the entire sequence space of all VOGs.

3.4. subVOG Covariance Models

We built covariance models for all structures found within subVOGs and, using cmsearch, found that in many cases, structures are conserved between different subVOGs and even between different VOGs. In most cases, this was due to a shared sequence domain. For example, the subVOG 64 from VOG00003 harbors four mRNA sequences from different nucleopolyhedroviruses, belonging to the family Baculoviridae. This subVOG was predicted to contain four conserved structures. One of these structures is a highly conserved stem-loop structure (Figure 11c). This structure can also be found in an mRNA of Heliothis virescens ascovirus 3e, belonging to the family Ascoviridae, which is part of VOG01276 (Figure 11d). The two structures are highly conserved with an SCI close to 1, although they are part of different VOGs and belong to mRNAs of different virus families. The alignment of the corresponding proteins revealed that these sequences share a common domain, but the sequence similarity is below the inclusion threshold of the VOG pipeline (Figure S3).

3.5. mRNA Stability and Length

It was shown for a number of eukaryotic and prokaryotic organisms that longer mRNAs exhibit more stable RNA structures, which allows for more efficient control of co-translational protein folding [45,46]. In our dataset of viral mRNA sequences, we also found a correlation between the free energy of the most stable 30-nucleotide segment of an mRNA (ΔGmin) and mRNA length (Pearson correlation coefficient −0.27; from here on referred to as Pearson's *r*), but no correlation between the average energy of all possible 30-nucleotide windows (ΔGmean) and mRNA length (Table 1, Figure 12a). We additionally calculated the free energy of the most and least stable local optimal segment found by RNALalifold as well as the mean energy of all found RNALalifold segments, and obtained Pearson's r values of −0.25, −0.07, and 0.29 respectively. The Pearson's r of folding energy and GC content lies between −0.5 for ΔGmax and −0.94 for ΔGmean (Table 1, Figure 12b). The number of bases that are within functional structures is positively correlated with the alignment length of subVOGS (Pearson's r 0.40, p-value $< 2.2^{-16}$), while this correlation becomes negative when considering the percentage of bases within structures (structural coverage) instead of the absolute value (Pearson's r −0.27, p-value $< 2.2^{-16}$) (Figure 13). In other words, longer mRNAs harbor more or longer structured regions, but at the same time, the percentage of positions in functional structures decreases with increasing length. The only explanation for this effect that we can think of is that there is a certain number of structured elements needed for regulatory functions, which is largely independent of the mRNA length. As expected (see Figure 8), there is a weak but significant negative correlation (Pearson's r −0.23, p-value $< 2.2^{-16}$) between structural coverage and the number of sequences in the MSA, with more taxonomically diverse alignments containing fewer conserved structures.

Table 1. Pearson correlation between alignment length or GC-content and the minimum (ΔGmin), maximum (ΔGmax), or mean (ΔGmean) folding energy of either all possible 30-nucleotide long-sequence windows or all local optimal structures found with RNALfold, of all mRNAs in our data set. *P*-values are given in parentheses.

Type of ΔG	Pearson Correlation Coefficient	
	ΔG vs. Sequence Length	**ΔG vs. GC-Content**
ΔGmin	−0.27 ($<2.2^{-16}$)	−0.73 ($<2.2^{-16}$)
ΔGmean	0.004 (0.1655)	−0.94 ($<2.2^{-16}$)
ΔGmax	0.17 ($<2.2^{-16}$)	−0.50 ($<2.2^{-16}$)
ΔGmin (RNALfold)	−0.24 ($<2.2^{-16}$)	−0.86 ($<2.2^{-16}$)
ΔGmean (RNALfold)	−0.16 ($<2.2^{-16}$)	−0.86 ($<2.2^{-16}$)
ΔGmax (RNALfold)	0.29 ($<2.2^{-16}$)	−0.07 ($<2.2^{-16}$)

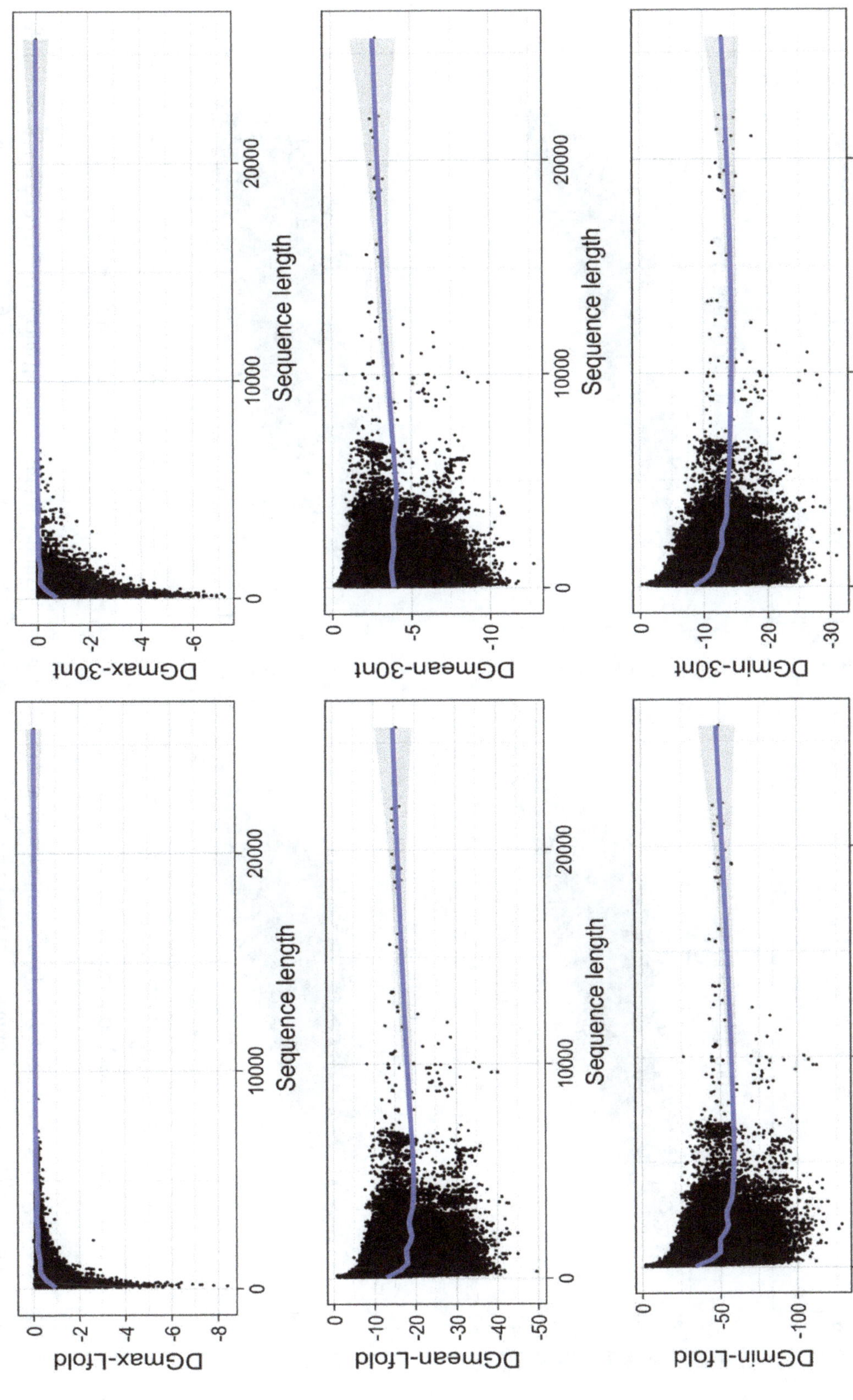

a)

Figure 12. *Cont.*

b)

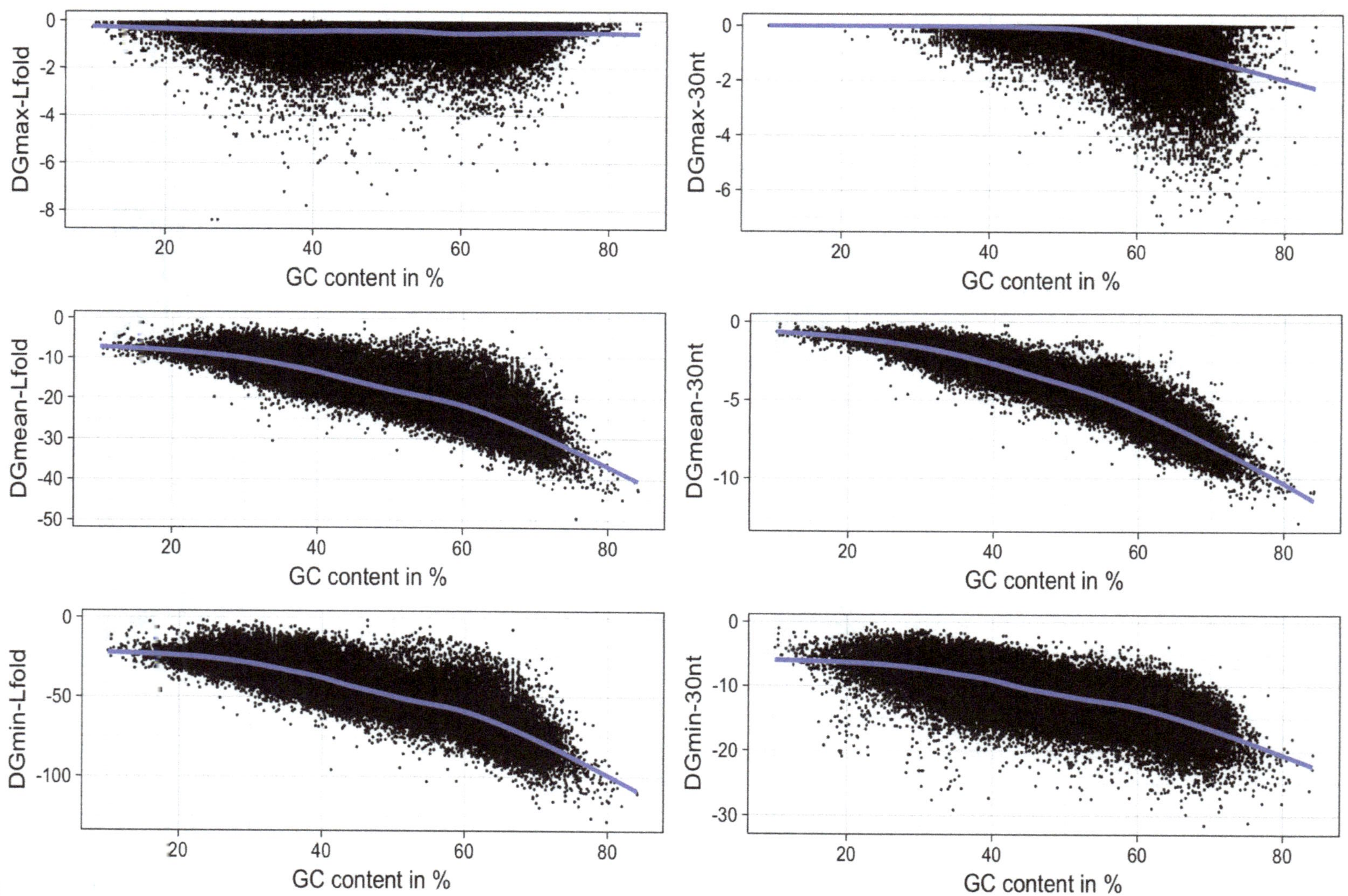

Figure 12. MRNA folding energy as a function of (**a**) sequence length and (**b**) GC-content. DGmin: Minimum folding energy of either all possible 30-nucleotide windows of a sequence or all found local optimal structures using RNALfold. DGmean and DGmax: Mean and maximum of all windows, respectively.

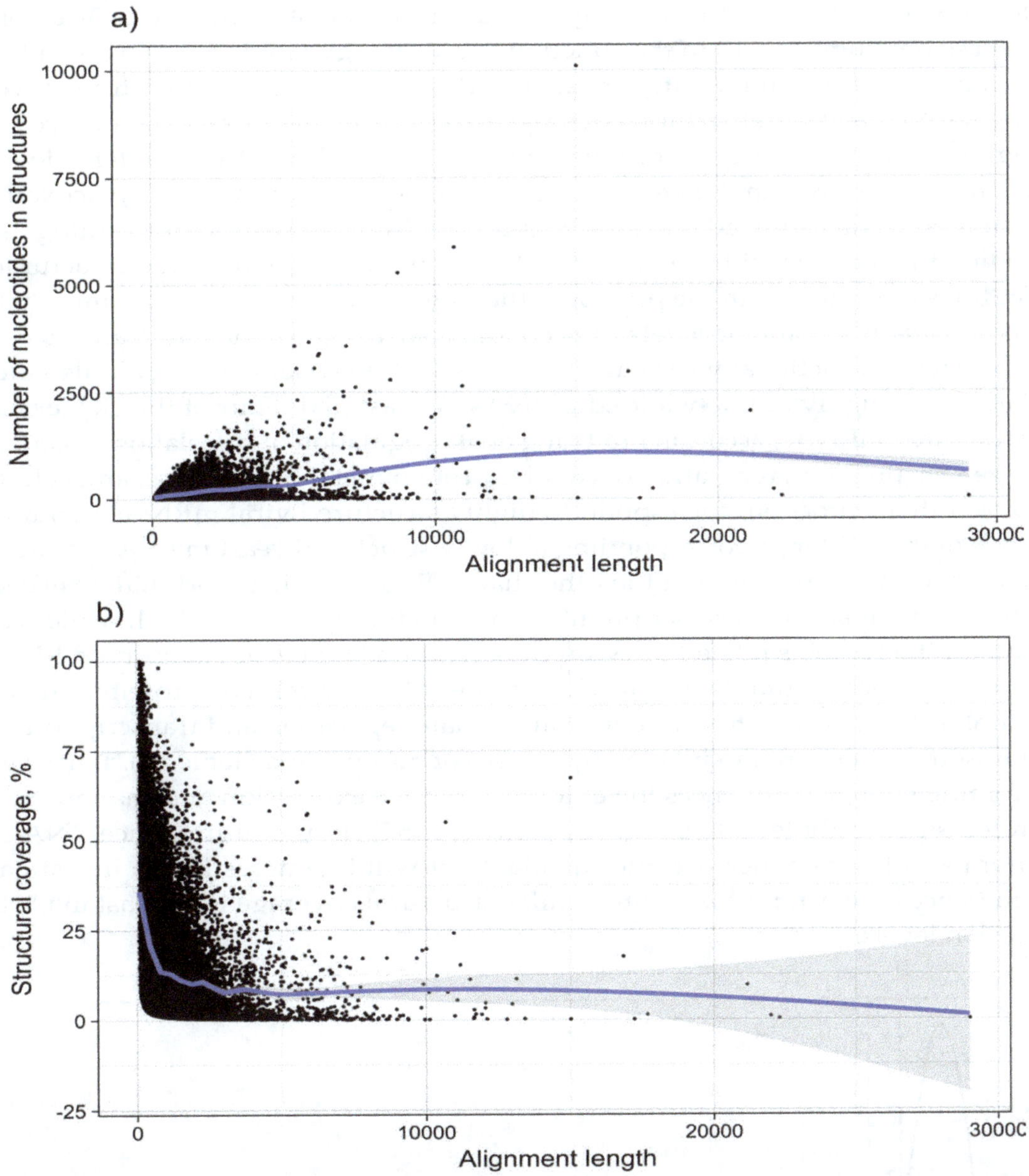

Figure 13. MRNA structure as a function of length. The graph shows the dependence of (**a**) the number of nucleotides within structures predicted to be functional, and (**b**) the structural coverage of the mRNAs in %, from the total length of mRNAs. Each point corresponds to one subVOG.

3.6. mRNA Structures and Protein Function

We analyzed the relationship between protein function and mRNA structure in viral subVOGs by comparing RNA structural coverage with gene ontology (GO) annotation. Using the QuickGO database, we identified a total of 814 VOG proteins that are manually or experimentally annotated (according to ECO evidence codes, as described in Materials and Methods) with GO terms, of which 727 are part of a subVOG, and thus harbor conserved structures according to our analysis. (For the sake of completeness, we also performed the same analysis for all GO annotated proteins, without regard for the annotation evidence codes, see Table S2). For each individual GO term, we only considered the structural coverage of mRNA sequences if that term was assigned to more than 50% of the proteins in a given subVOG. This resulted in 106 GO terms from the biological process sub-ontology and 17 terms from the molecular function sub-ontology. Note that no GO terms from the cellular component sub-ontology satisfied the criteria explained above.

Using Revigo, we derived 70 functionally similar groups of GO terms, with 57 belonging to the biological process ontology and 13 to the function sub-ontology (Table S1). The resulting GO term groups were subdivided into three categories, according to the average structural coverage of the corresponding subVOGs: Low structural coverage (up to 10%), medium structural coverage (up to 20%), and high structural coverage (more than 20%). We found that the standard deviation of the structural coverage values within the Revigo clusters was significantly smaller (Wilcoxon test p-value 1.068^{-10}), compared to randomized clusters (Figure 14). In other words, our findings suggest that mRNAs encoding the proteins with coherent functions tend to exhibit a similar structural coverage.

These findings are in line with the previous study by Vandivier et al. who found that transcripts in *Arabidopsis thaliana* with similar levels of secondary structure in their untranslated and coding regions tend to encode functionally similar proteins [48]. Likewise, Wang et al. also identified GO terms associated with highly or lowly folded mRNAs in yeast [55]. Four of the GO terms associated with highly structured mRNAs, according to Wang et al. (regulation of translation, posttranscriptional regulation of gene expression, regulation of cellular protein metabolic process, and cellular nitrogen compound biosynthetic process), correspond to highly structured viral mRNAs in our data. At the same time, none of the GO terms corresponding to lowly structured yeast mRNAs according to Wang et al. were enriched in our results. On the other hand, Fan Li et al. found that *Arabidopsis thaliana* mRNAs related to "regulation of transcription" were structurally unstable [56], while we found that mRNAs encoding the proteins related to "viral transcription" do harbor conserved RNA structures. We also found virus-specific trends not previously observed for cellular proteins, such as the high structure of viral mRNAs coding for proteins that regulate replication and transcription, suppression by viruses of host translation, or modulation by viruses of host process (Table S1). It has been reported that mRNA folding strength influences the efficiency of gene expression and that mRNAs encoding abundant proteins generally tend to be more structured [57]. In the future, once RNA-seq data for a sufficient number of viral genes becomes available, it will be interesting to investigate whether functional coherence between mRNAs with similar structural coverage is, at least in part, caused by similar expression levels.

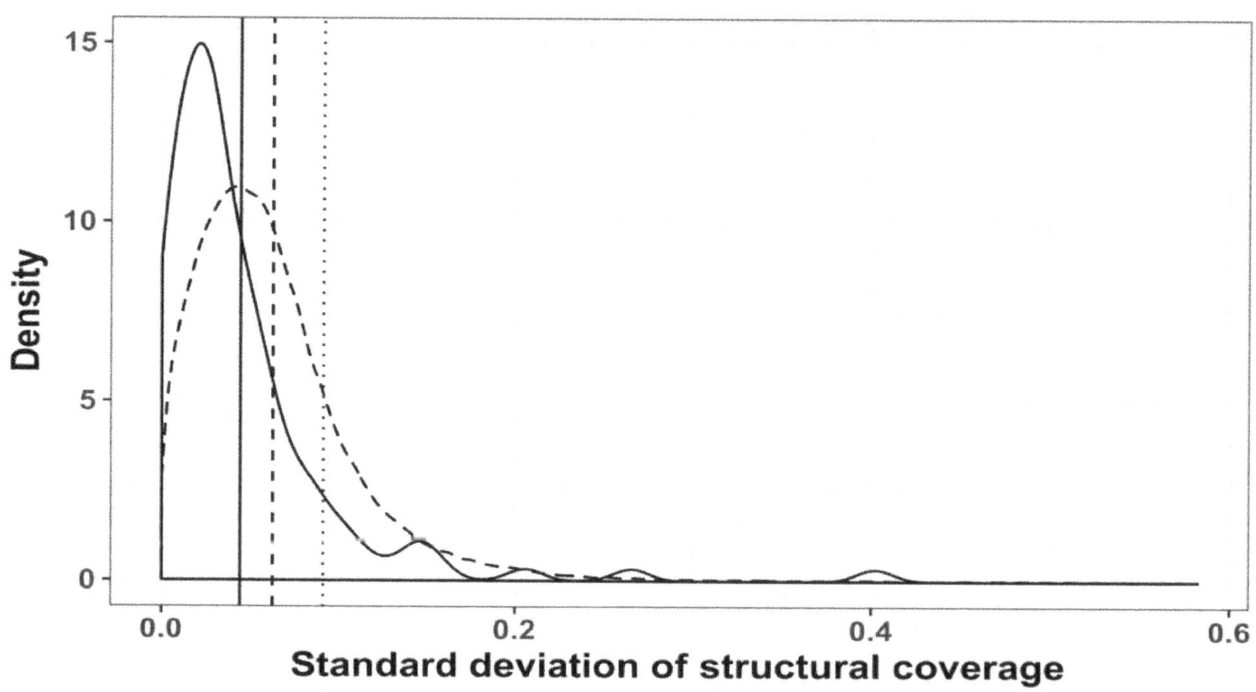

Figure 14. Distribution of standard deviations of mRNA structural coverage, mapped to GO-terms: Clustered with Revigo (solid line); randomized Revigo clusters (dashed line); not clustered (dotted line); vertical lines represent the mean of the corresponding dataset.

3.7. subVOG Online Resource

Structurally homogenous subVOGs can be accessed online (http://rnasiv.bio.wzw.tum.de) through two entry points: "Browse by VOG" and "Browse by taxonomy". The first option is a list of all VOGs, together with the consensus description of their constituent proteins. The list can be filtered with a keyword search and links to the corresponding subVOGs of each VOG are provided. The second option is an expandable taxonomic tree, based on the NCBI taxonomy [58], which allows navigation to the viral species of interest. For each species, mRNA sequences are provided, if available, interlinked to the corresponding subVOGs. Tree nodes containing only mRNAs that are not part of any subVOG are colored grey. Each subVOG contains at least two sequences that share at least one structural element predicted to be functional. If a species of interest is not contained in the subVOG database, the taxonomy tree makes it possible to find the taxonomically closest species. Web pages describing individual subVOGs contain four parts:

(i) General information, i.e., number of mRNAs in the subVOG, the number of proteins and species in the parent VOG, as well as a consensus functional description;
(ii) Information on conserved structures among the subVOG sequences. A plot outlining the SCI for each column of the subVOG MSA gives a brief overview over the structure of the subVOG members. Also provided is a table that shows a list of all structures found, including the corresponding values of SCI, mPID, and the GC content. The consensus structure can also be visualized by Forna, and a covariance model is provided, which can be used to search for similar structures. Additionally, the RNAz results for each individual structured region can be accessed, including structure visualization, dot plots, and the local structure-guided alignments;
(iii) The global MSA for the subVOG sequences. Alignment columns colored in blue correspond to the structured regions described in the previous section. The alignment is visualized with the javascript library MSAviewer [59], which is based on Jalview [60];
(iv) The list of subVOG members, including protein names, descriptions, and taxonomy. For each protein, a link to the RefSEQ entry is provided, as well as the amino acid and nucleotide sequences. The leftmost column of the list contains a checkbox for each subVOG member, which can be used to build a subset of members and analyze the RNA structures shared by these.

4. Discussion

In this work we set out to create a possibly complete census of conserved RNA secondary structures in the coding regions of viruses and to shed light on their biological role. Using sequence comparison and structure prediction methods, we derived structurally homogenous groups of viral mRNAs from subsets of viral orthologous groups (VOGs), which we call subVOGs. We identified a total of 147,087 conserved structures in 42,293 subVOGs, which we make accessible through our database RNASIV (RNA Structures in Viruses). On average, subVOGs contain three structured regions and their structural homogeneity decreases with increasing taxonomic diversity of the viruses and their hosts. We found that 63% of all subVOGs contain mRNAs from at least two genera and 21% from more than one taxonomic family. In line with the previous studies on cellular organisms, we confirm that, in viruses, longer mRNAs tend to contain more stable structures. However, the number of structures grows only slowly with length, which implies that there is a certain minimum amount of structures required to maintain regulatory functions and control protein folding. MRNAs annotated with similar GO terms tend to have a similar structural coverage, hinting at possible commonalities in the regulatory mechanisms of functionally related proteins. It is hoped that RNASIV will be a useful resource for exploring the structure–function relationships in viral mRNAs.

Supplementary Materials:
Figure S1: Virus lineages included in the VOGs; Figure S2: Mean pairwise sequence identity of VOG alignments as a function of VOG size; Figure S3: Sequence Alignment of a protein from Heliothis virescens ascovirus 3e and proteins belonging to the mRNAs of subVOG 64 of VOG00003; Table S1: Clustering of GO terms of subVOG

proteins and the average structural coverage of their corresponding mRNAs; Table S2: Clustering of GO terms of subVOG proteins and the average structural coverage of their corresponding mRNAs (regardless of GO evidence codes).

Author Contributions: Conceptualization, D.F., I.H., and T.R.; methodology, D.F., R.O., and M.K.; software, M.K.; validation, D.F. and M.K.; formal analysis, D.F. and M.K.; investigation, D.F., M.K., and T.R.; resources, I.H. and T.R.; data curation, M.K. and H.-J.H.; writing—original draft preparation, D.F. and M.K.; writing—review and editing, D.F. and M.K.; visualization, D.F. and M.K.; supervision, D.F.; project administration, D.F., I.H., and T.R.; funding acquisition, D.F., I.H., and T.R.

Acknowledgments: We are grateful to Jan Zaucha for his helpful comments.

References

1. Bevilacqua, P.C.; Blose, J.M. Structures, kinetics, thermodynamics, and biological functions of RNA hairpins. *Annu. Rev. Phys. Chem.* **2008**, *59*, 79–103. [CrossRef] [PubMed]
2. Serganov, A.; Patel, D.J. Ribozymes, riboswitches and beyond: Regulation of gene expression without proteins. *Nat. Rev. Genet.* **2007**, *8*, 776–790. [CrossRef] [PubMed]
3. Gray, N.K.; Hentze, M.W. Regulation of protein synthesis by mRNA structure. *Mol. Biol. Rep.* **1994**, *19*, 195–200. [CrossRef] [PubMed]
4. Kozak, M. Regulation of translation via mRNA structure in prokaryotes and eukaryotes. *Gene* **2005**, *361*, 13–37. [CrossRef]
5. Katz, L.; Burge, C.B. Widespread selection for local RNA secondary structure in coding regions of bacterial genes. *Genome Res.* **2003**, *13*, 2042–2051. [CrossRef]
6. Mortimer, S.A.; Kidwell, M.A.; Doudna, J.A. Insights into RNA structure and function from genome-wide studies. *Nat. Rev. Genet.* **2014**, *15*, 469–479. [CrossRef]
7. Kudla, G.; Murray, A.W.; Tollervey, D.; Plotkin, J.B. Coding-sequence determinants of gene expression in Escherichia coli. *Science* **2009**, *324*, 255–258. [CrossRef]
8. Duan, J.; Wainwright, M.S.; Comeron, J.M.; Saitou, N.; Sanders, A.R.; Gelernter, J.; Gejman, P.V. Synonymous mutations in the human dopamine receptor D2 (DRD2) affect mRNA stability and synthesis of the receptor. *Hum. Mol. Genet.* **2003**, *12*, 205–216. [CrossRef]
9. Ilyinskii, P.O.; Schmidt, T.; Lukashev, D.; Meriin, A.B.; Thoidis, G.; Frishman, D.; Shneider, A.M. Importance of mRNA secondary structural elements for the expression of influenza virus genes. *OMICS* **2009**, *13*, 421–430. [CrossRef]
10. Carlini, D.B.; Chen, Y.; Stephan, W. The relationship between third-codon position nucleotide content, codon bias, mRNA secondary structure and gene expression in the drosophilid alcohol dehydrogenase genes Adh and Adhr. *Genetics* **2001**, *159*, 623–633. [PubMed]
11. Nackley, A.G.; Shabalina, S.A.; Tchivileva, I.E.; Satterfield, K.; Korchynskyi, O.; Makarov, S.S.; Maixner, W.; Diatchenko, L. Human catechol-O-methyltransferase haplotypes modulate protein expression by altering mRNA secondary structure. *Science* **2006**, *314*, 1930–1933. [CrossRef]
12. Gu, W.; Zhou, T.; Wilke, C.O. A universal trend of reduced mRNA stability near the translation-initiation site in prokaryotes and eukaryotes. *PLoS Comput. Biol.* **2010**, *6*, e1000664. [CrossRef]
13. Del Campo, C.; Bartholomäus, A.; Fedyunin, I.; Ignatova, Z. Secondary Structure across the Bacterial Transcriptome Reveals Versatile Roles in mRNA Regulation and Function. *PLoS Genet.* **2015**, *11*, e1005613. [CrossRef]
14. Shabalina, S.A.; Ogurtsov, A.Y.; Spiridonov, N.A. A periodic pattern of mRNA secondary structure created by the genetic code. *Nucleic Acids Res.* **2006**, *34*, 2428–2437. [CrossRef]
15. Simmonds, P.; Tuplin, A.; Evans, D.J. Detection of genome-scale ordered RNA structure (GORS) in genomes of positive stranded RNA viruses: Implications for virus evolution and host persistence. *RNA* **2004**, *10*, 1337–1351. [CrossRef]
16. Meyer, I.M.; Miklós, I. Statistical evidence for conserved, local secondary structure in the coding regions of eukaryotic mRNAs and pre-mRNAs. *Nucleic Acids Res.* **2005**, *33*, 6338–6348. [CrossRef]
17. Olivier, C.; Poirier, G.; Gendron, P.; Boisgontier, A.; Major, F.; Chartrand, P. Identification of a conserved RNA motif essential for She2p recognition and mRNA localization to the yeast bud. *Mol. Cell. Biol.* **2005**, *25*, 4752–4766. [CrossRef]
18. Fricke, M.; Gerst, R.; Ibrahim, B.; Niepmann, M.; Marz, M. Global importance of RNA secondary structures in protein-coding sequences. *Bioinformatics* **2019**, *35*, 579–583. [CrossRef]

19. Thurner, C.; Witwer, C.; Hofacker, I.L.; Stadler, P.F. Conserved RNA secondary structures in Flaviviridae genomes. *J. Gen. Virol.* **2004**, *85*, 1113–1124. [CrossRef]
20. Fricke, M.; Dünnes, N.; Zayas, M.; Bartenschlager, R.; Niepmann, M.; Marz, M. Conserved RNA secondary structures and long-range interactions in hepatitis C viruses. *RNA* **2015**, *21*, 1219–1232. [CrossRef]
21. Pirakitikulr, N.; Kohlway, A.; Lindenbach, B.D.; Pyle, A.M. The Coding Region of the HCV Genome Contains a Network of Regulatory RNA Structures. *Mol. Cell* **2016**, *62*, 111–120. [CrossRef]
22. Simmonds, P.; Smith, D.B. Structural constraints on RNA virus evolution. *J. Virol.* **1999**, *73*, 5787–5794.
23. Moss, W.N.; Priore, S.F.; Turner, D.H. Identification of potential conserved RNA secondary structure throughout influenza A coding regions. *RNA* **2011**, *17*, 991–1011. [CrossRef]
24. Clyde, K.; Harris, E. RNA secondary structure in the coding region of dengue virus type 2 directs translation start codon selection and is required for viral replication. *J. Virol.* **2006**, *80*, 2170–2182. [CrossRef]
25. Goz, E.; Tuller, T. Widespread signatures of local mRNA folding structure selection in four Dengue virus serotypes. *BMC Genomics.* **2015**, *16* Suppl. 10, S4. [CrossRef]
26. Goz, E.; Tuller, T. Evidence of a Direct Evolutionary Selection for Strong Folding and Mutational Robustness Within HIV Coding Regions. *J. Comput. Biol.* **2016**, *23*, 641–650. [CrossRef]
27. Díez, J.; Jungfleisch, J. Translation control: Learning from viruses, again. *RNA Biol.* **2017**, *14*, 835–837. [CrossRef]
28. Pruitt, K.D.; Tatusova, T.; Brown, G.R.; Maglott, D.R. NCBI Reference Sequences (RefSeq): Current status, new features and genome annotation policy. *Nucleic Acids Res.* **2012**, *40*, D130–D135. [CrossRef]
29. Lorenz, R.; Bernhart, S.H.; Höner Zu Siederdissen, C.; Tafer, H.; Flamm, C.; Stadler, P.F.; Hofacker, I.L. ViennaRNA Package 2.0. *Algorithms Mol. Biol.* **2011**, *6*, 26. [CrossRef]
30. Gruber, A.R.; Findeiß, S.; Washietl, S.; Hofacker, I.L.; Stadler, P.F. RNAz 2.0: Improved noncoding RNA detection. *Pac. Symp Biocomput* **2010**, 69–79. [CrossRef]
31. O'Leary, N.A.; Wright, M.W.; Brister, J.R.; Ciufo, S.; Haddad, D.; McVeigh, R.; Rajput, B.; Robbertse, B.; Smith-White, B.; Ako-Adjei, D.; et al. Reference sequence (RefSeq) database at NCBI: Current status, taxonomic expansion, and functional annotation. *Nucleic Acids Res.* **2016**, *44*, D733–D745. [CrossRef]
32. Hulo, C.; de Castro, E.; Masson, P.; Bougueleret, L.; Bairoch, A.; Xenarios, I.; Le Mercier, P. ViralZone: A knowledge resource to understand virus diversity. *Nucleic Acids Res.* **2011**, *39*, D576–D582. [CrossRef]
33. Sievers, F.; Higgins, D.G. Clustal omega. *Curr. Protoc. Bioinformatics* **2014**, *48*, 3.13.1–3.13.16.
34. Eddy, S.R. A new generation of homology search tools based on probabilistic inference. *Genome Inform.* **2009**, *23*, 205–211.
35. Remmert, M.; Biegert, A.; Hauser, A.; Söding, J. HHblits: Lightning-fast iterative protein sequence searching by HMM-HMM alignment. *Nat. Methods* **2011**, *9*, 173–175. [CrossRef]
36. Enright, A.J.; Van Dongen, S.; Ouzounis, C.A. An efficient algorithm for large-scale detection of protein families. *Nucleic Acids Res.* **2002**, *30*, 1575–1584. [CrossRef]
37. UniProt Consortium UniProt: A hub for protein information. *Nucleic Acids Res.* **2015**, *43*, D204–D212. [CrossRef]
38. Altschul, S.F.; Gish, W.; Miller, W.; Myers, E.W.; Lipman, D.J. Basic local alignment search tool. *J. Mol. Biol.* **1990**, *215*, 403–410. [CrossRef]
39. Mihara, T.; Nishimura, Y.; Shimizu, Y.; Nishiyama, H.; Yoshikawa, G.; Uehara, H.; Hingamp, P.; Goto, S.; Ogata, H. Linking Virus Genomes with Host Taxonomy. *Viruses* **2016**, *8*, 66. [CrossRef]
40. Kazlauskas, D.; Krupovic, M.; Venclovas, Č. The logic of DNA replication in double-stranded DNA viruses: Insights from global analysis of viral genomes. *Nucleic Acids Res.* **2016**, *44*, 4551–4564. [CrossRef]
41. Saitou, N.; Nei, M. The neighbor-joining method: A new method for reconstructing phylogenetic trees. *Mol. Biol. Evol.* **1987**, *4*, 406–425.
42. Stajich, J.E.; Block, D.; Boulez, K.; Brenner, S.E.; Chervitz, S.A.; Dagdigian, C.; Fuellen, G.; Gilbert, J.G.R.; Korf, I.; Lapp, H.; et al. The Bioperl toolkit: Perl modules for the life sciences. *Genome Res.* **2002**, *12*, 1611–1618. [CrossRef]
43. Will, S.; Joshi, T.; Hofacker, I.L.; Stadler, P.F.; Backofen, R. LocARNA-P: Accurate boundary prediction and improved detection of structural RNAs. *RNA* **2012**, *18*, 900–914. [CrossRef]
44. Cui, X.; Lu, Z.; Wang, S.; Jing-Yan Wang, J.; Gao, X. CMsearch: Simultaneous exploration of protein sequence space and structure space improves not only protein homology detection but also protein structure prediction. *Bioinformatics* **2016**, *32*, i332–i340. [CrossRef]

45. Tuller, T.; Veksler-Lublinsky, I.; Gazit, N.; Kupiec, M.; Ruppin, E.; Ziv-Ukelson, M. Composite effects of gene determinants on the translation speed and density of ribosomes. *Genome Biol.* **2011**, *12*, R110. [CrossRef]
46. Faure, G.; Ogurtsov, A.Y.; Shabalina, S.A.; Koonin, E.V. Role of mRNA structure in the control of protein folding. *Nucleic Acids Res.* **2016**, *44*, 10898–10911. [CrossRef]
47. Ashburner, M.; Ball, C.A.; Blake, J.A.; Botstein, D.; Butler, H.; Cherry, J.M.; Davis, A.P.; Dolinski, K.; Dwight, S.S.; Eppig, J.T.; et al. Gene ontology: Tool for the unification of biology. The Gene Ontology Consortium. *Nat. Genet.* **2000**, *25*, 25–29. [CrossRef]
48. Vandivier, L.; Li, F.; Zheng, Q.; Willmann, M.; Chen, Y.; Gregory, B. Arabidopsis mRNA secondary structure correlates with protein function and domains. *Plant. Signal. Behav.* **2013**, *8*, e24301. [CrossRef]
49. Binns, D.; Dimmer, E.; Huntley, R.; Barrell, D.; O'Donovan, C.; Apweiler, R. QuickGO: A web-based tool for Gene Ontology searching. *Bioinformatics* **2009**, *25*, 3045–3046. [CrossRef]
50. Giglio, M.; Tauber, R.; Nadendla, S.; Munro, J.; Olley, D.; Ball, S.; Mitraka, E.; Schriml, L.M.; Gaudet, P.; Hobbs, E.T.; et al. ECO, the Evidence & Conclusion Ontology: Community standard for evidence information. *Nucleic Acids Res.* **2019**, *47*, D1186–D1194.
51. Supek, F.; Bošnjak, M.; Škunca, N.; Šmuc, T. REVIGO summarizes and visualizes long lists of gene ontology terms. *PLoS ONE* **2011**, *6*, e21800. [CrossRef] [PubMed]
52. Jiang, T.; Nogales, A.; Baker, S.F.; Martinez-Sobrido, L.; Turner, D.H. Mutations Designed by Ensemble Defect to Misfold Conserved RNA Structures of Influenza A Segments 7 and 8 Affect Splicing and Attenuate Viral Replication in Cell Culture. *PLoS ONE* **2016**, *11*, e0156906. [CrossRef]
53. Kutchko, K.M.; Madden, E.A.; Morrison, C.; Plante, K.S.; Sanders, W.; Vincent, H.A.; Cruz Cisneros, M.C.; Long, K.M.; Moorman, N.J.; Heise, M.T.; et al. Structural divergence creates new functional features in alphavirus genomes. *Nucleic Acids Res.* **2018**, *46*, 3657–3670. [CrossRef]
54. Weinberg, Z.; Breaker, R.R. R2R–software to speed the depiction of aesthetic consensus RNA secondary structures. *BMC Bioinformatics* **2011**, *12*, 3. [CrossRef] [PubMed]
55. Wang, X.; Li, P.; Gutenkunst, R.N. Systematic Effects Of mRNA Secondary Structure On Gene Expression And Molecular Function In Budding Yeast. *BioRxiv* **2017**. [CrossRef]
56. Li, F.; Zheng, Q.; Vandivier, L.E.; Willmann, M.R.; Chen, Y.; Gregory, B.D. Regulatory impact of RNA secondary structure across the Arabidopsis transcriptome. *Plant. Cell* **2012**, *24*, 4346–4359. [CrossRef]
57. Zur, H.; Tuller, T. Strong association between mRNA folding strength and protein abundance in *S. cerevisiae*. *EMBO Rep.* **2012**, *13*, 272–277. [CrossRef]
58. Federhen, S. The NCBI Taxonomy database. *Nucleic Acids Res.* **2012**, *40*, D136–D143. [CrossRef]
59. Yachdav, G.; Wilzbach, S.; Rauscher, B.; Sheridan, R.; Sillitoe, I.; Procter, J.; Lewis, S.E.; Rost, B.; Goldberg, T. MSAViewer: Interactive JavaScript visualization of multiple sequence alignments. *Bioinformatics* **2016**, *32*, 3501–3503. [CrossRef]
60. Waterhouse, A.M.; Procter, J.B.; Martin, D.M.A.; Clamp, M.; Barton, G.J. Jalview Version 2–a multiple sequence alignment editor and analysis workbench. *Bioinformatics* **2009**, *25*, 1189–1191. [CrossRef]

Proteogenomics Uncovers Critical Elements of Host Response in Bovine Soft Palate Epithelial Cells Following In Vitro Infection with Foot-And-Mouth Disease Virus

Florian Pfaff [1], Sara Hägglund [2], Martina Zoli [1], Sandra Blaise-Boisseau [3], Eve Laloy [3,4], Susanne Koethe [1], Daniela Zühlke [5], Katharina Riedel [5], Stephan Zientara [3], Labib Bakkali-Kassimi [3], Jean-François Valarcher [2], Dirk Höper [1], Martin Beer [1] and Michael Eschbaumer [1,*]

1. Institute of Diagnostic Virology, Friedrich-Loeffler-Institut, Federal Research Institute for Animal Health, 17493 Greifswald, Germany; Florian.Pfaff@fli.de (F.P.); martina.zoli2811@gmail.com (M.Z.); Susanne.Koethe@fli.de (S.K.); Dirk.Hoeper@fli.de (D.H.); Martin.Beer@fli.de (M.B.)
2. Swedish University of Agricultural Sciences, Host-pathogen interaction group, Division of Ruminant Medicine, 75007 Uppsala, Sweden; sara.hagglund@slu.se (S.H.); jean-francois.valarcher@slu.se (J.-F.V.)
3. Laboratoire de Santé Animale de Maisons-Alfort, UMR 1161 virologie, INRA, Ecole Nationale Vétérinaire d'Alfort, ANSES, Université Paris-Est, 94700 Maisons-Alfort, France; sandra.blaise-boisseau@anses.fr (S.B.-B.); eve.laloy@vet-alfort.fr (E.L.); stephan.zientara@vet-alfort.fr (S.Z.); Labib.BAKKALI-KASSIMI@anses.fr (L.B.-K.)
4. Biopôle EnvA, Ecole Nationale Vétérinaire d'Alfort, Université Paris-Est, 94700 Maisons-Alfort, France
5. Institute of Microbiology, Department for Microbial Physiology and Molecular Biology, University of Greifswald, 17489 Greifswald, Germany; daniela.zuehlke@uni-greifswald.de (D.Z.); riedela@uni-greifswald.de (K.R.)

* Correspondence: Michael.Eschbaumer@fli.de

Abstract: Foot-and-mouth disease (FMD) is the most devastating disease of cloven-hoofed livestock, with a crippling economic burden in endemic areas and immense costs associated with outbreaks in free countries. Foot-and-mouth disease virus (FMDV), a picornavirus, will spread rapidly in naïve populations, reaching morbidity rates of up to 100% in cattle. Even after recovery, over 50% of cattle remain subclinically infected and infectious virus can be recovered from the nasopharynx. The pathogen and host factors that contribute to FMDV persistence are currently not understood. Using for the first time primary bovine soft palate multilayers in combination with proteogenomics, we analyzed the transcriptional responses during acute and persistent FMDV infection. During the acute phase viral RNA and protein was detectable in large quantities and in response hundreds of interferon-stimulated genes (ISG) were overexpressed, mediating antiviral activity and apoptosis. Although the number of pro-apoptotic ISGs and the extent of their regulation decreased during persistence, some ISGs with antiviral activity were still highly expressed at that stage. This indicates a long-lasting but ultimately ineffective stimulation of ISGs during FMDV persistence. Furthermore, downregulation of relevant genes suggests an interference with the extracellular matrix that may contribute to the skewed virus-host equilibrium in soft palate epithelial cells.

Keywords: foot-and-mouth disease virus (FMDV); bovine soft palate; nasopharynx; transcriptomics; proteomics; bioinformatics; virus-host interaction; innate immune system; interferon-stimulated genes (ISG)

1. Introduction

Foot-and-mouth disease (FMD) is an acute and severe systemic vesicular disease of cloven-hoofed animals (*Artiodactyla*) with tremendous economic impact. The last FMD epizootic in the European Union in the United Kingdom, Ireland, France and the Netherlands in 2001 culminated in the slaughter of more than 6.5 million animals and an economic toll of over €5 billion [1]. The etiological agent is foot-and-mouth disease virus (FMDV), the type species of the genus *Aphthovirus* in the family *Picornaviridae* [2]. FMDV particles comprise a non-enveloped icosahedral capsid that surrounds a single-stranded positive-sense RNA genome with an approximate length of 8.4 kilobases [3]. FMD mainly affects livestock such as cattle, buffalo, pigs, goats, and sheep, but can also be transmitted to deer and wild boar. It is endemic in wild buffalo in Southern Africa [4,5]. Although more than 70 species are known to be susceptible to FMDV, its primary host seem to be buffalo and cattle, in which the disease causes very high morbidity, but only low mortality in adults [6]. During the onset of acute infection, cattle are highly febrile and small vesicles develop on the mucosal membranes of the muzzle, lips, and oral cavity, as well as on the coronary band and interdigital space, and the teats of the udder.

Usually, cattle clinically recover within 2–3 weeks if no secondary infection occurs. At this time, many have completely cleared the virus, however, about 50% of animals may remain subclinically infected for up to three years, depending on the species [7]. The World Organisation for Animal Health (OIE) defines an animal from which infectious FMDV can be recovered by probang sampling later than 28 days post infection (dpi) as persistently infected or a so-called "carrier" [7]. The fear of contagion from carrier animals has severe consequences for trade in live animals and animal products [8]. In vivo studies have shown that in cattle more than 50% of animals will become persistently infected [9], even if they had been vaccinated against FMDV and did not develop clinical disease [10–12]. The exact anatomical sites of persistence are still debated, but different tissues of the upper respiratory tract including the nasopharynx have been suggested. A recent study of the tissue-specific localization of FMDV in persistently infected steers identified the contiguous epithelia of the dorsal soft palate (SP) and the dorsal nasopharynx as the most likely sites of persistence [10]. Evidence for FMDV persistence in lymph nodes and germinal centers was also put forward although no viral replication could be detected [13].

The cellular factors that promote establishment of FMDV persistence and the viral strategies of immune avoidance in the bovine nasopharynx remain currently poorly understood, making it impossible to predict which animals will develop into carriers and which will clear the virus. Previous in vitro work that aimed to decipher cellular responses during persistent FMDV infection hardly reflected persistence in vivo as the used models were either based on immortalized non-bovine cell lines, such as hamster kidney (BHK) cells [14], or bovine cell lines, e.g., bovine kidney (MDBK, EBK) cells [15,16], that are not from the primary site of persistence. O'Donnell et al. [17] used a persistently infected bovine pharynx cell line to examine the gene expression of selected cellular cytokines by RT-qPCR, showing differences in the expression of key antiviral cytokines between acute and persistent infection.

The rationale of this study was therefore to use a novel air–liquid interphase cell culture model based on primary SP cells from cattle [18] that closely resembles the situation in vivo, together with an proteogenomics approach that combined transcriptomics by high-throughput sequencing, RT-qPCR, proteomics and bioinformatics. Our results provide detailed insights into the transcriptional responses of the SP in reaction to acute FMDV infection and reveal long-lasting changes of gene activation throughout persistence. The identified pathways and genes may give rise to further investigations leading to early detection of persistence in cattle and novel vaccines that prevent the carrier state.

2. Materials and Methods

A schematic visualization of the sample processing and analysis workflow can be found in the Supplementary Figure S1.

2.1. Ethics Statement

The tissues used for the study were collected from animals slaughtered for food production. The animals were being processed as part of the normal work of the abattoir, therefore no ethics approval was required.

2.2. Bovine Epithelial Cultures from Soft Palate

Multilayers of bovine dorsal soft palate cells were propagated at the air–liquid interface for 5 weeks before FMDV infection, as described previously [18]. Briefly, bovine dorsal soft palate tissue, collected immediately after slaughter, was dissected and digested at 4 °C overnight in incubation medium supplemented with protease XIV (Sigma-Aldrich, St. Louis, MO, USA). Epithelial cells were thereafter scraped off the underlying tissue, filtered, and incubated in cell culture flasks for 4 h at 37 °C and 5% CO_2. Cells that did not adhere to the plastic were centrifuged at 200× g for 10 min at room temperature, frozen, thawed, and propagated for three to five passages in cell culture flasks before being seeded in 12 mm diameter Corning®Transwell-COL collagen-coated PTFE membrane inserts with 3.0 µm pores (Sigma-Aldrich). The cell culture medium was removed from the upper compartment after five days of culture and changed in the lower compartment every two or three days. The average number of cells that constituted the upper layer of the multilayer was estimated at 750,000.

2.3. Experimental Design and FMDV Infection

After 5 weeks of culture on inserts without passage, cells were infected with a twice-plaque-purified viral clone (FMDV O Clone 2.2, "Cl 2.2") derived from the O/FRA/1/2001 strain that was further propagated on BHK-21 cells (four passages) [16], or negative cell lysate, as described previously [18]. Two experiments (experiment 1 and 2) were performed with SP cells that originated from two different animals (a male and female, respectively) and that were infected at a multiplicity of infection (MOI) of 0.01 (compare Supplementary Table S1). Briefly, the inserts were incubated for one hour with 500 µL of clarified cell lysate from infected or uninfected cell cultures. Following infection and thereafter at a maximum interval of 3 days, the upper compartments were washed with 500 µL cell culture medium containing 10% FCS and, similarly, but only from 2 dpi, the medium in the lower compartments was changed. For each experiment, at days 0, 1 and 28, SP cells from 2 inserts were lysed with 750 µL of TRIzol Reagent (Life Technologies, Carlsbad, CA, USA) and frozen separately at −80 °C for transcriptomic and proteomic analyses.

2.4. RNA and Protein Isolation

For isolation of high-quality total RNA and proteins, 150 µL of trichlormethane (Carl Roth, Karlsruhe, Germany) was added to the lysed cells in 750 µL TRIzol Reagent and the mixture was centrifuged in order to separate the RNA-containing aqueous phase from the DNA- and protein-containing organic phase. The aqueous phase was then mixed with an equal amount of 100% ethanol (Carl Roth) and total RNA was extracted using the RNeasy Mini Kit (Qiagen, Hilden, Germany) with on-column DNase digestion with the RNase-Free DNase Set (Qiagen), following the manufacturer's instructions. The quantity and quality of total RNA was subsequently analyzed using a NanoDrop 1000 spectrophotometer (Peqlab, Erlangen, Germany) and RNA 6000 Pico chips on an Agilent 2100 Bioanalyzer (Agilent Technologies, Böblingen, Germany). All samples were checked for contamination using the 260/280 and 260/230 nm ratios, as well as for RNA degradation using the RNA Integrity Number (RIN).

Polyadenylated mRNA was subsequently isolated from 1–3 µg of high-quality total RNA using the Dynabeads mRNA DIRECT Micro kit (Invitrogen, Carlsbad, CA, USA) following the manufacturer's instructions. Prior to isolation, the ERCC ExFold RNA Spike-In mix 1 (Invitrogen) was supplemented and used as an internal control for all following steps. The quality of the mRNA and the extent of ribosomal RNA contamination was assessed using RNA 6000 Pico chips on the Agilent 2100 Bioanalyzer (Agilent Technologies).

Proteins were extracted from the organic TRIzol phase following the manufacturer's instructions. Briefly, 0.3 mL of 100% ethanol (Carl Roth) were added and DNA was pelleted by centrifugation. To the supernatant, 1.5 mL of isopropanol (Carl Roth) were added and proteins were pelleted by centrifugation. Protein pellets were washed three times with 0.3 M guanidine hydrochloride (Carl Roth) in 95% ethanol (Carl Roth). After a final washing step using 95% ethanol, proteins were air-dried and the pellet was resuspended in freshly prepared 1% SDS (Carl Roth) by ultrasonication. Quality and quantity of the isolated proteins were checked using a 12% polyacrylamide gel (SDS-PAGE) and a colorimetric bicinchoninic acid (BCA) assay.

2.5. Library Preparation and Sequencing

For preparation of whole-transcriptome libraries the Ion Total RNA-Seq Kit v2 (Life Technologies) was used, following the manufacturer's instructions. Briefly, between 1 and 100 ng of the mRNA containing the ERCC spike-in control was treated with RNase III at 37 °C for 10 min. The fragmented mRNA was subsequently purified using the Magnetic Bead Cleanup Module (Life Technologies) and the resulting size distribution was assessed with the Agilent 2100 Bioanalyzer as described above. After hybridization and ligation of appropriate adapters, the fragmented mRNA was reverse transcribed into cDNA using SuperScript III enzyme. The cDNA was purified as described above and amplified for 14 cycles using Platinum PCR SuperMix High Fidelity along with appropriate primers for the generation of barcoded libraries. The resulting libraries were again purified using the method described above and the size distribution was assessed using the Agilent 2100 Bioanalyzer together with the DNA 7500 kit and chip (Agilent Technologies). All libraries were subsequently quantified using the KAPA Library Quantification Kit Ion Torrent (Kapa Biosystems, Wilmington, MA, USA) on a CFX96 Real-Time PCR Detection System (Bio-Rad Laboratories, München, Germany) and pooled at an equimolar ratio. For sequencing an Ion S5XL sequencing system (Life Technologies) along with the Ion 540 OT2 and Chip kit (Life Technologies) for the generation of up to 200 bp reads was used. Each library was sequenced in at least two independent sequencing runs.

2.6. Statistical Analysis of Differential Gene Expression

In order to detect problems and biases during mRNA isolation and library preparation, we used the ERCC_Analysis Plugin (version 5.8.0.1) provided in the Torrent Suite software (version 5.8.0). Only forward strand reads were selected for mapping and the minimum transcript count was set to 50. The raw reads from each sequencing library were quality checked using FastQC (version 0.11.7; Babraham Institute) with a focus on the read length distribution and adapter contamination. In order to quantify the expression of known bovine transcripts in these datasets we used Salmon (version 0.9.1) that uses a lightweight alignment method (quasi-mapping) for rapid transcript abundance estimation [19]. Briefly, we obtained the transcript reference of cattle (GCF000003055.6 Bos Taurus UMD 3.1.1) from NCBI and selected only transcripts that were featured as mRNAs, non-coding RNAs and miscellaneous RNAs (for details see Supplementary Table S2). A Salmon index was created using the option for a perfect hash, rather than a dense hash. Each sequencing library was then used as input for the major Salmon function "quant" using appropriate options for the library type (stranded single-end protocol with reads coming from the forward strand) and the mean read length. The number of bootstraps was set to 100 replicates. Subsequently a "tx2gene" table was prepared using the accessions of each RNA transcript from the aforementioned reference in the first column and the corresponding gene symbol in the second column (see also Supplementary Table S3). Using this table, the transcript abundancy datasets from Salmon "quant" were imported into R workspace (version 3.4.1; [20]) using the "tximport" package (version 1.6.0; [21]) as implemented in the Bioconductor library. For handling and manipulating R scripts, the software RStudio (version 1.0.153) was used. Technical replicates of samples (multiple sequencing of the same library from a single sample) were combined and the datasets were pre-filtered using only genes were more than four samples had raw gene counts greater than or equal to 100. In order to check the dataset for influence of treatment (infection, time and animal) and repeatability (replicates of same treatment) a principal components analysis (PCA) as well as a heatmap clustering was conducted using the regularized log transformed (rlog) read counts. DESeq2 (version 1.18.1; [22]) was then used to identify differentially expressed genes between the treatments based on the negative binomial distribution. The resulting *p*-values were adjusted with the Benjamini–Hochberg procedure and only genes with an adjusted *p*-value below 0.001 and an absolute fold change of >1 were considered significant. The logarithmic fold changes were further shrunken as recommended and described by Love et al. 2014 [22] in order to account for genes with low read counts. Significant differently expressed genes were annotated using the AnnotationDbi package (version 1.40.0) and used for further pathway enrichment analysis. Briefly, gene sets were analyzed using the "enrichPathway" function of the ReactomePA package (version 1.22.0; [23]) and the "enrichKEGG" function of the clusterProfiler package (version 3.6.0; [24]).

2.7. Quantitative Reverse Transcription PCR (RT-qPCR)

In order to confirm the results from the RNA sequencing experiment and to include samples from additional time points, a subset of six target genes (*ANKRD1*, *CASP7*, *IDO1*, *IFIH1*, *NCAM1*, *OAS2*) and two reference genes (*ACTB*, *GAPDH*) was selected for quantitative reverse transcription PCR (RT-qPCR) analysis. For each gene, two intron-spanning primer sets were designed using Primer3 (version 0.4.0; [25]) (for primer sequences see Supplementary Table S4). For RT-qPCR the QuantiTect Probe RT-PCR Kit (Qiagen) was used together with LightCycler 480 ResoLight Dye (Roche, Mannheim, Germany) according to the manufacturers' instructions. The following temperature profile was used on a Bio-Rad CFX96: 50 °C for 30 min, 95 °C for 15 min and 45 cycles of 94 °C for 15 s and 60 °C for 1 min. After each cycle, the fluorescence in the SYBR channel was detected and threshold cycle (Ct) values were deduced after each run. For each newly designed primer set the PCR efficiency was determined using an appropriate dilution series from an independent bovine RNA control. Using the ΔΔCt method, Ct values of target genes were first normalized by subtracting the Ct value of references genes ($\Delta Ct = Ct_{Target} - Ct_{Reference}$). Mean and standard deviation of ΔCt were calculated from two technical and two biological replicates for each treatment group. Subsequently, the ΔΔCt was calculated by subtracting the ΔCT of the control group from the ΔCt of the treatment group at different time points ($\Delta\Delta Ct = \Delta Ct_{Control} - \Delta Ct_{Treatment}$) and used for calculation of the fold change of each gene ($2^{-\Delta\Delta CT}$).

2.8. Protein Identification and Quantification, and Statistical Analysis of Differential Protein Expression

Thirty µg of the extracted proteins were separated by SDS-PAGE, stained with Colloidal Coomassie Brilliant Blue G-250 and afterwards lanes were cut into ten equidistant pieces. In-gel digestion using trypsin and purification of tryptic peptides using Ziptips (C18, Millipore) prior to MS analysis were done as described previously [26]. LC-MS/MS were conducted using an EASY-nLC II coupled to a LTQ Orbitrap-Velos mass spectrometer (Thermo Fisher Scientific, Waltham, MA, USA). Peptides were separated at a constant flow rate of 300 nL/min using a binary 76 min gradient from 5% B to 75% B 99.9% ACN, 0.1% acetic acid). Survey scans in the Orbitrap were recorded with a resolution of 60.000 in a m/z range of 300–1700. The 20 most intense peaks per scan cycle were selected for CID fragmentation in the LTQ. Ions with unknown charge state, as well as singly charged ions were excluded from fragmentation. Dynamic exclusion of precursor ions for 30 s was enabled. Internal calibration (lock mass 445.120025) was enabled as well. For protein identification, resulting spectra were searched against a database containing sequences of *Bos taurus* including reverse sequences and common laboratory contaminants (44,376 entries). Database searches using Sorcerer-SEQUEST (version v.27, rev.11; Sage-N Research, Inc., Milpitas, CA, USA) and Scaffold (version v.4.8.4; Proteome Software, Portland, OR, USA) were done as described earlier [27].

In order to identify differentially expressed proteins we imported the raw NSAF values into R workspace and created an ExpressionSet using the Biobase package (version 2.38.0; [28]). We then used the "Power Law Global Error Model" (PLGEM) as implemented in the "plgem" package (version 1.50.0; [29]) and first fitted it to the dataset using default settings and "FMDV infection" (samples from 24 h post infection (hpi) and 28 dpi) as fitting condition. After computation of observed and resampled signal-to-noise ratios, p-values for each detected protein were calculated.

2.9. Data Availability

The raw sequencing data along with deduced Salmon read count tables and substantial metadata are available at ArrayExpress (http://www.ebi.ac.uk/arrayexpress) under the accession number E-MTAB-7605. The mass spectrometry proteomics data have been deposited to the ProteomeXchange Consortium (http://www.proteomexchange.org) via the PRIDE [30] partner repository with the dataset identifier PXD012242

3. Results

3.1. A Cell Culture Model for FMDV Persistence

The establishment and full characterization of the primary SP cell culture model are presented by Hägglund et al. [18]. In summary, the primary SP cells formed into multilayers and showed typical features of stratified squamous epithelia, such as tight junctions and impermeability to cell culture media. The cultures were inoculated either with a viral clone derived from the O/FRA/1/2001 strain or negative cell lysate for control. Infection with FMDV at low MOI resulted in limited cytopathic effect, high viral loads and presence of detectable viral antigen. After 28 dpi, the multilayers were still intact, but FMDV antigen and genome remained detectable at very low levels and viable virus could be isolated—giving a clear indication of FMDV persistence.

3.2. RNA Sequencing and Exploratory Data Analysis

The polyadenylated RNA fraction from 21 primary bovine SP cell culture samples was sequenced. From these, 10 and 11 samples originated from a female and male bovine, respectively. Cells for analysis were harvested immediately before inoculation (0 hpi), 24 hpi and 28 dpi (Figure 1A). For each time point (0 hpi, 24 hpi, 28 dpi) and treatment (FMDV, control), a minimum of four biological replicates (two from each animal) was sequenced. The number of reads for each sample ranged from 21.8 to 28.0 million, with an average of 24.7 million. In total, 518.4 million reads were included in the analysis (Supplementary Table S1).

These reads were assigned to a cattle transcript reference and the raw gene count data was transformed with respect to library size and transcript length using an appropriate model (Supplementary Figure S2). A principal components analysis (PCA) based on the normalized gene counts revealed that the samples from non-infected cultures remained closely together independent of the time in culture, while FMDV-infected samples showed a clear time-dependent grouping (Figure 1B). Accordingly, the first principal component (50% of variance) was assumed to represent the transcriptional differences between the samples caused by FMDV infection. Furthermore, transcriptional differences between both animals were clearly visible and represented by principal component 2 (29% of variance). The PCA showed no batch-to-batch effects and biological replicates from the same animal were grouped tightly.

Unsupervised cluster analysis of the 45 most variable gene transcripts clearly separated the FMDV-infected samples from the non-infected controls (Figure 1C). In detail, the infected samples showed a strong positive deviation from the per gene mean count (Figure 1C, green and blue transcript cluster). Interestingly, samples from the acute and persistent phases of infection were clearly separated, because the activation of the aforementioned genes was reduced at 28 dpi (Figure 1C, blue cluster). As in the PCA, the non-infected samples were divided by the differential expression of another set of transcripts (Figure 1C, red transcript cluster) and grouped according to the donor animal they originated from. In summary, the explorative data analysis showed that the FMDV infection has a clear effect on gene expression that is distinct from the animal-dependent effects.

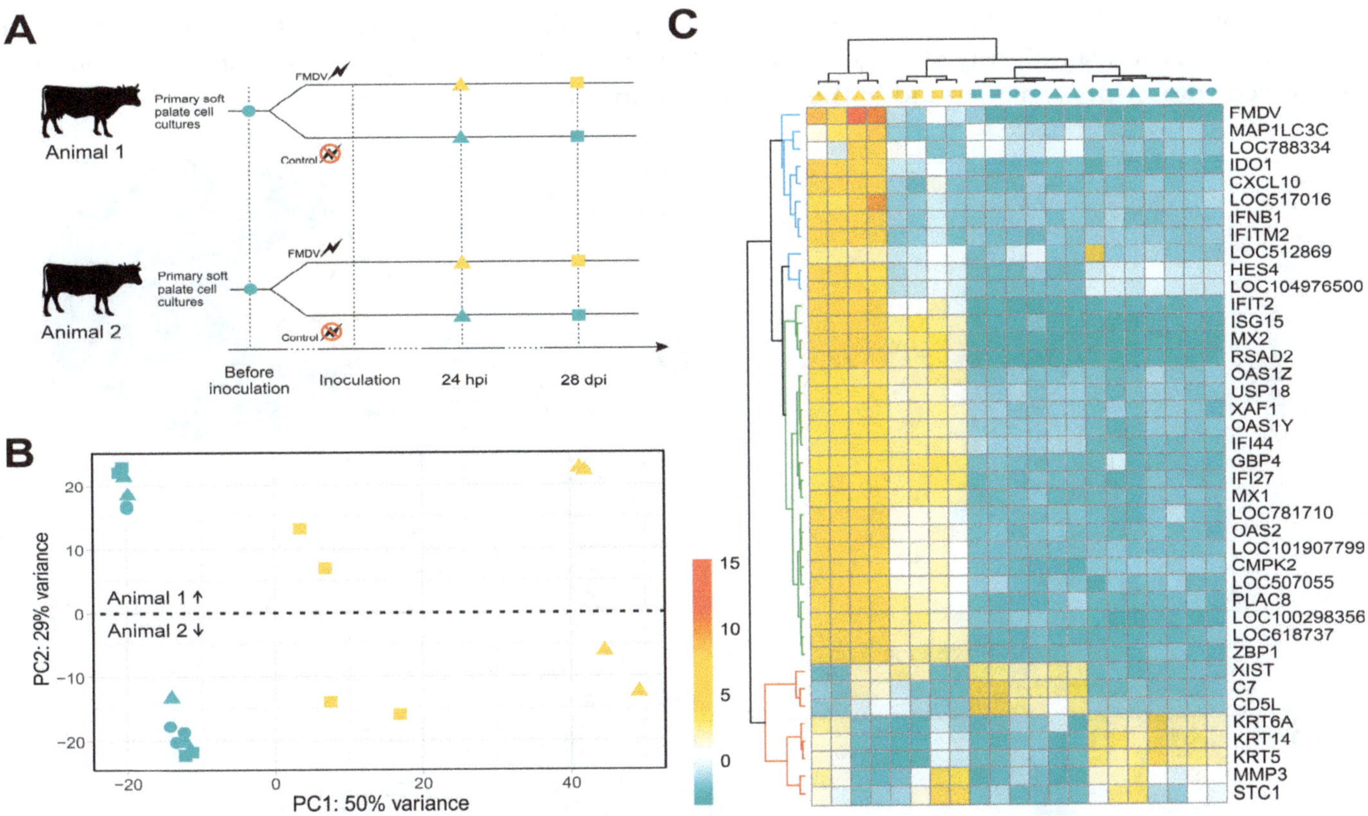

Figure 1. Experimental setup of RNA sequencing and exploratory data analysis. (**A**) Primary soft palate (SP) cell cultures were obtained from two animals. Baseline samples for RNA sequencing were collected immediately before inoculation (circle). Subsequently, the cell cultures were inoculated with foot-and-mouth disease virus (FMDV) (orange symbols) or mock-infected for use as controls (blue symbols). Cells were then harvested for sequencing at 24 h post infection (hpi) (triangles) or 28 days post infection (dpi) (squares), representing acute and persistent infection, respectively. (**B**) Principal components analysis based on normalized gene counts of the 1000 most variable genes. (**C**) The variance of normalized gene counts was calculated for each gene and the 45 genes with the highest variance were selected and visualized in a heat map. The color of the cells indicates the difference from the mean normalized gene count of the corresponding gene. Samples and genes were clustered according to these differences (trees). The transcript cluster highlighted in green is similarly activated by acute and persistent FMDV infection, while the blue cluster highlights transcripts that are only active during the acute phase of infection. The red cluster comprises transcripts whose abundance differs between donor animals used in this experiment.

3.3. Differential Expression during Acute and Persistent FMDV Infection and associated Pathways

In order to confirm the trends from the explorative data analysis and to address the transcriptional changes during acute and persistent infection in more detail, a differential expression analysis was conducted. The transcriptional response of cells from bovine SP tissue to FMDV infection was assessed at 24 hpi and 28 dpi, in contrast to non-infected controls from the same time points (Figure 2). A total of 312 and 73 gene transcripts were differentially expressed at 24 hpi and 28 dpi, respectively. With 305/312 (97.7%) for 24 hpi and 65/73 (89%) for 28 dpi, the majority of differentially expressed transcripts were up-regulated, while only 8/312 (2.5%) and 8/73 (11%) were down-regulated (Figure 2A,B). From a total of 324 differentially expressed transcripts at both time points, 249 and 10 transcripts were solely regulated during either acute or persistent phase, respectively. The expression of 63 transcripts was significantly up-regulated during both stages of infection (Figure 2C). The log2 fold change of these transcripts was much lower at 28 dpi when compared to 24 hpi (Figure 2D), with highest differences observed for *IFIT2*, LOC101907799 and *CMPK2*. In contrast, the log2 fold change for *IFI27*, *PARP9*, *IFI6*, and *OAS1Z* was comparable or nearly equal during both phases of infection. *PLAC8* was slightly stronger regulated at 28 dpi than at 24 hpi. A full list of differentially expressed transcripts can be found in the supplementary material (Supplementary Table S5).

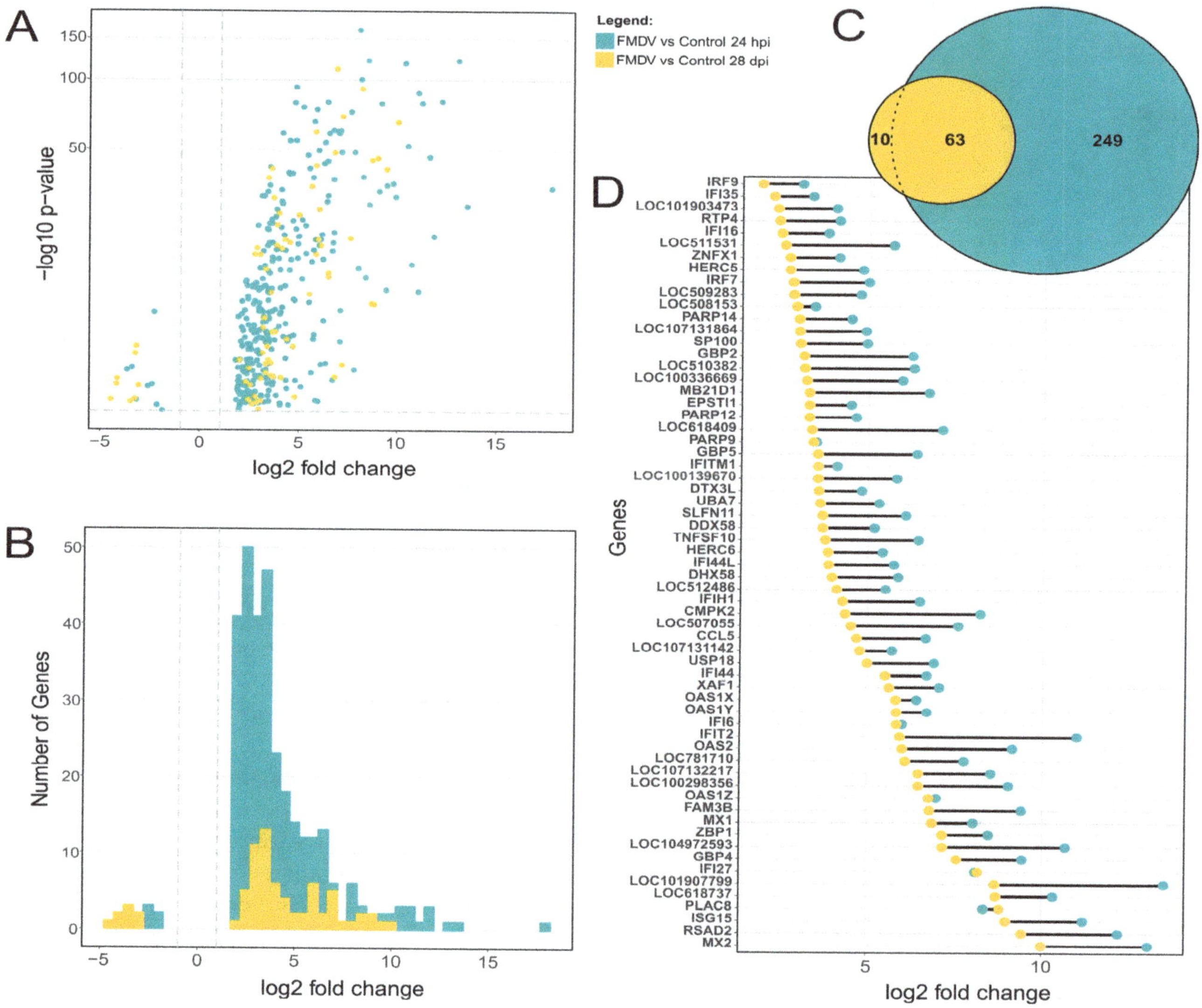

Figure 2. Differential gene expression of SP cells during FMDV infection. (**A**) Volcano plot showing the log2 fold change (x-axis) and the adjusted p-value (y-axis) for all differentially expressed genes during the acute (24 hpi, green) and persistent (28 dpi, orange) phase of FMDV infection. The log2 fold change and adjusted p-value are calculated relative to a non-infected control from the same time point. The grey dotted lines indicate the cutoff values: adjusted p-value < 0.001 and $|\text{log2 fold change}| > 1$. (**B**) The histogram summarizes the number of genes that have a certain log2 fold change. (**C**) The total number of differentially expressed genes is visualized in a Venn diagram. The overlap indicates the 63 genes that are differentially expressed during both phases of infection. (**D**) The log2 fold change of these 63 genes is compared for acute and persistent infection.

The significantly differently expressed genes during acute and persistent infection were further matched to specific metabolic and signaling pathways using the Reactome database [31] (Figure 3). During both infection stages, most of the regulated genes are associated with interferon signaling, in particular interferon α, β and γ signaling (Figure 3A and Table 1). The induction of these interferons appears to be mediated by the DDX58/IFIH1 pathway. Accordingly, a number of interferon-stimulated genes (ISG), such as the ubiquitin-like family member *ISG15*, that induce strong antiviral mechanisms of the innate immune system are highly activated. During acute infection, transcripts involved in the interleukin-1 (*IL1*) and -10 signaling pathways as well as the programmed cell death (apoptosis) pathway were enriched. However, it is of note that the *IL1* and *IL10* genes were not significantly differently expressed. The MHC I antigen presentation pathway, a part of both the innate and adaptive immune system, was also enriched at 24 hpi. During persistence, most of the genes belonging to the interleukin-1, -10, apoptosis and MHC I antigen presentation pathways were not significantly differentially expressed in comparison to control samples.

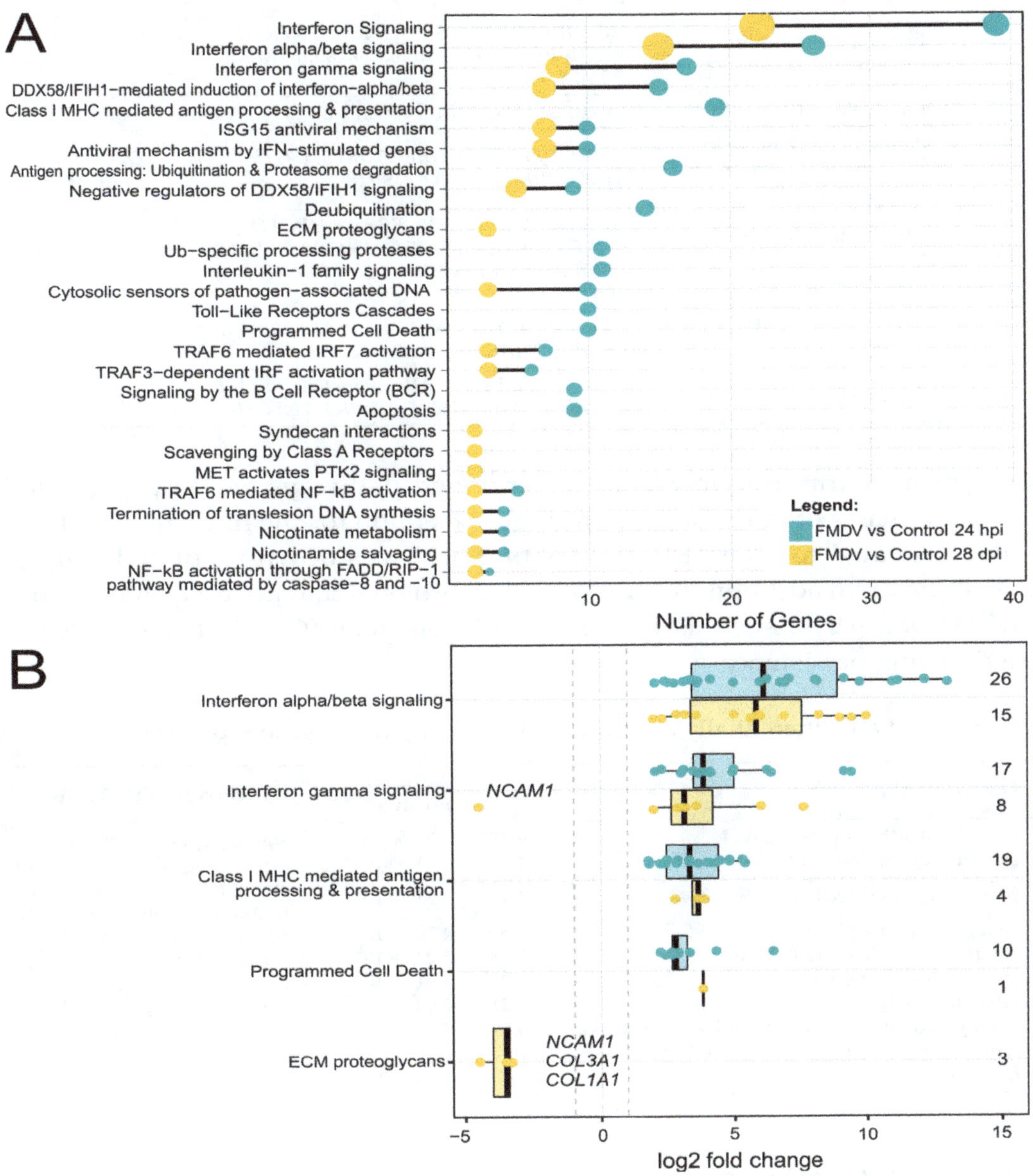

Figure 3. Pathway enrichment analyses of significantly differentially expressed genes during acute and persistent FMDV infection of bovine soft palate cells. (**A**) The number of genes that contribute to a significantly enriched Reactome pathway is shown for 24 hpi (blue dots) and 28 dpi (orange dots). Black lines connect pathways that are enriched in both datasets and the size of the circles represents the ratio of genes found for a certain pathway to the overall gene number. (**B**) The log2 fold change of genes that contribute to selected enriched Reactome pathways are highlighted (24 hpi—blue dots) and 28 dpi—orange dots). The grey dotted lines indicate a |log2 fold change| > 1. The numbers on the right correspond to the number of genes associated with each pathway.

Table 1. Selected enriched pathways and corresponding genes during acute and persistent FMDV infection.

Metabolic Complex	Pathway	24 hpi	28 dpi
Innate immune system	DDX58/IFIH1-mediated induction of interferon-alpha/beta	*DDX58, DHX58, HERC5, IFIH1, IFNB1, IRF1, IRF3, IRF7, ISG15, NFKBIA, NFKBIB, NLRC5, TNFAIP3, TRIM25, UBA7*	*DDX58, DHX58, HERC5, IFIH1, IRF7, ISG15, UBA7*
Cytokine signaling in immune system	Interferon alpha, beta signaling	*ADAR, GBP2, IFI27, IFI35, IFI6, IFIT2, IFIT3, IFITM1, IFITM2, IFNB1, IRF1, IRF3, IRF5, IRF7, IRF9, ISG15, MX1, MX2, OAS2, PSMB8, RNASEL, RSAD2, SOCS1, STAT2, USP18, XAF1*	*GBP2, IFI27, IFI35, IFI6, IFIT2, IFITM1, IRF7, IRF9, ISG15, MX1, MX2, OAS2, RSAD2, USP18, XAF1*
	Interferon gamma signaling	*GBP2, IFI27, IFI35, IFI6, IFIT2, IFITM1, IRF7, IRF9, ISG15, MX1, MX2, OAS2, RSAD2, USP18, XAF1*	*GBP2, GBP4, GBP5, IRF7, IRF9, NCAM1, OAS2, SP100*
	Interleukin-1 family signaling	*IL18, IL18BP, MAP3K8, NFKBIA, NFKBIB, PELI1, PSMB10, PSMB8, PSMB9, PSME2, PSMF1*	
	Interleukin-10 signaling	*CCL2, CCL5, CXCL10, CXCL2, CXCL8, IL18, IL6*	*CCL5*
	Antiviral mechanism by IFN-stimulated genes/ISG15 antiviral mechanism	*DDX58, EIF2AK2, HERC5, IRF3, ISG15, MX1, MX2, TRIM25, UBA7, USP18*	*DDX58, HERC5, ISG15, MX1, MX2, UBA7, USP18*
Adaptive immune system	Class I MHC mediated antigen processing and presentation	*AREL1, CTSS, DTX3L, ERAP2, HECTD2, HERC5, HERC6, PSMB10, PSMB8, PSMB9, PSME2, PSMF1, RBCK1, RNF114, RNF19B, SOCS1, TAP1, TRIM21, UBA7*	*DTX3L, HERC5, HERC6, UBA7*
	Activation of NF-kappaB in B cells	*NFKBIA, NFKBIB, NFKBIE, PSMB10, PSMB8, PSMB9, PSME2, PSMF1*	
Programmed cell death	Programmed cell death	*CASP7, CFLAR, PMAIP1, PSMB10, PSMB8, PSMB9, PSME2, PSMF1, RIPK3, TNFSF10*	*TNFSF10*

All of the genes contributing to the aforementioned pathways were up-regulated during the acute and persistent phases of infection, with a single exception for the nuclear cell adhesion molecule 1 (*NCAM1*) that is related to interferon γ signaling and was down-regulated during persistence (Figure 3B, and Table 2). In addition to *NCAM1*, other proteins that are associated with extracellular membrane (ECM) pathways, such as type I and III collagens (*COL1A1* and *COL3A1*) were also down-regulated during persistence.

Table 2. Genes that are differentially expressed only during the persistent phase of FMDV infection.

Gene	Description	LFC[†]	Adjusted p-value	Enriched DAVID Terms
NCAM1	neural cell adhesion molecule 1	−4.48	3.35×10^{-4}	signal peptide, secreted
ANKRD1	ankyrin repeat domain 1	−4.19	6.61×10^{-5}	positive regulation of apoptotic process
SFRP2	secreted frizzled related protein 2	−4.16	3.31×10^{-5}	signal peptide, secreted, positive regulation of apoptotic process
COL1A1	collagen type I alpha 1 chain	−3.50	4.16×10^{-4}	signal peptide, secreted
COL3A1	collagen type III alpha 1 chain	−3.28	3.23×10^{-4}	signal peptide, secreted
MYLK	myosin light chain kinase	−3.26	9.61×10^{-7}	
HTRA3	HtrA serine peptidase 3	−3.21	2.93×10^{-7}	signal peptide, secreted
ALDH1A2	aldehyde dehydrogenase 1 family member A2	−3.09	7.83×10^{-5}	positive regulation of apoptotic process
LY6E	lymphocyte antigen 6 family member E	3.45	2.59×10^{-6}	signal peptide
FBP1	fructose-bisphosphatase 1	4.09	2.94×10^{-5}	

[†]: log2 fold change.

3.4. RT-qPCR and Quantitative Proteomics

Results from the differential gene expression analysis based on RNA sequencing were confirmed using RT-qPCR. A panel of two reference (*ACTB, GAPDH*) and six target genes (*OAS2, IFIH1, NCAM1, ANKRD1, IDO1, CASP7*) were selected for analysis. These genes were chosen according to their apparent regulation: *OAS2, IFIH1* and *IDO1* were up-regulated during acute and persistent infection, while *NCAM1* and *ANKRD1* were only down-regulated during persistence. *CASP7* was only up-regulated during the acute phase of FMDV infection. The selected genes were also involved in relevant signaling pathways: *IFIH1* is part of the interferon induction cascade, while *OAS2* is an interferon-induced gene with antiviral activity. *CASP7* plays a key role in apoptosis and *IDO1* encodes a metabolically active protein that supports the immune system. *ANKRD1* and *NCAM1* were chosen as interesting target genes that were only regulated during persistence. The RT-qPCR analysis included samples from additional time points (2, 24, and 48 hpi; 7 and 28 dpi) for which log2 fold changes were then calculated using the delta-delta Ct method. For each time point and cell type four biological replicates (two replicates per animal) were used for analysis with the exception for 48 hpi, from which only two replicates derived from a single animal were available. The log2 fold-changes obtained by RT-qPCR of SP cells at the time points 24 hpi and 28 dpi confirm the results of the RNA sequencing (Figure 4A). All of the analyzed genes showed a time-dependent regulation during infection with FMDV (Figure 4B). The genes *IDO1, OAS2,* and *IFIH1* were strongly up-regulated from 2 to 24 hpi and were still up-regulated after 28 dpi. In contrast, *CASP7* was only up-regulated during acute infection. *ANKRD1* and *NCAM1* were down regulated as early as 48 hpi and throughout 28 dpi.

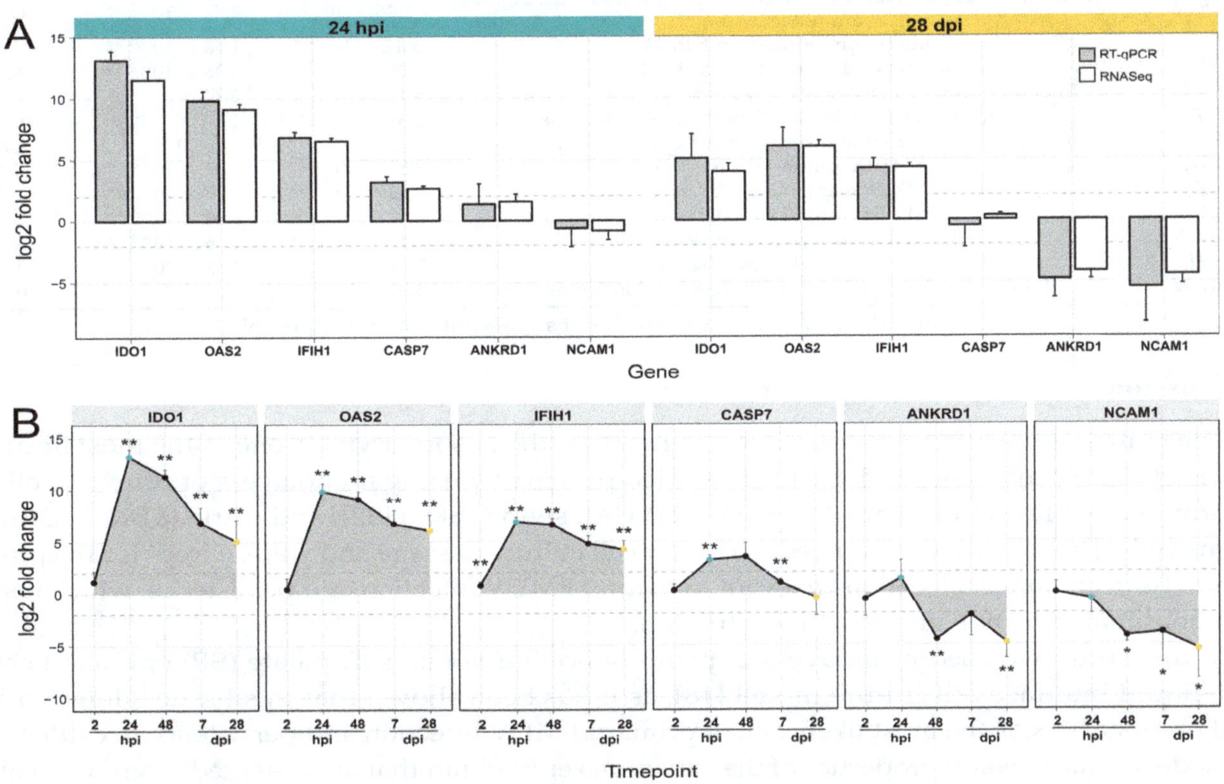

Figure 4. Expression analysis for selected genes by RT-qPCR at different time points. (**A**) Gene expression in FMDV infected soft palate cells compared to uninfected controls, at 24 hpi (blue panel) and 28 dpi (orange panel). The change in gene expression was analyzed for six selected target genes and two reference genes (not shown) using RT-qPCR. Log2 fold changes were calculated in comparison to a non-infected control from the same time points. Grey and white bars indicate the log2 fold change of gene expression using results from RT-qPCR and RNA sequencing (RNASeq), respectively. The grey dotted lines indicate a |log2 fold change| > 1. (**B**) The gene expression of the six selected genes was traced in a time course 2, 24, and 48 hpi; and 7 and 28 dpi. Statistically significant changes in comparison to control samples are highlighted with asterisks (p-value < 0.05: *; p-value < 0.01: **). Time points 24 hpi and 28 dpi are illustrated in blue and orange, respectively.

Additionally, a GeLC-MS/MS-based comparative proteome analysis was conducted to confirm the results of the RNA sequencing on the protein level, using two biological replicates for each time point and treatment from animal 2. Proteins with significantly different abundance in both experiments were identified and quantified by spectral-counting (Table 3). At the given thresholds, 11 and 8 proteins were differentially expressed when comparing FMDV-infected and control samples at 24 hpi and

28 dpi, respectively. Of these, 6 were differentially expressed at both time points. The Normalized Spectral Abundance Factor (NSAF), measured as the proportion of a single protein in relation to the detected proteins overall revealed an accumulation of the proteins *HERC6, IFI44, IFI44L, ISG15, MX1, MX2, OAS1X,* and *OAS1Y* in cells during persistent infection. In contrast, the proportion of *ATAD1, IFIT1,* and *IFIT2* to all other detected proteins was higher during acute than in persistent infection. FMDV polyprotein, *IFIT3* and *RSAD2* were not detected at the protein level at 28 dpi.

Table 3. Proteins with significantly different abundance at 24 hpi or 28 dpi. Only proteins that were also differentially expressed in the RNA sequencing experiment at the given time point are shown.

Protein	Description	24 hpi Mean NSAF	28 dpi Mean NSAF	$NSAF_{28\ dpi}/NSAF_{24\ hpi}$
ATAD1	ATPase Family, AAA Domain Containing 1	2.50×10^{-4} **	1.56×10^{-4}	0.6
CAG23917.14	FMDV polyprotein	3.16×10^{-4} **	$0^{\dagger}$	n.a.
HERC6	HECT And RLD Domain Containing E3 Ubiquitin Protein Ligase Family Member 6	2.46×10^{-5}	2.74×10^{-4} **	11.2
IFI44	Interferon Induced Protein 44	1.12×10^{-4} *	4.58×10^{-4} **	4.1
IFI44L	Interferon Induced Protein 44 like	8.32×10^{-5} *	6.98×10^{-4} **	8.4
IFIT1	Interferon-Induced Protein with Tetratricopeptide Repeats 1	5.41×10^{-4} **	7.62×10^{-5} *	0.1
IFIT2	Interferon Induced Protein With Tetratricopeptide Repeats 2	2.31×10^{-4} **	9.34×10^{-5} *	0.4
IFIT3	Interferon Induced Protein With Tetratricopeptide Repeats 3	3.78×10^{-4} **	$0^{\dagger}$	n.a.
ISG15	ISG15 Ubiquitin-Like Modifier	3.53×10^{-3} **	7.52×10^{-3} **	2.1
MX1	MX Dynamin Like GTPase 1	8.42×10^{-4} **	2.27×10^{-3} **	2.7
MX2	MX Dynamin Like GTPase 2	2.44×10^{-4} **	5.04×10^{-4} **	2.1
OAS1X	2′-5′-Oligoadenylate Synthetase 1 X	2.47×10^{-4} **	4.85×10^{-4} **	2.0
OAS1Y	2′-5′-Oligoadenylate Synthetase 1 Y	2.31×10^{-4} **	3.68×10^{-4} **	1.6
RSAD2	Radical S-Adenosyl Methionine Domain Containing 2	5.69×10^{-4} **	$0^{\dagger}$	n.a.

*: p-value < 0.01; **: p-value < 0.001; †: not significant; n.a. not applicable.

4. Discussion

The failure of the host response to clear virus from the nasopharynx is one of the most important features of FMDV infection, and has been studied previously using monolayers of primary cells and non-primary cell lines from hamsters (BHK-21) [14], swine (SK6) [32], and cattle (EBK, MDBK and pharynx cells) [15–17]. These cells needed regular cell culture passage and were, except for the pharynx cells, not from anatomical locations relevant to natural FMDV infection and persistence, which resulted in a suboptimal reflection of the in vivo situation.

In this study we used multilayers of primary bovine dorsal soft palate (SP) epithelial cells in an air–liquid interface cell culture model [18], as it has been shown that this tissue, along with the dorsal nasopharynx, is the most likely site of primary FMDV infection and persistence in cattle [10,33]. This model mimics several properties of the squamous epithelium that are observed in vivo and allows detailed studies of the target cells of primary FMDV infection. Specialized immune cells, such as B cells, T cells, and natural killer cells, which contribute to host responses in vivo, were depleted by the establishment of the cell culture in air [18], leading to improved standardization and a focus on the primary target cell. In vivo, neutralizing antibodies secreted by B cells are essential for resolving systemic FMDV infection, but have no effect on persistent infection in the nasopharynx [9]. Recent studies of gene expression in nasopharyngeal tissues of cattle have suggested an important role of cytotoxic T cells in the clearance of persistent FMDV infection [34,35]. The transcriptional changes analyzed in the present study, however, are limited to the response of SP epithelial cells themselves to infection with FMDV.

This response was analyzed using RNA sequencing, RT-qPCR and quantitative proteomics. Based on the results from exploratory RNA sequencing alone, infection with FMDV resulted in detectable changes in the transcriptome of the cells and the induced changes were distinct during the acute and persistent phases of infection. Despite that only a small proportion of cells were persistently infected, these influenced the transcriptome significantly. The non-infected controls did not show any comparable, time-dependent changes, indicating stable culture conditions without extensive cell differentiation or degradation. The observed changes induced by FMDV infection were also independent of the donor animals used for the preparation of the primary cell cultures, although transcriptional differences between the animals were detectable, e.g., sex-associated differences in *XIST* and keratin expression [36]. In summary, we observed genes that were highly regulated during both stages of infection (Figure 1C, green cluster), genes exclusively regulated during acute infection (Figure 1C, blue cluster) and animal-specific gene expression (Figure 1C, red cluster). During acute and persistent infection, comparably low numbers of genes were significantly differentially expressed and most of these were up-regulated. Interestingly, 63 genes were differentially expressed during both infection phases, indicating that FMDV infection induces long-lasting changes in the soft palate transcriptome. Furthermore, these genes were generally more up-regulated during acute infection. The transcriptional changes observed by RNA sequencing were supported at the protein level using quantitative proteomics. Although the number of detected differentially expressed proteins was much lower, it was confirmed that overexpression of ISGs leads to detectable levels of these proteins in the cells.

We found a strong activation of the innate immune response at 24 hpi that appears to be triggered by sensing of viral dsRNA over the cytosolic RNA sensors *IFIH1/MDA5* and *RIG-I/DDX58*, as the expression of MDA5 and RIG-I was significantly increased during acute infection. While RIG-I is sensing minus-strand RNA viruses by specific binding to 5′-triphosphate uncapped RNA genomes, MDA5 specifically senses plus-strand RNA viruses, such as FMDV, by binding to their dsRNA replication intermediates [37,38]. Although MDA5 is thought to be the main cellular detector of FMDV, it has been shown that RIG-I transcription is also elevated during FMDV infection [39]. After specific binding of MDA5 or RIG-I to their ligands they interact with TRIM25 and the mitochondrial protein MAVS that activate the transcription factors IRF3/IRF7 [40] as well as NF-κB [41], which is in accordance with the observed up-regulation of *IRF3, IRF7, NFKBIA, NFKBIB* and *TRIM25* (Table 1). Activated IRF3/IRF7 and NF-κB complexes thereafter induce the expression of type I/III interferons (IFNs) and proinflammatory cytokines, respectively [42]. We found that during acute infection especially IFN-β was expressed at elevated levels with significant overexpression of *IFNB1*, while IFN-α and IFN-λ were only expressed at very low levels (Supplementary Figure S3). However, we observed expression of several ISGs with antiviral activity, such as *IRF1, ISG15, MX1/2, OAS1/2,* and *RSAD2* (see Table 1). This indicates that while IFN-α transcription may be limited, IFN-β alone is able to induce a potent antiviral response in bovine SP cells. It is known that the viral leader proteinase L^{pro} of FMDV can inhibit IFN-β transcription and protein translation, thereby blocking the cells innate immune response [43]. However, at least in this model L^{pro} does not induce a full blocking of IFN-β transcription. A specific blocking of IFN-α transcription has only been shown for swine dendritic cell populations from blood and skin during acute FMDV infection [44]. The induction of IFN-β and associated ISGs was furthermore coincident with high viral genome copy numbers and presence of viral proteins (Table 3) as has been described before [45].

In contrast, no interferon expression was observed during the persistent phase of infection, although the aforementioned virus sensors MDA5/RIG-I and their associated transcription factor *IRF7* were still highly expressed (see *IFIH1* in Figure 3B). A possible factor reducing the transcription of interferons during persistence is the down-regulation of *ANKRD1*, as *ANKRD1* is directly involved in the signal transduction of *IRF3* and *IRF7* by binding IRF7/IRF3 complexes and thereby enhances the expression of type I/III interferons [46]. *ANKRD1* was initially observed in human dermal endothelial cells, where it is induced by inflammatory cytokines [47] and is substantially involved in the fibroblast mediated wound healing [48]. In agreement with our results, the expression of *ANKRD1* has previously been shown to be down-regulated in vivo during persistent FMDV infection of bovine nasopharyngeal tissue [35]. Gene silencing of *ANKRD1* in cells infected with herpes simplex virus resulted in increased viral load and reduced IFN-β (*IFNB1*) and IFN-λ (IL29) expression [46]. Therefore, the strong downregulation of *ANKRD1* as early as 48 hpi may be involved in the decrease of IFN-β expression during persistence.

Although no interferon expression was detectable at 28 dpi, the expression of many ISGs with known antiviral activity, such as *MX1/2* and *OAS1/2*, was significantly increased (see Figures 2D and 3B). Furthermore, based on the proteome data, the translation of these genes was stable, as their encoded proteins accumulated in the cells (Table 3). This indicates long-lasting induction of ISGs during persistent infection, with viral protein below the detection limit and low FMDV genome copy numbers. A comparable pattern of highly expressed ISGs as observed during FMDV persistence (*ISG15, MX1, OAS1/2,* and *USP18*) has been identified in liver samples from chronic infections with hepatitis C virus (HCV) [49,50]. Interestingly, the elevated expression of *ISG15/USP18* during chronic HCV infection, as we also observed during FMDV persistence, correlated with decreased responses to IFN-α treatment. Furthermore, the strong expression of *OAS1/2, MX1/2* and *IRF7*, but not of *IRF3* at 28 dpi is in accordance with increased interferon receptor (*IFNR*) signaling [51,52]. Observations from persistent lymphocytic choriomeningitis virus (LCMV) infections revealed that persistence is driven by chronic *IFNR* signaling, characterized by similar ISG activation patterns as observed in the present study [52]. The blockade of *IFNR* by antibodies abolished the expression of *IL-10* and *PD-L1*, two immunosuppressive T-cell exhaustion factors expressed by dendritic cells, and ultimately led to clearance of persistent LCMV infection by activation of IFN-γ expressing CD4+ T cells [52]. In this study, we did not observe upregulation of *IL-10* or *PD-L1*, as no dendritic cells were present in the SP cultures; however, an overexpression of T-cell exhaustion factors was previously found in nasopharyngeal tissues of FMDV carrier animals [35] and therefore may play a role in FMDV persistence in vivo.

The immunosuppressive factors identified in that study also included transforming growth factor β (TGF-β). While an overexpression of TGF-β itself was not evident in our data, lymphocyte antigen 6 family member E (*LY6E*) was one of the two genes that were up-regulated during the persistent phase only. *LY6E*, an ISG, has been implicated in the TGF-β-mediated escape from immune surveillance in many forms of cancer [53]. *LY6E* has been previously shown to promote viral infection [54] and it is essential for clathrin-mediated endocytosis of virus particles [55], a pathway that is also used by FMDV [56].

During acute infection, the interferon-mediated induction of ISGs appears to trigger apoptosis, as indicated by pathway analysis. Interestingly, genes associated with apoptosis, such as *CASP7*, were only expressed during acute infection and their expression waned with increasing time post infection (Figure 3B). The apparent absence of apoptotic processes during persistent infection is in accordance with recent findings that indicated an inhibition of apoptotic pathways in nasopharyngeal tissues from FMDV carrier animals [34,35]. Furthermore, we observed a specific down-regulation of pro-apoptotic genes during persistence, such as *ALDH1A2, ANKRD1,* and *SFRP2* (Table 2). Similarly, overexpression of fructose-bisphosphatase 1, the other gene that was only up-regulated during the persistent phase, inhibits many forms of apoptosis by increasing total cellular glutathione [57].

Another interesting observation was the downregulation of *NCAM1* (CD56), a member of the immunoglobulin superfamily, as early as 48 hpi and its further decrease until 28 dpi. Besides its role as a differentiation marker for natural killer cells, *NCAM1* is involved in cell binding and migration [58]. Downregulation of *NCAM1* is associated with decreased cell adhesion capacity, enhanced tumor cell invasiveness and is triggered in other viral infections [59]. The expression of *HTRA3*, a serine protease involved in remodeling of the extracellular matrix (ECM), as well as the expression of the ECM components *COL1A1* and *COL3A1* were significantly down-regulated during persistent infection. The role of these changes during FMDV persistence is currently unknown but lends itself to some speculation. The interaction of cells with the ECM plays a key role in epithelial maturation [60], which in turn is critical for the life cycle of some persistent viruses such as papillomaviruses [61]. Even though FMDV and papillomaviruses are biologically very different, certain features of FMDV persistence in vivo are conspicuously similar to what is observed in papillomavirus infection—particularly the difference between the distribution of viral genome, which is concentrated in the basal stratum germinativum [62], and viral antigen, which is concentrated in the superficial layers of the soft palate epithelium [63]. This segregation helps papillomaviruses evade the immune system as high levels of viral replication and protein synthesis occur only in terminally differentiated cells that are not subject to immune surveillance [64]. Whether a similar mechanism is involved in the maintenance of FMDV persistence remains to be investigated. However, it is difficult to faithfully recreate the complex epithelial structure and ECM interactions in vitro and investigations of the role of the ECM in persistent FMDV infection have to be performed with ex vivo tissue samples.

5. Conclusions

In conclusion, our study independently confirms earlier findings of a polygenic inhibition of apoptosis during persistent FMDV infection, which has been put forward as one of the principal mechanisms for the maintenance of persistence [34,35]. Another proposed mechanism, Th2 polarization and T-cell exhaustion, was not directly represented in our data, because the SP culture model does not include specialized immune cells. Nevertheless, we demonstrated the utility of state-of-the-art proteogenomics for the analysis of transcriptional signatures of acute and persistent FMDV infection in a near-natural in vitro system. We will proceed to apply this technology to tissue samples collected from carrier animals to obtain the first comprehensive picture of the transcriptomic and proteomic alterations associated with FMDV persistence in the natural host. Unraveling the cellular mechanisms of FMDV persistence may ultimately give rise to improved diagnostics and prevention of the FMDV carrier state.

Supplementary Materials:
Figure S1: Schematic visualization of the proteogenomic workflow, Figure S2: Comparison of different transformation methods for the raw read counts per gene, Figure S3: Comparison of interferon expression, Table S1: Summary of primary soft palate cell culture samples, Table S2: Summary of excluded and included RNA reference transcripts, Table S3: Replaced gene aliases of the *Bos taurus* reference genome, Table S4: Sequences of primers used for the RT-qPCR analysis, Table S5: Differentially expressed genes, File S1: Raw read counts for all significantly differently expressed genes.

Author Contributions: Writing—Original Draft, F.P. and M.E.; Methodology, S.H., F.P. and M.Z.; Investigation, F.P., S.H., M.Z., S.B., E.L., S.K., D.Z. and J.V.; Formal Analysis, Visualization, and Data Curation, F.P.; Supervision, S.Z., L.B., J.V., K.R., D.H., M.B. and M.E.; Conceptualization, S.Z., L.B., J.V., D.H., M.B. and M.E.; Project Administration, S.Z.

Acknowledgments: We thank Cells for Life Platform, partly funded by the Infrastructure Committee at SLU, Sweden, for providing facilities and equipment. We acknowledge Patrick Zitzow and Jenny Lorke of the Institute of Diagnostic Virology at the Friedrich-Loeffler-Institut for their excellent technical assistance. We are grateful to Dörte Becher and Claudia Hirschfeld from the Institute of Microbiology at the University of Greifswald for helping with the mass spectrometry analyses.

References

1. Knight-Jones, T.J.; Rushton, J. The economic impacts of foot and mouth disease—What are they, how big are they and where do they occur? *Prev. Vet. Med.* **2013**, *112*, 161–173. [CrossRef] [PubMed]
2. Zell, R.; Delwart, E.; Gorbalenya, A.E.; Hovi, T.; King, A.M.Q.; Knowles, N.J.; Lindberg, A.M.; Pallansch, M.A.; Palmenberg, A.C.; Reuter, G.; et al. ICTV Virus Taxonomy Profile: Picornaviridae. *J. Gen. Virol.* **2017**, *98*, 2421–2422. [CrossRef] [PubMed]
3. Forss, S.; Strebel, K.; Beck, E.; Schaller, H. Nucleotide sequence and genome organization of foot-and-mouth disease virus. *Nucleic Acids Res.* **1984**, *12*, 6587–6601. [CrossRef] [PubMed]
4. Alexandersen, S.; Mowat, N. Foot-and-mouth disease: Host range and pathogenesis. *Curr. Top. Microbiol. Immunol.* **2005**, *288*, 9–42. [CrossRef] [PubMed]
5. Thomson, G.R.; Vosloo, W.; Bastos, A.D. Foot and mouth disease in wildlife. *Virus Res.* **2003**, *91*, 145–161. [CrossRef]
6. Grubman, M.J.; Baxt, B. Foot-and-mouth disease. *Clin. Microbiol. Rev.* **2004**, *17*, 465–493. [CrossRef] [PubMed]
7. World Organisation for Animal Health (OIE). Foot and mouth disease (infection with foot and mouth disease virus). In *Manual of Diagnostic Tests and Vaccines for Terrestrial Animals (Terrestrial Manual)*; OIE: Paris, France, 2018; 2.1.8.

8. Arzt, J.; Belsham, G.J.; Lohse, L.; Botner, A.; Stenfeldt, C. Transmission of Foot-and-Mouth Disease from Persistently Infected Carrier Cattle to Naive Cattle via Transfer of Oropharyngeal Fluid. *mSphere* **2018**, *3*. [CrossRef]
9. Alexandersen, S.; Zhang, Z.; Donaldson, A.I. Aspects of the persistence of foot-and-mouth disease virus in animals—The carrier problem. *Microbes Infect.* **2002**, *4*, 1099–1110. [CrossRef]
10. Pacheco, J.M.; Smoliga, G.R.; O'Donnell, V.; Brito, B.P.; Stenfeldt, C.; Rodriguez, L.L.; Arzt, J. Persistent Foot-and-Mouth Disease Virus Infection in the Nasopharynx of Cattle; Tissue-Specific Distribution and Local Cytokine Expression. *PLoS ONE* **2015**, *10*, e0125698. [CrossRef]
11. Cox, S.J.; Voyce, C.; Parida, S.; Reid, S.M.; Hamblin, P.A.; Paton, D.J.; Barnett, P.V. Protection against direct-contact challenge following emergency FMD vaccination of cattle and the effect on virus excretion from the oropharynx. *Vaccine* **2005**, *23*, 1106–1113. [CrossRef]
12. Cox, S.J.; Voyce, C.; Parida, S.; Reid, S.M.; Hamblin, P.A.; Hutchings, G.; Paton, D.J.; Barnett, P.V. Effect of emergency FMD vaccine antigen payload on protection, sub-clinical infection and persistence following direct contact challenge of cattle. *Vaccine* **2006**, *24*, 3184–3190. [CrossRef] [PubMed]
13. Juleff, N.; Windsor, M.; Reid, E.; Seago, J.; Zhang, Z.; Monaghan, P.; Morrison, I.W.; Charleston, B. Foot-and-mouth disease virus persists in the light zone of germinal centres. *PLoS ONE* **2008**, *3*, e3434. [CrossRef] [PubMed]
14. Han, L.; Xin, X.; Wang, H.; Li, J.; Hao, Y.; Wang, M.; Zheng, C.; Shen, C. Cellular response to persistent foot-and-mouth disease virus infection is linked to specific types of alterations in the host cell transcriptome. *Sci. Rep.* **2018**, *8*, 5074. [CrossRef] [PubMed]
15. Zhu, J.; Weiss, M.; Grubman, M.J.; de los Santos, T. Differential gene expression in bovine cells infected with wild type and leaderless foot-and-mouth disease virus. *Virology* **2010**, *404*, 32–40. [CrossRef]
16. Kopliku, L.; Relmy, A.; Romey, A.; Gorna, K.; Zientara, S.; Bakkali-Kassimi, L.; Blaise-Boisseau, S. Establishment of persistent foot-and-mouth disease virus (FMDV) infection in MDBK cells. *Arch. Virol.* **2015**, *160*, 2503–2516. [CrossRef] [PubMed]
17. O'Donnell, V.; Pacheco, J.M.; Larocco, M.; Gladue, D.P.; Pauszek, S.J.; Smoliga, G.; Krug, P.W.; Baxt, B.; Borca, M.V.; Rodriguez, L. Virus-host interactions in persistently FMDV-infected cells derived from bovine pharynx. *Virology* **2014**, *468–470*, 185–196. [CrossRef] [PubMed]
18. Hägglund, S.; Laloy, E.; Näslund, K.; Pfaff, F.; Eschbaumer, M.; Romey, A.; Relmy, A.; Rikberg, A.; Svensson, A.; Huet, H.; et al. Model of persistent foot-and–mouth disease virus infection in multilayered cells derived from bovine dorsal soft palate. *Transbound. Emerg. Dis.* **2018**, under review.
19. Patro, R.; Duggal, G.; Love, M.I.; Irizarry, R.A.; Kingsford, C. Salmon provides fast and bias-aware quantification of transcript expression. *Nat. Methods* **2017**, *14*, 417–419. [CrossRef]
20. R Core Team. *R: A Language and Environment for Statistical Computing*; R Foundation for Statistical Computing: Vienna, Austria, 2017.
21. Soneson, C.; Love, M.I.; Robinson, M.D. Differential analyses for RNA-seq: Transcript-level estimates improve gene-level inferences. *F1000Research* **2015**, *4*, 1521. [CrossRef]
22. Love, M.I.; Huber, W.; Anders, S. Moderated estimation of fold change and dispersion for RNA-seq data with DESeq2. *Genome Biol.* **2014**, *15*, 550. [CrossRef]
23. Yu, G.; He, Q.Y. ReactomePA: An R/Bioconductor package for reactome pathway analysis and visualization. *Mol. Biosyst.* **2016**, *12*, 477–479. [CrossRef] [PubMed]
24. Yu, G.; Wang, L.G.; Han, Y.; He, Q.Y. clusterProfiler: An R package for comparing biological themes among gene clusters. *Omics* **2012**, *16*, 284–287. [CrossRef] [PubMed]
25. Untergasser, A.; Cutcutache, I.; Koressaar, T.; Ye, J.; Faircloth, B.C.; Remm, M.; Rozen, S.G. Primer3—New capabilities and interfaces. *Nucleic Acids Res.* **2012**, *40*, e115. [CrossRef] [PubMed]
26. Eymann, C.; Lassek, C.; Wegner, U.; Bernhardt, J.; Fritsch, O.A.; Fuchs, S.; Otto, A.; Albrecht, D.; Schiefelbein, U.; Cernava, T.; et al. Symbiotic Interplay of Fungi, Algae, and Bacteria within the Lung Lichen *Lobaria pulmonaria* L. Hoffm. as Assessed by State-of-the-Art Metaproteomics. *J. Proteome Res.* **2017**, *16*, 2160–2173. [CrossRef] [PubMed]
27. López-Mondéjar, R.; Zühlke, D.; Větrovský, T.; Becher, D.; Riedel, K.; Baldrian, P. Decoding the complete arsenal for cellulose and hemicellulose deconstruction in the highly efficient cellulose decomposer Paenibacillus O199. *Biotechnol. Biofuels* **2016**, *9*, 104. [CrossRef] [PubMed]

28. Huber, W.; Carey, V.J.; Gentleman, R.; Anders, S.; Carlson, M.; Carvalho, B.S.; Bravo, H.C.; Davis, S.; Gatto, L.; Girke, T.; et al. Orchestrating high-throughput genomic analysis with Bioconductor. *Nat. Methods* **2015**, *12*, 115–121. [CrossRef] [PubMed]
29. Pavelka, N.; Pelizzola, M.; Vizzardelli, C.; Capozzoli, M.; Splendiani, A.; Granucci, F.; Ricciardi-Castagnoli, P. A power law global error model for the identification of differentially expressed genes in microarray data. *BMC Bioinform.* **2004**, *5*, 203. [CrossRef]
30. Vizcaino, J.A.; Csordas, A.; del-Toro, N.; Dianes, J.A.; Griss, J.; Lavidas, I.; Mayer, G.; Perez-Riverol, Y.; Reisinger, F.; Ternent, T.; et al. 2016 update of the PRIDE database and its related tools. *Nucleic Acids Res.* **2016**, *44*, D447–D456. [CrossRef]
31. Fabregat, A.; Jupe, S.; Matthews, L.; Sidiropoulos, K.; Gillespie, M.; Garapati, P.; Haw, R.; Jassal, B.; Korninger, F.; May, B.; et al. The Reactome Pathway Knowledgebase. *Nucleic Acids Res.* **2018**, *46*, D649–D655. [CrossRef]
32. Ni, Z.; Yang, F.; Cao, W.; Zhang, X.; Jin, Y.; Mao, R.; Du, X.; Li, W.; Guo, J.; Liu, X.; et al. Differential gene expression in porcine SK6 cells infected with wild-type and SAP domain-mutant foot-and-mouth disease virus. *Virol. Sin.* **2016**, *31*, 249–257. [CrossRef]
33. Arzt, J.; Juleff, N.; Zhang, Z.; Rodriguez, L.L. The pathogenesis of foot-and-mouth disease I: Viral pathways in cattle. *Transbound. Emerg. Dis.* **2011**, *58*, 291–304. [CrossRef] [PubMed]
34. Stenfeldt, C.; Eschbaumer, M.; Smoliga, G.R.; Rodriguez, L.L.; Zhu, J.; Arzt, J. Clearance of a persistent picornavirus infection is associated with enhanced pro-apoptotic and cellular immune responses. *Sci. Rep.* **2017**, *7*, 17800. [CrossRef] [PubMed]
35. Eschbaumer, M.; Stenfeldt, C.; Smoliga, G.R.; Pacheco, J.M.; Rodriguez, L.L.; Li, R.W.; Zhu, J.; Arzt, J. Transcriptomic Analysis of Persistent Infection with Foot-and-Mouth Disease Virus in Cattle Suggests Impairment of Apoptosis and Cell-Mediated Immunity in the Nasopharynx. *PLoS ONE* **2016**, *11*, e0162750. [CrossRef] [PubMed]
36. Brown, C.J.; Hendrich, B.D.; Rupert, J.L.; Lafreniere, R.G.; Xing, Y.; Lawrence, J.; Willard, H.F. The human XIST gene: Analysis of a 17 kb inactive X-specific RNA that contains conserved repeats and is highly localized within the nucleus. *Cell* **1992**, *71*, 527–542. [CrossRef]
37. Feng, Q.; Hato, S.V.; Langereis, M.A.; Zoll, J.; Virgen-Slane, R.; Peisley, A.; Hur, S.; Semler, B.L.; van Rij, R.P.; van Kuppeveld, F.J. MDA5 detects the double-stranded RNA replicative form in picornavirus-infected cells. *Cell Rep.* **2012**, *2*, 1187–1196. [CrossRef]
38. Kato, H.; Takeuchi, O.; Sato, S.; Yoneyama, M.; Yamamoto, M.; Matsui, K.; Uematsu, S.; Jung, A.; Kawai, T.; Ishii, K.J.; et al. Differential roles of MDA5 and RIG-I helicases in the recognition of RNA viruses. *Nature* **2006**, *441*, 101–105. [CrossRef]
39. Zhu, Z.; Wang, G.; Yang, F.; Cao, W.; Mao, R.; Du, X.; Zhang, X.; Li, C.; Li, D.; Zhang, K.; et al. Foot-and-Mouth Disease Virus Viroporin 2B Antagonizes RIG-I-Mediated Antiviral Effects by Inhibition of Its Protein Expression. *J. Virol.* **2016**, *90*, 11106–11121. [CrossRef]
40. Lin, R.; Heylbroeck, C.; Pitha, P.M.; Hiscott, J. Virus-dependent phosphorylation of the IRF-3 transcription factor regulates nuclear translocation, transactivation potential, and proteasome-mediated degradation. *Mol. Cell. Biol.* **1998**, *18*, 2986–2996. [CrossRef]
41. Liu, S.; Chen, J.; Cai, X.; Wu, J.; Chen, X.; Wu, Y.T.; Sun, L.; Chen, Z.J. MAVS recruits multiple ubiquitin E3 ligases to activate antiviral signaling cascades. *Elife* **2013**, *2*, e00785. [CrossRef]
42. Belgnaoui, S.M.; Paz, S.; Hiscott, J. Orchestrating the interferon antiviral response through the mitochondrial antiviral signaling (MAVS) adapter. *Curr. Opin. Immunol.* **2011**, *23*, 564–572. [CrossRef]
43. De Los Santos, T.; de Avila Botton, S.; Weiblen, R.; Grubman, M.J. The leader proteinase of foot-and-mouth disease virus inhibits the induction of beta interferon mRNA and blocks the host innate immune response. *J. Virol.* **2006**, *80*, 1906–1914. [CrossRef] [PubMed]
44. Nfon, C.K.; Ferman, G.S.; Toka, F.N.; Gregg, D.A.; Golde, W.T. Interferon-alpha production by swine dendritic cells is inhibited during acute infection with foot-and-mouth disease virus. *Viral. Immunol.* **2008**, *21*, 68–77. [CrossRef] [PubMed]
45. Eschbaumer, M.; Stenfeldt, C.; Rekant, S.I.; Pacheco, J.M.; Hartwig, E.J.; Smoliga, G.R.; Kenney, M.A.; Golde, W.T.; Rodriguez, L.L.; Arzt, J. Systemic immune response and virus persistence after foot-and-mouth disease virus infection of naive cattle and cattle vaccinated with a homologous adenovirus-vectored vaccine. *BMC Vet. Res.* **2016**, *12*, 205. [CrossRef] [PubMed]

46. Bin, L.; Li, X.; Richers, B.; Streib, J.E.; Hu, J.W.; Taylor, P.; Leung, D.Y.M. Ankyrin repeat domain 1 regulates innate immune responses against herpes simplex virus 1: A potential role in eczema herpeticum. *J. Allergy Clin. Immunol.* **2018**. [CrossRef] [PubMed]
47. Chu, W.; Burns, D.K.; Swerlick, R.A.; Presky, D.H. Identification and characterization of a novel cytokine-inducible nuclear protein from human endothelial cells. *J. Biol. Chem.* **1995**, *270*, 10236–10245. [CrossRef]
48. Samaras, S.E.; Almodovar-Garcia, K.; Wu, N.; Yu, F.; Davidson, J.M. Global deletion of Ankrd1 results in a wound-healing phenotype associated with dermal fibroblast dysfunction. *Am. J. Pathol.* **2015**, *185*, 96–109. [CrossRef] [PubMed]
49. Chen, L.; Borozan, I.; Feld, J.; Sun, J.; Tannis, L.L.; Coltescu, C.; Heathcote, J.; Edwards, A.M.; McGilvray, I.D. Hepatic gene expression discriminates responders and nonresponders in treatment of chronic hepatitis C viral infection. *Gastroenterology* **2005**, *128*, 1437–1444. [CrossRef] [PubMed]
50. MacQuillan, G.C.; Mamotte, C.; Reed, W.D.; Jeffrey, G.P.; Allan, J.E. Upregulation of endogenous intrahepatic interferon stimulated genes during chronic hepatitis C virus infection. *J. Med. Virol.* **2003**, *70*, 219–227. [CrossRef]
51. Au, W.C.; Moore, P.A.; Lowther, W.; Juang, Y.T.; Pitha, P.M. Identification of a member of the interferon regulatory factor family that binds to the interferon-stimulated response element and activates expression of interferon-induced genes. *Proc. Natl. Acad. Sci. USA* **1995**, *92*, 11657–11661. [CrossRef]
52. Wilson, E.B.; Yamada, D.H.; Elsaesser, H.; Herskovitz, J.; Deng, J.; Cheng, G.; Aronow, B.J.; Karp, C.L.; Brooks, D.G. Blockade of chronic type I interferon signaling to control persistent LCMV infection. *Science* **2013**, *340*, 202–207. [CrossRef]
53. AlHossiny, M.; Luo, L.; Frazier, W.R.; Steiner, N.; Gusev, Y.; Kallakury, B.; Glasgow, E.; Creswell, K.; Madhavan, S.; Kumar, R.; et al. Ly6E/K Signaling to TGFbeta Promotes Breast Cancer Progression, Immune Escape, and Drug Resistance. *Cancer Res.* **2016**, *76*, 3376–3386. [CrossRef] [PubMed]
54. Schoggins, J.W.; Wilson, S.J.; Panis, M.; Murphy, M.Y.; Jones, C.T.; Bieniasz, P.; Rice, C.M. A diverse range of gene products are effectors of the type I interferon antiviral response. *Nature* **2011**, *472*, 481–485. [CrossRef] [PubMed]
55. Hackett, B.A.; Cherry, S. Flavivirus internalization is regulated by a size-dependent endocytic pathway. *Proc. Natl. Acad. Sci. USA* **2018**, *115*, 4246–4251. [CrossRef] [PubMed]
56. Han, S.C.; Guo, H.C.; Sun, S.Q.; Jin, Y.; Wei, Y.Q.; Feng, X.; Yao, X.P.; Cao, S.Z.; Liu, D.X.; Liu, X.T. Productive Entry of Foot-and-Mouth Disease Virus via Macropinocytosis Independent of Phosphatidylinositol 3-Kinase. *Sci. Rep.* **2016**, *6*, 19294. [CrossRef] [PubMed]
57. Voehringer, D.W.; Hirschberg, D.L.; Xiao, J.; Lu, Q.; Roederer, M.; Lock, C.B.; Herzenberg, L.A.; Steinman, L.; Herzenberg, L.A. Gene microarray identification of redox and mitochondrial elements that control resistance or sensitivity to apoptosis. *Proc. Natl. Acad. Sci. USA* **2000**, *97*, 2680–2685. [CrossRef] [PubMed]
58. Moos, M.; Tacke, R.; Scherer, H.; Teplow, D.; Früh, K.; Schachner, M. Neural adhesion molecule L1 as a member of the immunoglobulin superfamily with binding domains similar to fibronectin. *Nature* **1988**, *334*, 701–703. [CrossRef] [PubMed]
59. Blaheta, R.A.; Beecken, W.D.; Engl, T.; Jonas, D.; Oppermann, E.; Hundemer, M.; Doerr, H.W.; Scholz, M.; Cinatl, J. Human cytomegalovirus infection of tumor cells downregulates NCAM (CD56): A novel mechanism for virus-induced tumor invasiveness. *Neoplasia* **2004**, *6*, 323–331. [CrossRef]
60. Bonnans, C.; Chou, J.; Werb, Z. Remodelling the extracellular matrix in development and disease. *Nat. Rev. Mol. Cell Biol.* **2014**, *15*, 786–801. [CrossRef]
61. Feller, L.; Khammissa, R.A.; Wood, N.H.; Lemmer, J. Epithelial maturation and molecular biology of oral HPV. *Infect. Agent Cancer* **2009**, *4*, 16. [CrossRef]
62. Prato Murphy, M.L.; Forsyth, M.A.; Belsham, G.J.; Salt, J.S. Localization of foot-and-mouth disease virus RNA by in situ hybridization within bovine tissues. *Virus Res.* **1999**, *62*, 67–76. [CrossRef]
63. Stenfeldt, C.; Eschbaumer, M.; Rekant, S.I.; Pacheco, J.M.; Smoliga, G.R.; Hartwig, E.J.; Rodriguez, L.L.; Arzt, J. The Foot-and-Mouth Disease Carrier State Divergence in Cattle. *J. Virol.* **2016**, *90*, 6344–6364. [CrossRef] [PubMed]
64. McBride, A.A. Mechanisms and strategies of papillomavirus replication. *Biol. Chem.* **2017**, *398*, 919–927. [CrossRef] [PubMed]

10

Interpreting Viral Deep Sequencing Data with GLUE

Joshua B. Singer [1,*], Emma C. Thomson [1], Joseph Hughes [1], Elihu Aranday-Cortes [1], John McLauchlan [1], Ana da Silva Filipe [1], Lily Tong [1], Carmen F. Manso [2], Robert J. Gifford [1], David L. Robertson[1], Eleanor Barnes [3], M. Azim Ansari [3], Jean L. Mbisa [2], David F. Bibby [2]Daniel Bradshaw [2] and David Smith [3]

1 MRC-University of Glasgow Centre for Virus Research, Glasgow G61 1QH, UK; Emma.Thomson@glasgow.ac.uk (E.C.T.); joseph.hughes@glasgow.ac.uk (J.H.); Elihu.Aranday-Cortes@glasgow.ac.uk (E.A.-C.); John.McLauchlan@glasgow.ac.uk (J.M.); Ana.daSilvaFilipe@glasgow.ac.uk (A.d.S.F.); lily.tong@glasgow.ac.uk (L.T.); Robert.Gifford@glasgow.ac.uk (R.J.G.); David.L.Robertson@glasgow.ac.uk (D.L.R.)

2 Virus Reference Department, National Infection Service, Public Health England, Colindale, London NW9 5EQ, UK; Carmen.Manso@phe.gov.uk (C.F.M.); Tamyo.Mbisa@phe.gov.uk (J.L.M.); David.Bibby@phe.gov.uk (D.F.B.); Daniel.Bradshaw@phe.gov.uk (D.B.)

3 Peter Medawar Building for Pathogen Research, Nuffield Department of Medicine, University of Oxford, Oxford OX1 3SY, UK; ellie.barnes@ndm.ox.ac.uk (E.B.); ansari@well.ox.ac.uk (M.A.A.); david.smith@stcatz.ox.ac.uk (D.S.)

* Correspondence: josh.singer@glasgow.ac.uk

Abstract: Using deep sequencing technologies such as Illumina's platform, it is possible to obtain reads from the viral RNA population revealing the viral genome diversity within a single host. A range of software tools and pipelines can transform raw deep sequencing reads into Sequence Alignment Mapping (SAM) files. We propose that interpretation tools should process these SAM files, directly translating individual reads to amino acids in order to extract statistics of interest such as the proportion of different amino acid residues at specific sites. This preserves per-read linkage between nucleotide variants at different positions within a codon location. The samReporter is a subsystem of the GLUE software toolkit which follows this direct read translation approach in its processing of SAM files. We test samReporter on a deep sequencing dataset obtained from a cohort of 241 UK HCV patients for whom prior treatment with direct-acting antivirals has failed; deep sequencing and resistance testing have been suggested to be of clinical use in this context. We compared the polymorphism interpretation results of the samReporter against an approach that does not preserve per-read linkage. We found that the samReporter was able to properly interpret the sequence data at resistance-associated locations in nine patients where the alternative approach was equivocal. In three cases, the samReporter confirmed that resistance or an atypical substitution was present at NS5A position 30. In three further cases, it confirmed that the sofosbuvir-resistant NS5B substitution S282T was absent. This suggests the direct read translation approach implemented is of value for interpreting viral deep sequencing data.

Keywords: deep sequencing; virus genomics; hepatitis C virus; variant calling; sequence interpretation; drug resistance; bioinformatics

1. Introduction

For some virus species, their highly error-prone replication mechanism produces a population of related genomic variants of the virus within a single infected host individual [1]. Sequencing systems such as Illumina's platform produce short, relatively accurate nucleotide sections of viral genome, often generating thousands of reads for a given genomic location from a single sample [2]. Such deep sequencing technologies therefore offer methods for understanding the nature of viral intra-host diversity. Whole genome and deep sequencing of virus genomes has been widely applied in basic virology research but has also found applications in clinical contexts such as the detection of drug resistance [3].

Various bioinformatics stages must be applied in the interpretation of viral deep sequencing data. Reads unrelated to the virus genome are removed and low-quality reads removed or trimmed. Following this, we must then construct an alignment: how the reads are arranged relative to each other within the virus genome, accounting for sequence homology. Reference-based alignment or mapping methods such as Bowtie 2 [4], BWA [5], MOSAIK [6], Stampy [7] or Tanoti [8] use one or more reference sequences to guide the alignment of reads. In contrast, de novo assembly approaches such as SPAdes [9] and VICUNA [10] use associations derived purely from the read data itself to propose large genome fragments, avoiding the biases arising from the choice of reference sequence. A drawback of de novo methods is that they may not accurately capture the full genomic structure or diversity, thus, for well-known viruses with high levels of genomic diversity, combinations of de novo assembly and reference-based alignment methods, such as shiver [11], are often used. One aspect common to almost all recent methods in this area is that they output their results in the form of a sequence alignment mapping (SAM) file. The SAM format [12] integrates nucleotide, read quality and alignment data in a single file. It was standardised at an early point in the adoption of deep sequencing, allowing diverse methods to be compared with each other and integrated into processing pipelines for a broad set of applications.

A range of variant-calling methods have been developed to analyse genomic heterogeneity within deep sequencing data. Error rates in short read technologies such as Illumina are low but it can be challenging to distinguish errors from real single nucleotide variants (SNVs) occurring at frequencies comparable to the error rate. Therefore, variant-calling methods such as LoFreq [13] and V-Phaser [14,15] apply statistical techniques to the aligned read data to identify probable SNVs occurring even at very low frequency in the presence of sequencing errors. The focus on low frequency SNVs is critical in applications such as cancer genomics where somatic deviations from the consensus are both rare and of high consequence [16]. However, virus bioinformatics has distinct priorities from fields focused on eukaryote or bacterial organisms with higher replication fidelity [17]. Since viruses typically have a low replication fidelity, there is a higher level of diversity within an infected host and the viral population can be expected to contain many variants.

For reasons of clarity within the research community, virus genome locations can be defined in terms of a standard virus strain with a well-established "master reference" genome. In hepatitis C virus (HCV) for example, codons within viral proteins are numbered relative to the H77 strain [18]. Polymorphisms at these standardised locations are reported with phenotypic associations established experimentally or in clinical trials. The advent of deep sequencing data prompts questions such as: What are the relative proportions of different amino acid residues at a given genome location? What proportion of reads support the presence of a certain sequence motif? What proportion of reads indicate a deletion? However, it is challenging to answer these questions within deep sequencing data, since read alignments do not in general use the reference coordinate space, and a mapping between the two spaces must be established and applied.

The genomes of the virus population may contain multiple nucleotide base variants at different positions within a single codon location. Both Verbist et al. [19] and Döring et al. [20] pointed out that linkage between nucleotide positions is lost when variants are called as SNVs. This linkage must be retained within datasets in order to accurately predict the amino acid residues arising from protein translation. Suppose for example we observe significant levels of both adenine (A) and thymine (T) at the first position of a particular codon location. At the second position, we observe cytosine (C) and guanine (G). With cytosine at the third position, what amino acids are the genomes in the virus population coding for? Without retaining linkage, these observations are consistent with a mix of Threonine (codon ACC) and Cysteine (TGC), or alternatively with purely Serine (AGC/TCC), or with any combination of these amino acids. By retaining linkage, we can accurately select between these interpretations. Haplotype reconstruction methods aim to determine linkage by associating sub-populations of reads as haplotypes. Schirmer et al. [21] found that this was exceptionally unreliable for viral deep sequencing data. However, full haplotype reconstruction may not be necessary for practical applications where the variants of interest are linked within the span of a single read or read pair.

We present a subsystem of the GLUE software package [22] called samReporter, focused on the analysis of aligned deep sequencing viral genome data. It directly processes the SAM file format produced by most methods, and can also process the more compact Binary Alignment Map (BAM) format. The samReporter can be instantiated within an existing GLUE project containing reference sequences and alignments for a given virus. This allows the software to establish a reading frame for coding region reads within a SAM file, and map between the read alignment coordinate space and standardised genome locations. In turn, this facilitates the scanning of reads directly for different classes of sequence pattern such as codons, amino acid residues, indels and motifs. This approach of scanning reads directly has the advantage of retaining linkage and we can report how often combinations of variations appear together on the same viral RNA, certainly within a codon location but also further, at least as far as paired-end read data allows.

We demonstrate the benefits of applying the GLUE samReporter to hepatitis C virus (HCV) deep sequencing data. HCV is a positive-sense single-stranded RNA virus of the family Flaviviridae. Its genome of about 9000 bases codes for a single polyprotein that produces 10 mature viral proteins. HCV affects over 100 million people worldwide and can cause liver disease and cirrhosis. The infection can be treated with a range of direct-acting antiviral (DAA) drugs which inhibit three of the mature proteins: NS3, NS5A and NS5B. Such therapies produce a sustained virological response (SVR) in the vast majority of patients, clearing the virus in around 95% of cases [23]. Notwithstanding this therapeutic success, HCV is proving difficult to treat in certain categories of patients, including "retreatment" patients: those for whom prior DAA treatment has failed. It has been shown both in vitro and in vivo that certain resistance-associated substitutions (RASs) in the viral genome confer resistance to DAA drugs [24].

Vermehren et al. suggested that retreatment patients have RASs in multiple drug target genes and that therefore "genomic resistance testing may be useful to select the optimal combination and treatment duration" for subsequent rounds of drug therapy [23]. Recent guidance suggests that, if deep sequencing is used, observing a RAS in 15% of the virus population may be clinically relevant [25]. Tools aimed at HCV resistance testing such as geno2pheno[ngs-freq] [20] suggest frequencies of 2%, 10% and 15%. Thus, while HCV RAS testing can benefit from deep sequencing methods, moderately-low rather than ultra-low frequencies are of most interest. RAS testing for retreatment patients therefore provides a good case study for a deep sequencing data interpretation system.

2. Results

We analysed viral genome diversity within a group of 241 HCV retreatment patients sampled within the United Kingdom. A range of genotypes were represented: Gt1 $n = 115$, Gt2 $n = 5$, Gt3 $n = 104$, Gt4 $n = 14$ and Gt6 $n = 3$. The five most frequent subtypes: were 1a $n = 98$, 1b $n = 13$, 3a $n = 95$, 3b $n = 6$ and 4r $n = 6$. Fourteen other subtypes were represented each by three or fewer patients. In three cases, a subtype could not be assigned.

The samReporter scans aligned reads directly, retaining linkage within reads with the intention of more accurate detection of specific variants. To test the benefits of this approach, we also contrived an alternative method that attempts to capture within-host diversity without retaining linkage.

Besides the four concrete bases A, C, G and T/U, IUPAC notation for nucleic acids, used in the FASTA sequence file format, contains 11 ambiguity codes, covering all possible combinations of more than one base [26]. For example, code S can represent a combination of C and G. Any method which calls SNVs can encode these variants within a FASTA file. This aspect of the encoding is used to capture minor nucleotide variants. Web-based HCV drug resistance interpretation systems such as HCV-GLUE [27] and geno2pheno[hcv] [28] do attempt to interpret ambiguity codes if they appear in the input data. We produced FASTA files with ambiguity codes for each sample using the samReporter `nucleotide-consensus` command, which produces one IUPAC code for each nucleotide position in the SAM reference coordinate space. Read bases with a Phred quality score of less than 25 were excluded. A "concrete" base (A, C, G or T) was encoded at a given position if it appeared both in at least five individual reads and in 5% of the quality-filtered reads at that location. IUPAC ambiguity codes are then used if multiple concrete bases are to be encoded. We found that FASTA files for all but two samples contained at least one ambiguity code. On average, files contained ambiguity codes which represent two bases at 1.02% of nucleotide positions (std. dev. 1.3%) and codes which represent three bases at 0.0161% of nucleotide positions (std. dev. 0.0475%).

A triplet of concrete bases, i.e. a codon, specifies a single amino acid. If ambiguity codes occur within nucleotide data for a given codon location, multiple distinct codons are present in the underlying data at that location; the precise composition is unknown. For a given ambiguous triplet (possibly containing ambiguity codes), there is set of "possible" amino acids comprising any residues coded by one or more of the possible codons. For example, for the ambiguous triplet YTM, the set of possible amino acids is Leucine (L) and Phenylalanine (F) because the set of codons and their corresponding amino acids are CTA (L), CTC (L), TTA (L) and TTC (F). Additionally, there is a (possibly empty) subset of "definite" amino acid residues, i.e. those that must be coded by at least some of the underlying codons, whatever the composition, under reasonable assumptions. For the ambiguous triplet YTM, every combination of codons which produces the ambiguity codes contains at least some codons for Leucine (L); this is the single definite amino acid. In general, if there is a single ambiguity code encoding two bases within the triplet, there will be one or two definite amino acids, and these will also be the only possible amino acids.

The FASTA files were analysed for "ambiguous" codon locations where the definite and possible amino acid sets were different. This typically occurs when there are two ambiguity codes within a triplet. Such locations present a challenge for drug resistance interpretation systems based on FASTA file inputs. Whereas amino acid residues in the definite set can be inferred to be present in the virus population, the status of amino acids in the possible set but not the definite set cannot be established clearly from FASTA data. We excluded from the analysis degenerate locations i.e. those where the possible set, excluding stops, contained more than five amino acids or where the read depth for the whole codon location was less than 10.

In total, 435 ambiguous locations were found in patients, within all ten viral proteins: Core $n = 62$, E1 $n = 24$, E2 $n = 205$, p7 $n = 5$, NS2 $n = 14$, NS3 $n = 30$, NS4A $n = 1$, NS4B $n = 9$, NS5A $n = 38$ and NS5B $n = 47$. The full set of ambiguous locations is given in the Supplementary Materials. Scaling by the length of each region, this implies that the E2 and Core proteins had a higher rate of ambiguous locations. The drug target proteins NS3, NS5A and NS5B have rates in the lower part of the range.

The current version of HCV-GLUE [27] documents 44 locations associated in the literature with resistance to six DAA drugs in current use: 18 in NS3, 15 in NS5A and 11 in NS5B. These are listed in the Supplementary Materials. Within the FASTA data, we found 10 ambiguous resistance-associated locations in nine patients, six in NS5A and four in NS5B. We resolved these locations using the GLUE samReporter, calculating the frequencies of codons and amino acids by directly analysing reads. Codons were excluded if any Phred base quality was below 25. Amino acid residues were deemed to be present if 5% or more of filtered read codons at the location coded for the residue. The 10 locations are shown in Table 1. HCV-GLUE classifies an amino acid as typical at a location for a given subtype if 10% or more of the GenBank sequences of that subtype contain the residue, these are also shown in the table. In all cases except one (R25, NS5B position 159) the definite residues set (not shown) was empty.

Table 1. Ambiguous resistance-associated locations resolved using GLUE samReporter.

Sequencing Facility	Sample ID	Subtype	Virus Protein	Codon Location	Ambiguous Triplet	Typical Residue (s)	Possible Residues Set	Confirmed Residues Set
Glasgow	HCV294	3b	NS5B	282	WSY	S	CST	S
Glasgow	HCV300	3a	NS5A	30	RMG	A	AEKT	AK
PHE	R127	1a	NS5A	24	RSG	K	AGRT	GT
PHE	R164	3a	NS5A	30	RMG	A	AEKT	AK
PHE	R25	4r	NS5B	159	YTM	L	FL	L
PHE	R25	4r	NS5B	282	WSC	S	CST	S
PHE	R36	4r	NS5B	282	WSC	S	CST	S
PHE	R67	1a	NS5A	30	YAW	Q	HQY	QY
PHE	R91	1a	NS5A	28	RYG	M	AMTV	MV
Oxford	7444	3a	NS5A	62	SYA	ST	ALPV	AL

In the cases shown in Table 1, samReporter was able to eliminate many possible residues. Whereas the possible set contained up to four residues, samReporter confirmed that only one or two were actually present at the 5% level. Two resistance locations occur three times each and merit a discussion. For NS5A position 30, in subtype 1a the typical residue is Glutamine (Q). In sample R67, samReporter found Tyrosine (Y) at around 74%, which has not been documented as a RAS but is atypical for the subtype. In samples HCV300 and R164 (subtype 3a) the typical residue is Alanine (A); samReporter found Lysine (K), a well-documented RAS, at levels of 92% and 82% respectively. Thus, in these three cases samReporter confirmed the presence of a RAS or atypical substitution. Substitutions at NS5B position 282 have been strongly associated with resistance to sofosbuvir, particularly the substitution of the typical Serine (S) with Threonine (T). In contrast with NS5A position 30, samReporter was able to eliminate this resistant residue and the other possible atypical residue Cysteine (C); in these three ambiguous cases, only Serine is present, but is actually coded by significantly distinct codons in each case. For sample HCV294, the codons were TCT at 54%, AGT at 25% and AGC at 20%, for sample R25, AGC at 85% and TCC at 15%, and, for sample R36, TCC at 92% and AGC at 7%. The effect of the presence of these diverse codons is to create ambiguity at the nucleotide level. One possible explanation for the codon diversity is that Threonine codons became frequent in the viral population during sofosbuvir treatment, and that following the end of the treatment course the descendants of these virions reverted to Serine, but now coded using diverse alternative codons.

The HCV296 sample is typical in the sense that the size of the BAM file (24.6 MB) was closest to the mean for this dataset, it contains ≈282,000 paired-end reads with an average depth across the HCV polyprotein of ≈3700. To evaluate performance, we ran some samReporter commands on this file using a 2014 MacBook Pro with a 2.5 GHz Intel Core i7 processor and 16 GB of RAM. The samReporter was configured to use up to four CPU cores. The `amino-acid` command was run to translate reads for the whole polyprotein, producing amino acid residue frequencies at each codon location, without any read filters. Using the auto-align feature with a known target reference sequence, this command took 7.5 s. Using the max-likelihood-placer feature, the command took 29.2 s, with most of the extra time spent in the RAxML-EPA step. See Section 3.2 for details of the samReporter design.

3. Materials and Methods

3.1. Sequencing Data

Deep whole genome HCV sequencing data was derived from blood samples collected from 241 patients resident in the United Kingdom, who had not achieved virological clearance after previous courses of antiviral therapy. Sequencing was performed using target enrichment on Illumina sequencers at three different institutes: the MRC-University of Glasgow Centre for Virus Research (n = 56), the University of Oxford Nuffield Department of Medicine (n = 25) and the Virus Reference Department at Public Health England (PHE) (n = 160).

The Glasgow library preparation protocol was as follows. RNA was isolated from 200 μL of plasma using the RNAdvance Blood extraction kit (Beckman Coulter, Brea, CA, United States) and collected in 27 μL of water. Following conversion of RNA to double-stranded DNA, libraries were prepared for Illumina sequencing using the KAPA DNA LTP Library Preparation Kit (Roche, Basel, Switzerland), and NEBNext Multiplex Oligos for Illumina (New England Biolabs, Ipswich, MA, United States). Libraries were quantified using Qubit dsDNA HS Assay Kit (Invitrogen, Carlsbad, CA, United States) and size distribution assessed using Agilent TapeStation with D1K High Sensitivity Kit (Agilent, Santa Clara, CA, United States); libraries were normalised according to viral load and mass. A 500 ng aliquot of the pooled library was enriched using SeqCap EZ Developer Probes (Roche), following the manufacturer's protocol. Following a 14 cycle post-enrichment PCR, the cleaned pool was sequenced with 151-base paired-end reads on a NextSeq cartridge (Illumina, San Diego, CA, United States).

The Oxford libraries were prepared for Illumina sequencing using the NEBNext Ultra Directional RNA Library Prep Kit (New England Biolabs) with 8 μL of RNA extracted from plasma using NUCLISENS easyMAG (bioMérieux, Marcy-l'Étoile, France) and previously published modifications of the manufacturer's guidelines (v2.0) [29]: omission of heat fragmentation, omission of Actinomycin D at first-strand reverse transcription, library amplification for 15 PCR cycles using custom indexed primers [30], and post-PCR clean-up with 0.85x volume Ampure XP (Beckman Coulter). Libraries were quantified using Quant-iT PicoGreen dsDNA Assay Kit (Invitrogen) and size distribution analysed using Agilent TapeStation D1K High Sensitivity kit. A 500 ng aliquot of the pooled libraries (96 plex) was enriched using the xGen Lockdown protocol (Rapid Protocol for DNA Probe Hybridization and Target Capture Using an Illumina TruSeq or Ion Torrent Library v1.0, Integrated DNA Technologies, Coralville, IA, United States) with equimolar-pooled 120 nt DNA oligonucleotide probes (Integrated DNA Technologies) followed by a 12-cycle on-bead post-enrichment PCR. The cleaned post-enrichment ve-Seq library was quantified by qPCR with the KAPA SYBR FAST qPCR Kit (Roche) and sequenced with 150b paired-end reads on a single run of the Illumina MiSeq.

The PHE library preparation protocol is the laboratory component of a pipeline aimed at clinical use; a manuscript describing the full pipeline is in preparation. RNA was extracted from 350 μL of plasma using the NUCLISENS easyMAG system (bioMérieux). Total eluates were subjected to Turbo DNAse treatment (Thermo Fisher, Waltham, MA, United States) followed by library preparation using KAPA RNA HyperPrep kit (Roche). Libraries were pooled based upon DNA concentration and HCV quantity, assessed using the Quant-iT kit on the Glomax platform (Promega, Madison, WI, United States) and the Qiagen QuantiTect kit with primers and probes from Davalieva et al. [31]. Pools were enriched by hybridisation to a biotinylated probe set (Integrated DNA Technologies, described by Bonsall et al. [32]) followed by further PCR cycles depending upon HCV quantity. The two pools were pooled, again by concentration and HCV quantity. The final pool was quantified using the KAPA SYBR FASTA qPCR kit (Roche) on a PRISM 7500 (Applied BioSystems, Foster City, CA, United States) before being sequenced on a MiSeq using Reagent kit v2 (Illumina).

The Illumina read data were processed into SAM files using different bioinformatics pipelines at the different institutions. At Glasgow, reads were trimmed and filtered using TrimGalore [33] with quality threshold 30 and minimum read length 75. The most appropriate HCV reference sequence was identified via a *k*-mer-based approach, using *k*-mers unique to each genotype [34]. SAM files were generated by mapping against the best-matching HCV reference using Tanoti [8]. At Oxford, de-multiplexed sequence read-pairs were trimmed of low-quality bases using QUASR v7.01 [35] and adapter sequences with CutAdapt version 1.7.1 [36] and subsequently discarded if either read had fewer than 50 bases in its remaining sequence or if both reads matched the human reference sequence using Bowtie version 2.2.4 [4]. Remaining reads were mapped using BWA mem [5] and Stampy [7] against a database of reference sequences, both to choose an appropriate reference and to select those reads which formed a majority population for de novo assembly using VICUNA [10] and finishing with V-FAT [37]. The reads were then mapped back to this assembly using MOSAIK [6]. At PHE, human reads were filtered out from trimmed FASTQ files using SMALT [38], remaining reads were then assembled using VICUNA de novo assembly [10]. Contigs were matched to HCV reference genomes using BLAST [39] and gaps filled using LASTZ [40] to generate a draft assembly. Reads were then mapped to the draft assembly with BWA [5].

The deep sequencing data used in this study has been deposited in the NCBI Sequence Read Archive (http://www.ncbi.nlm.nih.gov/sra), under BioProject accession number PRJNA527067 and experiment accession numbers SRX5528430 to SRX5528670.

3.2. GLUE samReporter Design

The GLUE samReporter aims to provide a convenient tool for interpreting viral deep sequencing data. As part of the wider GLUE system [22], it can be used interactively in the command line interpreter or within bioinformatics scripts. Instantiated within a GLUE project for a specific category of viruses such as HCV-GLUE, it can take advantage of certain data objects within that project.

When interpreting viral deep sequencing data, one obstacle is mapping the SAM file coordinate space to a standard codon numbering system. Within the HCV-GLUE project the H77 strain (RefSeq accession NC_004102) is defined as the "master" reference sequence object. The precursor polyprotein and the 10 mature proteins are defined as coding features and their locations are specified on the H77 sequence. A wider set of reference sequence objects is also defined within HCV-GLUE; there are currently over 200 of these, based on the ICTV HCV resource [41]. HCV-GLUE then specifies an unconstrained "master" alignment object containing all these reference sequences, which is used to map their locations to those on the H77 sequence. HCV-GLUE also contains a reference phylogeny of the same set of sequences, computed using RAxML [42].

SAM files for HCV typically map each read to a single coordinate space. To interpret individual reads the samReporter must infer sequence homology (i.e., pairwise alignment) between this SAM file coordinate space and one of the reference sequences defined within GLUE—the "target" reference. The simplest method, to specify that the SAM file coordinate space is identical to that of a specific target reference sequence, is appropriate if one of the project's reference sequences was used for the SAM file coordinate space. A more flexible "auto-align" method allows GLUE to generate a codon-aware pairwise alignment between the consensus of the SAM file and a selected target reference, using techniques based on BLAST+ [39]. This is appropriate if the SAM virus strain is closely related to the target reference, but importantly, it allows the method producing the SAM file, which may have a de novo element, to construct a coordinate space appropriate for the viral reads. The final and most general and robust method is "max-likelihood-placer". This allows GLUE to select the target reference itself, by feeding the consensus of the SAM file into the first two stages of the GLUE genotyping pipeline. This consists of incorporating it into the master alignment using MAFFT [43], placing it in the reference phylogeny using RAxML-EPA [42] and selecting as the target the reference sequence with the lowest patristic distance from the SAM consensus. The auto-align method is then used to generate the homology. The master alignment will also typically act as the "linking" alignment, providing a mapping between the target reference and the master reference. The result of this process is then a chain of pairwise homology relationships, as shown in Figure 1, from each individual read to the SAM file coordinate space, to the target reference sequence and finally, via the linking alignment, to the master reference sequence.

The samReporter offers a range of GLUE commands for interpreting SAM files (Table 2). These each accept similar arguments for specifying the coordinate homology and genome region. The "variation scan" command scans each read for the presence or absence of sequence patterns defined by GLUE *Variation* objects [22]. If paired-end read data are supplied, the reads in each pair are processed together. *Variation* objects can encapsulate insertions, deletions, regular expressions and combinations at the nucleotide or amino acid level. In HCV drug resistance this capability may become important. For example, the Magellan-1 trial of the drug pibrentasvir found that the combination of a Methionine at NS5A position 28 with the deletion of the residue at NS5A position 32 was associated with resistance to the drug for HCV subtype 1b [44]. It remains to be seen whether such deletions and combinations occurring as minority variants are clinically relevant but if so, the samReporter offers a means of detecting these.

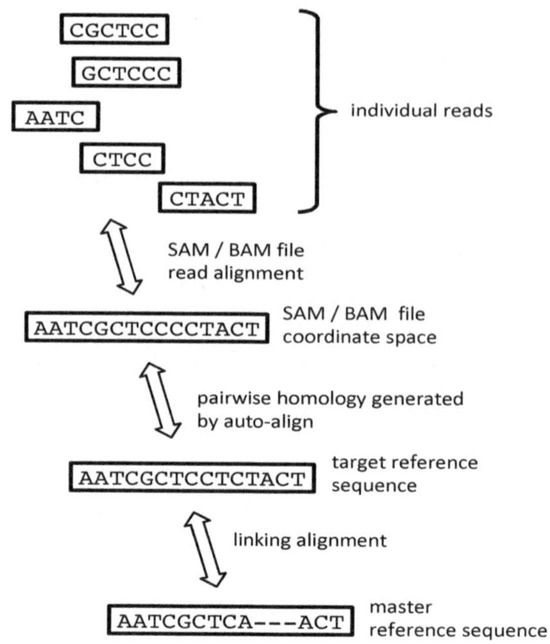

Figure 1. The chain of pairwise homology relationships between reads and the master reference sequence (H77 for HCV), established during the operation of GLUE samReporter.

Table 2. GLUE samReporter commands.

Command	Description
`nucleotide`	Generate a table of nucleotide frequencies within a specific genome region.
`depth`	Generate a table of read depths within a specific genome region.
`nucleotide-consensus`	Generate a FASTA consensus file, optionally using ambiguity codes.
`amino-acid`	Generate a table of amino acid residue frequencies within a specific protein-coding region.
`codon-triplets`	Generate a table of codon frequencies within a specific protein-coding region.
`variation scan`	Scan for the presence or absence of GLUE *Variations* within reads.
`export nucleotide-alignment`	Export a specific part of the SAM alignment as a FASTA file.

The commands also allow simple, optional filtering based on Phred base qualities, MAPQ mapping quality and depth. In command outputs, codon numbering is based on the system proposed by Kuiken et al. [18]; nucleotide coordinates both within the SAM file and the mapped location on the master reference are also given. Individual input files may be processed more quickly using parallelisation of command operations across multiple processors. Finally, for paired-end read data, regions where paired reads overlap are not counted twice in command outputs. As part of the GLUE engine, the SAM reporter is implemented in Java, using the Htsjdk library [45] to interpret the SAM format.

The samReporter is delivered as part of the GLUE software package. This study used GLUE version 1.1.33, HCV-GLUE project version 0.1.51 with PHE-HCV-DRUG-RESISTANCE extension version 0.1.21. GLUE is licensed under the open source GNU Affero General Public License version 3.0. and may be installed on Mac OSX, Windows or Linux systems. Documentation specific to the samReporter may be found at: http://tools.glue.cvr.ac.uk/#/deepSequencingData. Documentation for other aspects of GLUE and links to the source code repository can be found on other pages within the same web site.

4. Discussion

The results show that, within virus genomes of HCV retreatment patients, linkage between nucleotide variants within a codon location is not a purely theoretical issue. In a small number of cases among UK retreatment patients, such linkage did occur at sites critical for drug resistance. Approaches that do not preserve linkage, such as those encoding variants as ambiguity codes, cannot correctly resolve these cases. How would such a system deal with amino acid residues in the possible set but not in the definite set? If the system is configured not to report such residues, the result is false negative detection of a substitution at NS5A position 30 in three patients, obscuring A30K RAS in two cases. Conversely, if the system reports these residues, the result is false positive detection of the NS5B RAS S282T in three other patients. As far as we are aware, the current study is novel in terms of quantifying the effect of such linkage on resistance detection in real HCV patient data.

The current HCV-GLUE database [27] documents many RASs that combine substitutions at locations within the span of a typical Illumina read. The samReporter can report the presence or absence of these on any read (or read pair for paired-end data) that covers the relevant locations. Future work might consider whether detection of these "combination" RASs at a minority level are of clinical relevance. It would also be of interest to incorporate existing low frequency variant-calling mechanisms from the literature into GLUE. The samReporter and GLUE generally are intended to be useful in both research and clinical contexts. However, even once a SAM file has been generated from a sample, the samReporter only represents one part of the process in terms of drug resistance analysis. The HCV-GLUE system is currently being developed to provide a comprehensive drug resistance report, using samReporter to interpret a SAM file.

Other software, for example the VirVarSeq system [19], calls variants at the codon level but is focused on very low frequency variants. DiversiTools [46] provides frequencies of amino acids on a per read basis but does not link to a standardised coordinate system as is available in the GLUE framework. The geno2pheno[ngs-freq] system [20] directly interprets drug resistance in deep sequencing data. Users must transform their data into a table of nucleotide or codon frequencies and a web-based system then performs interpretation on this table, using a user-defined frequency threshold. In comparison with samReporter, this design facilitates fast transfer over a network since the frequency table is much more compact than a typical SAM or BAM file. However, some information is necessarily lost in the processing, for example the codon frequency table cannot encode linkage beyond a codon location which would be required for example to detect combination RASs.

While the current study applied samReporter to HCV, it can also be used to analyse deep sequencing data for other viruses. In many simple cases, the prerequisites would simply be a nucleotide alignment of alternative target reference sequences and a master reference sequence with coding region annotations. In more complex cases, for example where virus genomes contain ambisense genomes or RNA editing, GLUE and samReporter would need to take account of this. The GLUE samReporter shows that a simple, pragmatic software design can conveniently answer some common questions concerning within-host variation in viral deep sequencing data.

Author Contributions: Conceptualisation, J.B.S., E.C.T., J.L.M., J.H., J.M., E.B., D.L.R. and R.J.G.; methodology, J.B.S., J.H., D.S., D.F.B. and M.A.A.; software, formal analysis, and validation, JBS; data curation, E.A.-C., E.C.T., D.B., D.S., D.F.B. and M.A.A.; resources, A.d.S.F., L.T., C.F.M., D.F.B., E.A.C. and M.A.A.; writing—original draft preparation, J.B.S.; writing—review and editing, J.H., E.C.T., J.L.M., R.J.G., E.B., D.L.R. and J.M.; and supervision, R.J.G., D.L.R., E.C.T., J.M., E.B. and J.L.M.

Acknowledgments: The authors would like to thank the participants and clinicians at NHS sites who contributed samples, and Will Irving and Emma Hudson for their contributions, particularly for sample collection and project management.

Abbreviations

The following abbreviations are used in this manuscript:

BAM	Binary alignment mapping
DAA	Direct-acting Antiviral
DNA	Deoxyribonucleic Acid
ICTV	International Committee for the Taxonomy of Viruses
HCV	Hepatitis C virus
RAS	Resistance-associated Substitution
RNA	Ribonucleic Acid
SAM	Sequence alignment mapping
SNV	Single nucleotide variant
SVR	Sustained virological response

References

1. Holmes, E.C. *The Evolution and Emergence of RNA Viruses*; Oxford University Press, Oxford, UK, 2009.
2. Goodwin, S.; McPherson, J.D.; McCombie, W.R. Coming of age: Ten years of next-generation sequencing technologies. *Nat. Rev. Genet.* **2016**, *17*, 333–351. [CrossRef] [PubMed]
3. Houldcroft, C.J.; Beale, M.A.; Breuer, J. Clinical and biological insights from viral genome sequencing. *Nat. Rev. Microbiol.* **2017**, *15*, 183–192. [CrossRef] [PubMed]
4. Langmead, B.; Salzberg, S.L. Fast gapped-read alignment with Bowtie 2. *Nat. Methods* **2012**, *9*, 357–359. [CrossRef]
5. Li, H.; Durbin, R. Fast and accurate short read alignment with Burrows–Wheeler transform. *Bioinformatics* **2009**, *25*, 1754–1760. [CrossRef]
6. Lee, W.P.; Stromberg, M.P.; Ward, A.; Stewart, C.; Garrison, E.P.; Marth, G.T. MOSAIK: A Hash-Based Algorithm for Accurate Next-Generation Sequencing Short-Read Mapping. *PLoS ONE* **2014**, *9*, e90581. [CrossRef]
7. Lunter, G.; Goodson, M. Stampy: A statistical algorithm for sensitive and fast mapping of Illumina sequence reads. *Genome Res.* **2011**, *21*, 936–939. [CrossRef]
8. Tanoti: A BLAST-Guided Reference-Based Short Read Aligner. Available online: http://www.bioinformatics.cvr.ac.uk/tanoti.php (accessed on 24 February 2019).
9. Bankevich, A.; Nurk, S.; Antipov, D.; Gurevich, A.A.; Dvorkin, M.; Kulikov, A.S.; Lesin, V.M.; Nikolenko, S.I.; Pham, S.; Prjibelski, A.D.; et al. SPAdes: A New Genome Assembly Algorithm and Its Applications to Single-Cell Sequencing. *J. Comput. Biol.* **2012**, *19*, 455–477. [CrossRef]
10. Yang, X.; Charlebois, P.; Gnerre, S.; Coole, M.G.; Lennon, N.J.; Levin, J.Z.; Qu, J.; Ryan, E.M.; Zody, M.C.; Henn, M.R. De novo assembly of highly diverse viral populations. *BMC Genom.* **2012**, *13*, 475. [CrossRef]
11. Wymant, C.; Fraser, C.; Hall, M.; Golubchik, T.; Bannert, N.; Fellay, J.; Fransen, K.; Porter, K.; Gourlay, A.; Grabowski, M.K.; et al. Easy and accurate reconstruction of whole HIV genomes from short-read sequence data with shiver. *Virus Evol.* **2018**, *4*, vey007. [CrossRef]
12. Li, H.; Wysoker, A.; Handsaker, B.; Marth, G.; Abecasis, G.; Ruan, J.; Homer, N.; Durbin, R.; Fennell, T.; The 1000 Genome Project Data Processing Subgroup. The Sequence Alignment/Map format and SAMtools. *Bioinformatics* **2009**, *25*, 2078–2079. [CrossRef] [PubMed]
13. Wilm, A.; Aw, P.P.K.; Bertrand, D.; Yeo, G.H.T.; Ong, S.H.; Wong, C.H.; Khor, C.C.; Petric, R.; Hibberd, M.L.; Nagarajan, N. LoFreq: A sequence-quality aware, ultra-sensitive variant caller for uncovering cell-population heterogeneity from high-throughput sequencing datasets. *Nucleic Acids Res.* **2012**, *40*, 11189–11201. [CrossRef]
14. Macalalad, A.R.; Zody, M.C.; Charlebois, P.; Lennon, N.J.; Newman, R.M.; Malboeuf, C.M.; Ryan, E.M.; Boutwell, C.L.; Power, K.A.; Brackney, D.E.; et al. Highly Sensitive and Specific Detection of Rare Variants in Mixed Viral Populations from Massively Parallel Sequence Data. *PLoS Comput. Biol.* **2012**, *8*, e1002417. [CrossRef] [PubMed]

15. Yang, X.; Patrick, C.; Macalalad, A.; Henn, M.R.; Zody, M.C. V-Phaser 2: Variant inference for viral populations. *BMC Genom.* **2013**, *14*, 674. [CrossRef]
16. Sandmann, S.; de Graaf, A.O.; Karimi, M.; van der Reijden, B.A.; Hellström-Lindberg, E.; Jansen, J.H.; Dugas, M. Evaluating Variant Calling Tools for Non-Matched Next-Generation Sequencing Data. *Sci. Rep.* **2017**, *7*, 43169. [CrossRef]
17. Ibrahim, B.; McMahon, D.P.; Hufsky, F.; Beer, M.; Deng, L.; Mercier, P.L.; Palmarini, M.; Thiel, V.; Marz, M. A new era of virus bioinformatics. *Virus Res.* **2018**, *251*, 86–90. [CrossRef] [PubMed]
18. Kuiken, C.; Combet, C.; Bukh, J.; Shin-I, T.; Deleage, G.; Mizokami, M.; Richardson, R.; Sablon, E.; Yusim, K.; Pawlotsky, J.M.; et al. A comprehensive system for consistent numbering of HCV sequences, proteins and epitopes. *Hepatology* **2006**, *44*, 1355–1361. [CrossRef]
19. Verbist, B.M.; Aerssens, J.; Reumers, J.; Thys, K.; Van der Borght, K.; Clement, L.; Thas, O.; Talloen, W.; Wetzels, Y. VirVarSeq: A low-frequency virus variant detection pipeline for Illumina sequencing using adaptive base-calling accuracy filtering. *Bioinformatics* **2014**, *31*, 94–101. [CrossRef] [PubMed]
20. Döring, M.; Büch, J.; Friedrich, G.; Pironti, A.; Kalaghatgi, P.; Knops, E.; Heger, E.; Obermeier, M.; Däumer, M.; Thielen, A.; et al. geno2pheno[ngs-freq]: a genotypic interpretation system for identifying viral drug resistance using next-generation sequencing data. *Nucleic Acids Res.* **2018**, *46*, W271–W277. [CrossRef] [PubMed]
21. Schirmer, M.; Sloan, W.T.; Quince, C. Benchmarking of viral haplotype reconstruction programmes: An overview of the capacities and limitations of currently available programmes. *Brief. Bioinform.* **2012**, *15*, 431–442. [CrossRef]
22. Singer, J.B.; Thomson, E.C.; McLauchlan, J.; Hughes, J.; Gifford, R.J. GLUE: A flexible software system for virus sequence data. *BMC Bioinform.* **2018**, *19*, 532. [CrossRef] [PubMed]
23. Vermehren, J.; Park, J.S.; Jacobson, I.M.; Zeuzem, S. Challenges and perspectives of direct antivirals for the treatment of hepatitis C virus infection. *J. Hepatol.* **2018**, *69*, 1178–1187. [CrossRef] [PubMed]
24. Sorbo, M.C.; Cento, V.; Maio, V.C.D.; Howe, A.Y.; Garcia, F.; Perno, C.F.; Ceccherini-Silberstein, F. Hepatitis C virus drug resistance associated substitutions and their clinical relevance: Update 2018. *Drug Resist. Updates* **2018**, *37*, 17–39. [CrossRef] [PubMed]
25. European Association for the Study of the Liver. Recommendations on Treatment of Hepatitis C 2016. *J. Hepatol.* **2017**, *66*, 153–194. [CrossRef] [PubMed]
26. IUPAC-IUB Commission on Biochemical Nomenclature (CBN). Abbreviations and symbols for nucleic acids, polynucleotides and their constituents. Recommendations 1970. *Biochem. J.* **1970**, *120*, 449–454. [CrossRef]
27. HCV-GLUE: A Sequence Data Resource for Hepatitis C Virus. Available online: http://hcv.glue.cvr.ac.uk (accessed on 24 February 2019).
28. Kalaghatgi, P.; Sikorski, A.M.; Knops, E.; Rupp, D.; Sierra, S.; Heger, E.; Neumann-Fraune, M.; Beggel, B.; Walker, A.; Timm, J.; et al. Geno2pheno[HCV]—A Web-based Interpretation System to Support Hepatitis C Treatment Decisions in the Era of Direct-Acting Antiviral Agents. *PLoS ONE* **2016**, *11*, e0155869. [CrossRef]
29. Batty, E.M.; Wong, T.H.N.; Trebes, A.; Argoud, K.; Attar, M.; Buck, D.; Ip, C.L.C.; Golubchik, T.; Cule, M.; Bowden, R.; et al. A Modified RNA-Seq Approach for Whole Genome Sequencing of RNA Viruses from Faecal and Blood Samples. *PLoS ONE* **2013**, *8*, e66129. [CrossRef] [PubMed]
30. Lamble, S.; Batty, E.; Attar, M.; Buck, D.; Bowden, R.; Lunter, G.; Crook, D.; El-Fahmawi, B.; Piazza, P. Improved workflows for high throughput library preparation using the transposome-based Nextera system. *BMC Biotechnol.* **2013**, *13*, 104. [CrossRef] [PubMed]
31. Davalieva, K.; Kiprijanovska, S.; Plaseska-Karanfilska, D. Fast, reliable and low cost user-developed protocol for detection, quantification and genotyping of hepatitis C virus. *J. Virol. Methods* **2014**, *196*, 104–112. [CrossRef]
32. Bonsall, D.; Ansari, M.; Ip, C.; Trebes, A.; Brown, A.; Klenerman, P.; Buck, D.; STOP-HCV Consortium.; Piazza, P.; Barnes, E.; Bowden, R. ve-SEQ: Robust, unbiased enrichment for streamlined detection and whole-genome sequencing of HCV and other highly diverse pathogens [version 1; referees: 2 approved, 1 approved with reservations]. *F1000Research* **2015**, *4*, 1062. [CrossRef] [PubMed]
33. TrimGalore: A Wrapper around Cutadapt and FastQC to Consistently Apply Adapter and Quality Trimming to FastQ Files, with Extra Functionality for RRBS Data. Available online: https://github.com/FelixKrueger/TrimGalore (accessed on 24 February 2019).

34. Sreenu, V.B. (MRC-University of Glasgow Centre for Virus Research, Glasgow, UK). Personal Communication, 2019.
35. Gaidatzis, D.; Lerch, A.; Hahne, F.; Stadler, M.B. QuasR: Quantification and annotation of short reads in R. *Bioinformatics* **2014**, *31*, 1130–1132. [CrossRef]
36. Martin, M. Cutadapt removes adapter sequences from high-throughput sequencing reads. *EMBnet J.* **2011**, *17*, 10–12. [CrossRef]
37. V-FAT: A Post-Assembly Pipeline for the Finishing and Annotation of Viral Genomes. Available online: https://www.broadinstitute.org/viral-genomics/v-fat (accessed on 24 February 2019).
38. SMALT: A Mapper for DNA Sequencing Reads. Available online: https://www.sanger.ac.uk/science/tools/smalt-0 (accessed on 24 February 2019).
39. Camacho, C.; Coulouris, G.; Avagyan, V.; Ma, N.; Papadopoulos, J.; Bealer, K.; Madden, T.L. BLAST+: Architecture and applications. *BMC Bioinform.* **2009**, *10*, 421. [CrossRef] [PubMed]
40. Harris, R.S. Improved Pairwise Alignment of Genomic DNA. Ph.D. Thesis, Pennsylvania State University, State College, PA, USA, 2007.
41. Smith, D.B.; Bukh, J.; Kuiken, C.; Muerhoff, A.S.; Rice, C.M.; Stapleton, J.T.; Simmonds, P. Expanded classification of hepatitis C virus into 7 genotypes and 67 subtypes: Updated criteria and genotype assignment web resource. *Hepatology* **2014**, *59*, 318–327. [CrossRef]
42. Stamatakis, A. RAxML version 8: A tool for phylogenetic analysis and post-analysis of large phylogenies. *Bioinformatics* **2014**, *30*, 1312–1313. [CrossRef] [PubMed]
43. Katoh, K.; Standley, D.M. MAFFT Multiple Sequence Alignment Software Version 7: Improvements in Performance and Usability. *Mol. Biol. Evol.* **2013**, *30*, 772–780. [CrossRef] [PubMed]
44. Poordad, F.; Pol, S.; Asatryan, A.; Buti, M.; Shaw, D.; Hézode, C.; Felizarta, F.; Reindollar, R.W.; Gordon, S.C.; Pianko, S.; et al. Glecaprevir/Pibrentasvir in patients with hepatitis C virus genotype 1 or 4 and past direct-acting antiviral treatment failure. *Hepatology* **2018**, *67*, 1253–1260. [CrossRef] [PubMed]
45. A Java API for High-Throughput Sequencing Data (HTS) Formats. Available online: http://samtools.github.io/htsjdk/ (accessed on 24 February 2019).
46. DiversiTools: Tool for Analysing Viral Diversity from HTS. Available online: http://josephhughes.github.io/DiversiTools/ (accessed on 24 February 2019).

11

Virus-Incorporated Biomimetic Nanocomposites for Tissue Regeneration

Iruthayapandi Selestin Raja [1,†], Chuntae Kim [2,†], Su-Jin Song [3], Yong Cheol Shin [4], Moon Sung Kang [3], Suong-Hyu Hyon [5], Jin-Woo Oh [2,*] and Dong-Wook Han [3,*]

1 Monocrystalline Bank Research Institute, Pusan National University, Busan 46241, Korea
2 Department of Nanofusion Technology, College of Nanoscience & Nanotechnology, Pusan National University, Busan 46241, Korea
3 Department of Cogno-Mechatronics Engineering, College of Nanoscience & Nanotechnology, Pusan National University, Busan 46241, Korea
4 Department of Medical Engineering, Yonsei University, College of Medicine, Seoul 03722, Korea
5 Joint Faculty of Veterinary Medicine, Kagoshima University, Kagoshima 890-8580, Japan
* Correspondence: ojw@pusan.ac.kr (J.-W.O.); nanohan@pusan.ac.kr (D.-W.H.)
† These authors contributed equally to this work.

Abstract: Owing to the astonishing properties of non-harmful viruses, tissue regeneration using virus-based biomimetic materials has been an emerging trend recently. The selective peptide expression and enrichment of the desired peptide on the surface, monodispersion, self-assembly, and ease of genetic and chemical modification properties have allowed viruses to take a long stride in biomedical applications. Researchers have published many reviews so far describing unusual properties of virus-based nanoparticles, phage display, modification, and possible biomedical applications, including biosensors, bioimaging, tissue regeneration, and drug delivery, however the integration of the virus into different biomaterials for the application of tissue regeneration is not yet discussed in detail. This review will focus on various morphologies of virus-incorporated biomimetic nanocomposites in tissue regeneration and highlight the progress, challenges, and future directions in this area.

Keywords: virus; tissue regeneration; biomimetic nanocomposites; phage display

---

1. Introduction

1.1. Emerging Trends in Tissue Regeneration

Tissue engineering is a part of the regenerative medicine field, which emphasizes the fabrication of various functional biological constructs to reduce the increased demand for donor organs [1]. The shortage of organ donors and the increased number of people undergoing transplantation have necessitated the development of effective biomimetic materials adopting advanced technologies [2]. The aim of the field is to harness nature's ability to treat the damaged tissues., ensuring biocompatibility and supporting cellular biological events. When muscles are damaged by incidents such as illness, accidents, and microbial invasion, they lose integrity for healthy functioning at the cellular level and subsequently follow a cascade of biochemical events, including hemostasis, inflammation, proliferation, and maturation, to restore integrity [3]. However, the first ever immediate response after an injury is the production of reactive oxygen species by NADPH oxidase before inflammatory reaction [4]. Moreover, the time consumption to retain normal function in the dysfunctional organ is dependent on one's age and the seriousness of the damage. When tissue fails in the ability for self-regeneration, especially in a pathological condition, the external application of a scaffold becomes inevitable [5].

Stainless steel was first implanted as an artificial hip in 1929, which paved the way for the researchers to design biomimetic materials to replace body organs [6]. However, tissue and organ failure were considered as major medical complications until 1954, as there were not enough remedies to treat dysfunctional organs. The Nobel Laureate in medicine, Joseph Murray, transplanted a healthy kidney to a genetically identical twin brother successfully [1]. Following this historical incident, the researchers began to utilize cell structures to treat damaged tissues. These structures are autografts, isografts, allografts, and xenografts, which are biological tissues harvested from a patient's organ, a genetically identical twin, a genetically non-identical individual of the same species, and a donor of a different species, respectively [7]. The autograft has some advantages over other forms of grafts, with no rejection concerns. However, it requires a secondary procedure for the reconstructive surgery and faces donor site complication.

Meanwhile, isografts, allografts, and xenografts express slight or significant degrees of rejection as the Major Histocompatibility Complex (MHC) recognizes them as foreign antigens causing cytotoxic T-cell-mediated demolition. Unlike autografts, other forms of grafts eliminate the need for the secondary procedure without the donor site complications [8]. Though researchers are involved in finding therapeutic techniques, such as immunosuppressive medications, infectious prophylaxis, and DNA-based tissue typing, graft-versus-host disease (GVHD) remains a hurdle limiting the use of grafts in tissue regeneration [9].

The extracellular matrix (ECM) is an important component in tissues, which has the intrinsic cues to provide sites with all their cellular activities, including migration, orientation, shape, plasticity, and cell–matrix and cell–cell interactions [10]. The ECM is a complex three-dimensional network comprising proteins, such as collagen, elastin, and laminin, proteoglycans, and glycoproteins [11]. It acts as a glue and platform to bind cells together with the connective tissues. The decellularized extracellular matrix (DECM), a water-insoluble matrix, can be prepared from any organ or tissue by removing cellular components from ECM. In recent years, DECM has been used as a scaffold to hold the cells together and maintain integrity during wound healing processes [12]. Over the past two decades, the researchers have been inclined towards the preparation of artificial biomimetic grafts from the naturally available biomacromolecules, synthetic polymers, and ceramic materials, to replace tissue grafts and DECM. The remarkable features of artificial grafts are that they are economically inexpensive, their ease of preparation and storage, biocompatibility, and the tuning of the physicochemical and biological properties [13]. The most widely studied natural polymers include collagen, chitosan, hyaluronic acid, carboxymethyl cellulose, alginate, and silk fibroin, which have been reported to trigger a less immunogenic reaction in the human body, while the synthetic polymers, such as polycaprolactone, polyvinyl alcohol, nylon, and polyphosphazene, provide longer shelf life. To obtain combinatorial properties, nowadays researchers utilize a blend of polymers and their derivatives [14,15]. Ceramic and bioglass materials exhibit osteoconductive properties, showing remarkable mechanical strength and resistance to deformation, and hence, they have been widely used for bone tissue regeneration [16]. While designing a synthetic biomimetic scaffold, it is a requisite to create a benign microenvironment in a way that mimics the ECM niche for the application of effective tissue regeneration.

In the past ten years, researchers have explored a variety of biomimetic nanocomposites by incorporating bioactive molecules, such as growth factors, cytokines, genes, antibiotics, and anti-inflammatory drugs within the scaffold to increase tissue regeneration potential [17,18]. As the direct administration of therapeutic medications is not effective, leading to overdose toxicity and short time exposure to biological tissue, these biomimetic nanocomposites were much appreciated by chemists, biologists, and nanotechnologists [19]. Mostly, the bioactive molecules are cross-linked with the primary polymeric component by physical, chemical, and enzymatical treatment with or without the aid of cross-linkers and their release from the matrix is adjusted for either immediate, sustained, or slow release, according to the desired site of the target tissue. The drug release rate from the biomaterial majorly involves simple diffusion, erosion, or degradation of the matrix and swelling of the polymeric scaffold [5]. After the advent of nanotechnology, the nanoparticles were entrapped into the scaffold to exploit their medicinal properties. A composite of collagen and silver nanoparticles has been used to augment the burn tissue repair process [20]. Mieszawska et al. studied a composite film composed of silk and nanoclay to serve as a supportive biomaterial to improve bone tissue regeneration [21]. The in vivo effects of reduced graphene oxide and hydroxyapatite nanocomposite powders were investigated on critical-sized calvarial defects in a rabbit model and it was reported that the nanocomposite stimulated osteogenesis and enhanced bone formation without inflammatory responses [22]. Our research group also studied the influence of graphene oxide dispersed into a polylactic-co-glycolic acid (PLGA) electrospun nanofiber towards stimulation of myogenesis and enhanced vascular tissue regeneration [23,24]. The nanoparticles with diameters in the range of 50–700 nm acted as therapeutic drug carriers to pass through the capillaries into cells, facilitating the regeneration of new tissues [5,25].

Cell-laden scaffolds have also received attention among researchers aiming to achieve a tissue-imitating engineered graft. Kizilel et al. encapsulated pancreatic islets into nano-thin polyethylene glycol coating for enhanced insulin secretion [26]. Yoon et al. fabricated a three-dimensional layered structure using the blend of collagen epidermal keratinocytes and dermal fibroblasts to progress migration and proliferation of keratinocytes and fibroblasts during the skin repair process [27]. Within this context, microbe-based biomimetic materials have appeared as an emerging trend in tissue regeneration in recent times. Virus-based biomaterials have many biomedical applications, including cancer markers, antibacterial materials, drug carriers, and tissue regeneration [28]. Not only do they encapsulate and release the therapeutic agents to the target site, but the morphology of biomaterials also plays a pivotal role in altering physicochemical and biological properties. In this review, we have focused on summarizing the impacts of various virus-incorporated biomimetic nanocomposites with different morphologies, such as nanoparticles, nanofibers, hydrogels, and organic-inorganic hybrids, in the field of tissue regeneration. The same has been schematically shown in Figure 1. The nanoparticles that have a large surface area to volume have proven their effectiveness with long-term functionality and stability in the biological milieu. The nanoparticles can diffuse across the cell membrane and interact with cellular biomacromolecules residing inside the cell [20]. Remarkably, the hydrogel provides wettability and cell migration, while the nanofibrous matrix ensures air permeability and mechanical properties in tissue regeneration [29,30].

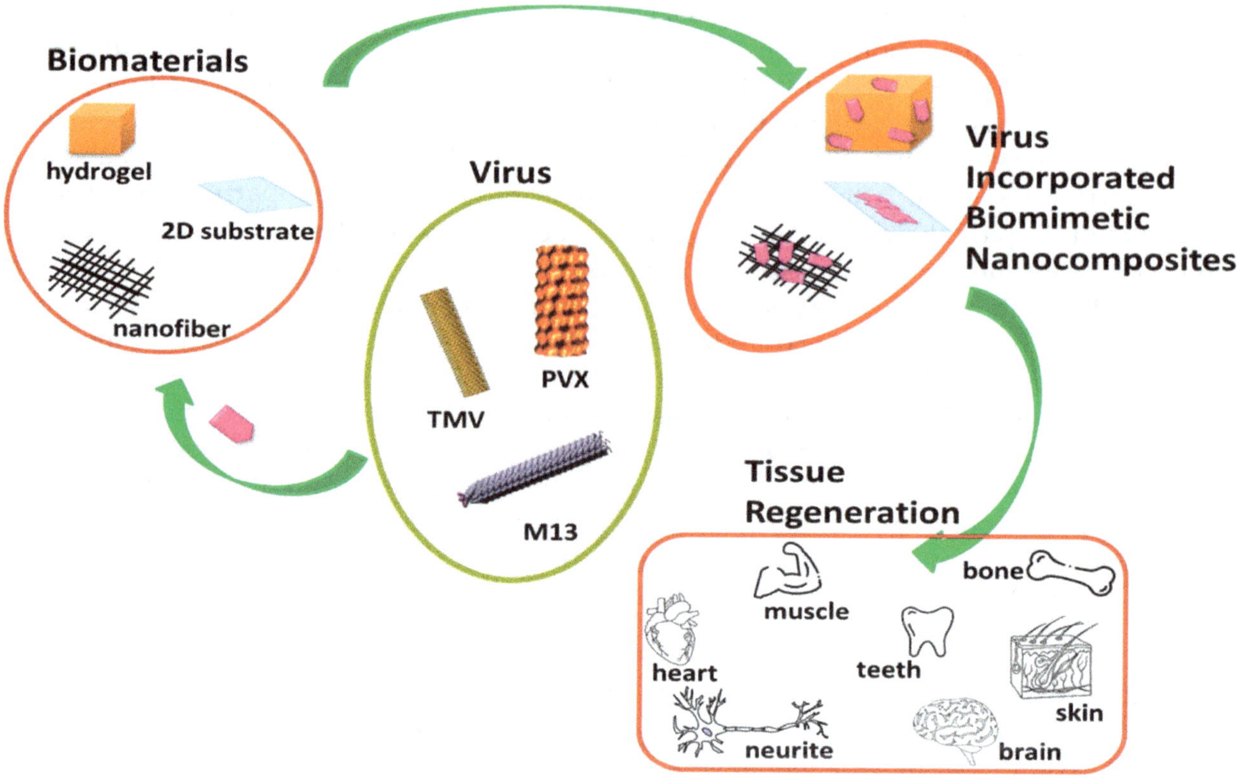

Figure 1. A compendium of tissue regeneration functionality of various virus incorporated biomimetic nanocomposites. Plant-based viruses Tobacco Mosaic Virus (TMV) and Potato Virus X (PVX) and bacteriophage M13 were shown. Reproduced with permission from [31]. Copyright Elsevier, 2010.

1.2. Remarkable Properties of Medicinally Valuable Viruses

Not all viruses cause infectious diseases in the human body. Viruses can be classified as lytic, temperate, or lysogenic based on the level of adverse effects produced in its host [32]. During infection, lytic phages kill the host bacteria, triggering the release of progeny. Lysogenic phages do not affect the host cell and infection occurs with replication of the phage genome but not the host bacterial genome. Temperate phages reside in host bacteria for amplification with no lysis, however some phages, including the λ phage, exceptionally, have both categories and thrive following either lytic or lysogenic cycles. The lytic phages, including T1–T7, contain a head and flexible tail but lack filaments. The T7 phage belongs to the Podoviridae family and structurally has an icosahedral head and a short tail. They were reported to lyse the host cell within a minute by secreting the lysozyme enzyme [33]. Professionally, T4 phages have found applications in food preservation, antibiotics, detection of bacteria, DNA and protein packing systems, and DNA-based vaccines. The literature reports revealed that Podovirus P22 assisted the assembly of cadmium sulfide nanocrystals to improve photosensitization in tissue imaging [34]. The filamentous phages, including Ff, f1, M13, N1, and Ike, are the examples of temperate phages. As they can act as a template for the synthesis of nanomaterials, the general applications of temperate phages are huge compared to lytic phages [33].

Virus-based biomimetic materials are generally derived from plant viruses and bacteriophages, as they rarely generate harmful side effects in human beings. Generally, bacteriophages (excluding fd and M13) are categorized into filamentous type of viruses, follow a non-lytic mode to infect and thrive in bacteria. It has been reported that these phages do not consist of mammalian promoter sequences in their genome, and hence, do not instigate dreadful human diseases [35]. In the human body, bacteriophages are present abundantly in the gut, bladder, and oral cavity, functioning to shape bacterial metabolisms and populations of microbial communities. It has been described that the potential role of phages increases from childhood to adulthood [36]. The monodispersed phages can self-assemble themselves into hierarchically ordered structures, such as rope-like bundles and liquid crystals. The protein surface can be modified either by covalent and non-covalent interactions or genetic alterations [35]. These unique properties have prompted the researchers to take a long stride in utilizing the phage-based biomaterials towards a wide range of biomedical applications, including biomedical imaging, drug delivery, biosensors, tissue regeneration, energy, and catalysis [37].

Owing to economically inexpensive, large scale production, ease of manipulation, and stability against a wide range of pH and temperature, a variety of phage-based biomimetic nanocomposites have been constructed for the application of effective tissue regeneration [38]. As far as morphology is concerned, a typical phage has a diameter of 68 A° and a length in the range of 800–2000 nm. The circular single-stranded DNA (ssDNA) of the phage encodes 10 genes containing 5000–8000 nucleotides, which encode a highly ordered major coat protein (p8) located around the center of phage, two minor coat proteins (p7 and p9) at one end, and two others (p3 and p6) at another terminal portion of the phage. The helical arrays of major coat proteins assemble to form the capsid shell. Generally, minor coat proteins display larger sized peptides than the major coat protein (p8) [39]. The major coat protein, p8, of M13 phage, has different segments, such as the N-terminal amphipathic, hydrophobic transmembrane (TM), and DNA binding segments. The small residues (Gly, Ala, and Ser) present on these segments have been reported to be involved in helix–helix axial and lateral interactions, which facilitate extrusion of the virion from the membrane during assembly, and hence have been known as conserved regions in the DNA sequence. Fiber diffraction and spectroscopic data show that M13 differs from fd at the 12th residue, where M13 replaces Asp of fd with Asn [40]. Filamentous phages are defined as non-enveloped bacterial viruses, having some properties in common, namely, life cycles, organization, and morphology.

The ssDNA has a left-handed helix structure possessing strong interactions with the positively charged inner surface of the capsid shell. The diffraction pattern studies classified filamentous phages into two distinct groups. Class I symmetry group consists of fd, M13, If1, and IKe, which are consistent with 5-fold symmetry. Class II symmetry group includes Pf3, Pf1, and Xf, wherein the helices are arranged with a rising per monomer of about 3.0 A° [39]. The aligned solid-state NMR studies proved that fd has O-P-O phosphate linkages in an ordered manner, whereas Pf1 did not possess such linkages [41]. According to NMR studies, phage fd has strong electrostatic interactions between the negatively charged phosphate backbone of the ssDNA nucleotide and two of four positively charged amino acid residues present at the C-terminal portion of the major coat protein, which is attributed to stabilization of the DNA core structure. The literature reports revealed that M13 and IKe showed similarity in π–π interactions between the residues of Tyr9 of one p8 and Tyr29 of an adjacent p8 [42].

Infection of E. coli by phage is initiated by the attachment of N-terminal amino acids of p3, which is present in the specialized threadlike appendage, F pilus. Subsequently, the coat protein of the phage dissolves onto the envelope of the host, which allows the only ssDNA into the cytoplasm. The host machinery synthesizes a complementary DNA strand with the involvement of two virally encoded proteins, p2 and p10, which leads to the formation of a double-stranded replicative form. The replicative form acts as a template to transcript phage genes for the synthesis of progeny ssDNAs. These progeny phage particles discharge from the bacterial cell envelope through the membrane pore complex, acquire coat proteins from the membrane, and appear as mature virions. The fact is that the infected cells undergo division at a slower rate than the uninfected cells [39].

In recent times, the researchers have sought to explore multifunctional phage-based biomaterials by precisely adjusting the surface chemistry of phage nanofibers. Covalent, non-covalent, and genetic modifications of phage coat proteins have been described comprehensively by the researchers. The genetic modification of phage coat proteins would display various foreign peptides with different functional groups at the side wall and the two termini of the phage. The endogenous amino acids of phage coat proteins are genetically combined with the foreign amino acid sequence in order to form a hybrid fusion protein, which is incorporated into phage particles and released from the cell subsequently. As a result, the foreign peptide is displayed on the surface of the phage coat protein [35]. The phage display is generally specified after N-terminal modification in its respective coat proteins. For example, if the N-terminus of p3 of the phage undergoes modification, the resulting phage is designated as a p3 display. When two or more coat proteins are controlled for modification in the same phage, then they can be known as double display, and so on [43]. In the phage coat protein, the carboxylates of aspartic and glutamic acid residues, the amine of lysine, and the phenol of tyrosine are the majorly available functional entities for the chemical modification. Introducing aldehyde into the reactive amine group has been involved in a wide range of bio-conjugation reactions, whereas the cross-linkage of p-azidophenylalanine has provided an azide handle on the phage surface, which can be easily modified for further reactions [44,45]. The EDC treatment has been helpful in cross-linking the reactive carboxylate groups with amine-functionalized moieties in phage proteins [46]. Strong nucleophile selenocysteine has been successfully genetically incorporated into phage protein using an opal stop, codon suppressing mRNA [47].

The phage-display library, with a heterogeneous mixture of phages carrying different foreign DNA insert, was created for selective binding of phage proteins with the target ligands, such as polymers, proteins, organic and inorganic crystals, small molecules, such as trinitrotoluene, and cells [48–51]. Among the phage-display libraries, the reports of p3 and p6 libraries have been well documented in research publications. Conventionally, research studies have adopted the biopanning method to find extensive use of phage particles in tissue regeneration. Biopanning is a typical technique to form a population of enriched phage-displayed peptides and specifically identify a target binding peptide [52]. According to this selection procedure, initially, a phage-display random library is incubated with the targets. Subsequently, the non-bound phage particles are eliminated with the help of detergent solubilized buffer. The target-bound phage particles are then eluted using a specialized buffer maintaining acidic pH around 2.2, and the amplification process is followed by infection of host bacteria. The resulting amplified phages form a newly enriched sub-library with more specificity to interact with the targets. The procedure is repeated several times until a only few desired peptides are predominantly available in the sub-library [53]. In the subsequent section, we will investigate the contribution of plant virus and phage-based biomimetic nanocomposites in the field of tissue regeneration.

2. Different Morphologies of Virus-Incorporated Biomimetic Nanocomposites in Tissue Regeneration

2.1. Virus-Based Nanoparticles

Many plants and phage-based viral nanoparticles have been employed so far for tissue regeneration. Plant viral nanoparticles are mono-dispersed, meta-stable, and structurally uniformed [54]. Li revealed that when the virus-based nanoparticle is more robust, the functional nanostructure is more stable, but at the same time they might be harmful to the encapsulated cargo [55].

Though the unmodified TMV nanoparticles have the potential to accelerate osteogenic differentiation in adult stem cells, the lack of affinity to the mammalian cell surface diminishes the cell adhesion property. Hence, the researchers opt for either genetic or chemical modification in viral nanoparticles in order to increase the cell binding capacity and find versatile biomedical applications. Sitasuwan et al. [62] modified the surface of a TMV nanoparticle by coupling azide-derivatized Arg-Gly-Asp-(RGD) tripeptide with tyrosine residues through Cu (I) catalyzed azide-alkyne cycloaddition reaction. When incorporated into the artificial scaffold, the RGD peptides overexpressed on ECM increase initial cell attachment by binding integrin receptors. The spacing between RGD motifs alter biological events, such as fibroblast adhesion and spreading (<440 nm), focal adhesion assembly (<140 nm), and induction of stress fiber formation (<60 nm). Owing to lack of mammalian cell infectivity, cost-effectiveness, and highly uniform size, plant viral nanoparticles have gained attention among nano and biomedical researchers. The Tobacco Mosaic Virus (TMV) constitutes a rod-like shaped nanoparticle with a diameter of 18 nm and length of 300 nm. The TMV nanoparticle consists of a capsid with 2130 identical coat protein subunits, which are responsible for assembling into a helical structure around the ssRNA. When each subunit is modified, the resulting

TMV is a polyvalent nanoparticle. The TMV can withstand temperatures up to 60 °C and can be stable in a pH range of 2–10. The TEM micrograph of wild type TMV is shown in Figure 2a [56].

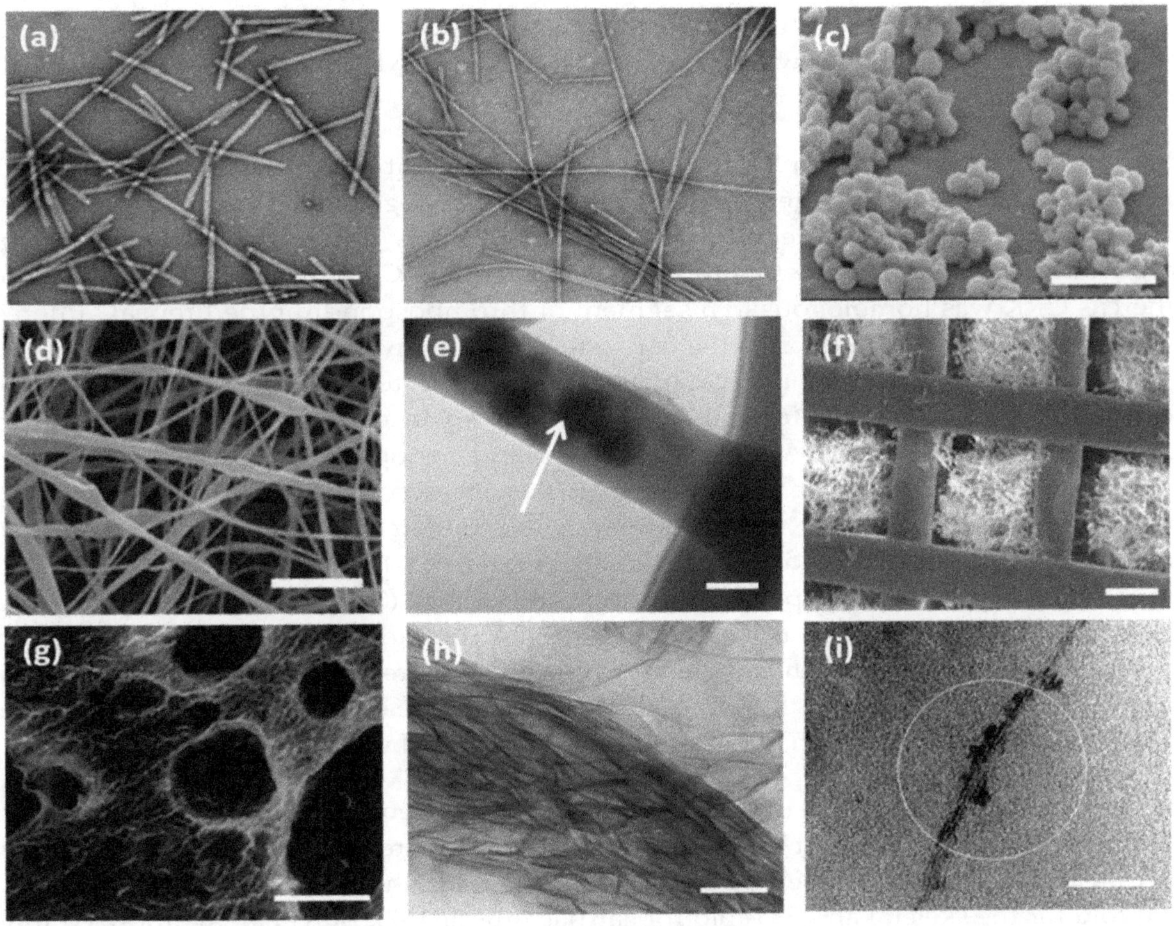

Figure 2. TEM micrographs of wild type Tobacco Mosaic Virus (TMV) nanoparticles, scale bar 200 nm (**a**) and TMV/PANI/PSS nanofiber (**b**) generated by flow assembly method, scale bar 500 nm. Reproduced with permission from [56]. Copyright American Chemical Society, 2015. (**c**) SEM micrograph of freeze-dried capsules of T4 bacteriophage/alginate water in oil emulsion in chloroform, scale bar 5 µm. (**d**) SEM and TEM micrographs (**e**) of PEO electrospun nanofiber containing T4 bacteriophage/alginate have been shown scale bars of 10 µm and 100 nm, respectively. The arrow indicates the presence of T4/alginate into the nanofiber. Reproduced with permission from [57]. Copyright John Wiley and Sons, 2013. (**f**) SEM image of the 3D printed bioceramic bone scaffold consisting of biphasic calcium phosphate with pores filled with a matrix of chitosan and RGD phage, scale bar 200 µm. Reproduced with permission from [58]. Copyright John Wiley and Sons, 2014. (**g**) SEM micrograph of porous alginate hydrogel containing TMV particles displaying interconnected channels and macropores, scale bar 100 µm. Reproduced with permission from [59]. Copyright American Chemical Society, 2012. (**h**) The TEM image of a mineralized E8-displaying phage bundle formed after 90 h incubation in the solution containing calcium and phosphate ions, scale bar 100 nm. Reproduced with permission from [60]. Copyright John Wiley and Sons, 2010. (**i**) The TEM image of the HAP-fd phage bundle with scale bar 100 nm. The circle indicates the presence of hydroxyapatite nanoparticles in the fibrous structure. Reproduced with permission from [61]. Copyright American Chemical Society, 2010.

In the human body, bone tissue regenerates to a greater extent when compared to other types of tissues. However, the regeneration process is complicated in the case of tumor resection, hip implant revision, and major fractures [63]. Pi et al. constructed a cartilage targeting gene delivery nanocomposite system by conjugating polyethylenimine (PEI) with M13 phage-displayed chondrocyte-affinity peptide

(CAP), DWRVIIPPRPSA, which was isolated after two rounds of biopanning. During incubation, the phages expressing CAP showed higher affinity towards rabbit chondrocytes at 265.5-fold when compared to unmodified phages. They reported that the CAP-conjugated PEI particles had no species specificity in binding chondrocytes of rabbit and humans. Furthermore, most of the particles were found to enter chondrocytes without being trapped in ECM, which acknowledges their larger transfection efficiency [64].

T7 viral nanoparticles were explored to display two different functional peptides CARSKNKDC (CAR) and CRKDKC (CRK) to target the microvasculature of regenerating wound tissue, including skin and tendon [25]. Skin disintegration may occur in many ways, such as bruising, abrasion, hacking, burning, stabbing, and laceration. It was observed that CAR was similar to heparin-binding sites, whereas CRK was homologous to a segment of thrombospondin type I repeat. Interestingly, CAR displayed a dominant function in the early stages of skin wound healing, while CRK showed preferences in the later stages of the same process. As the terminal residues contain cysteine, the screened peptides had more feasibility to be involved in disulfide bond formation to form a molecular cycle structure. The CAR-expressing T7 phage nanoparticles had been found to appear in wound sites 100–140-fold more efficiently than the non-recombinant phage nanoparticle [65]. The biomedical application of siRNAs is minimal owing to their low absorption across the stratum corneum, a horny outer layer of skin. Hsu et al. [66] explored M13 phage (from Ph.D-C7C library) viral nanoparticle-expressing skin penetrating and cell entering (SPACE) peptide with the sequence of AC-KTGSHNQ-CG in order to reach therapeutic macromolecules, including siRNAs, into the skin-associated cells. The in vitro physicochemical studies explored that the various macromolecules, including siRNA, penetrated across the stratum corneum into the epidermis layer of skin through the macropinocytosis pathway when the molecules were conjugated with SPACE.

A muscle binding M13 phage nanoparticle with peptide sequence ASSLNIA was identified to possess more excellent selectivity (at least five-times more) compared to the control phage nanoparticle. While investigating overall muscle selectivity on different organs, the muscle binding affinity was found to be 9–20-fold for the skeletal and 5–9-fold for cardiac muscle [67]. Sun et al. synthesized functional multivalent M13 phage (Ph.D.-7^{TM} display library) nanoparticles to express RIYKGVIQA and SEEL sequences, which are found in Nogo-66, a neurite outgrowth inhibitory protein. They selectively bound negative growth regulatory protein 1 (NgR1) with electrostatic forces of repeated leucine residues, enhancing neural differentiation of pc12 cells. Hence, this specific engineered viral nanoparticle has been appreciated for its potential use in neurite tissue regeneration, including spinal cord injury, optic nerve injury, ischemic stroke, and neurodegenerative diseases [68]. Collett et al. suggested that hepatitis C virus-based nanoparticles could act as a quadrivalent vaccine to trigger humoral and cellular immune responses. They explored biophysical, biochemical, and biomechanical properties of nanoparticles using Atomic Force Microscopy and observed that glycosylation occurred on the surface of the nanoparticle with ordered packing of the core [69]. The literature reports revealed that Sendai virus vectors displaying cardiac transcription factors could efficiently reprogram both mouse and human fibroblasts into induced cardiomyocyte-like cells in vitro. In addition, they could reduce scar formation, maintaining cardiac function in myocardial infarction affected animals [70].

The phosphate tailored TMV nanoparticle was demonstrated to induce expression of osteospecific genes of rat bone marrow stem cells (BMSCs), including osteocalcin and osteopontin, when compared to unmodified TMV nanoparticles. As shown in Figure 3d–f, the enhanced cell attachment and spreading of BMSCs were observed in phosphate grafted TMV (TMV-Phos) coated Ti substrates more than TMV coated substrates after 14 days of incubation in cell culture [73].

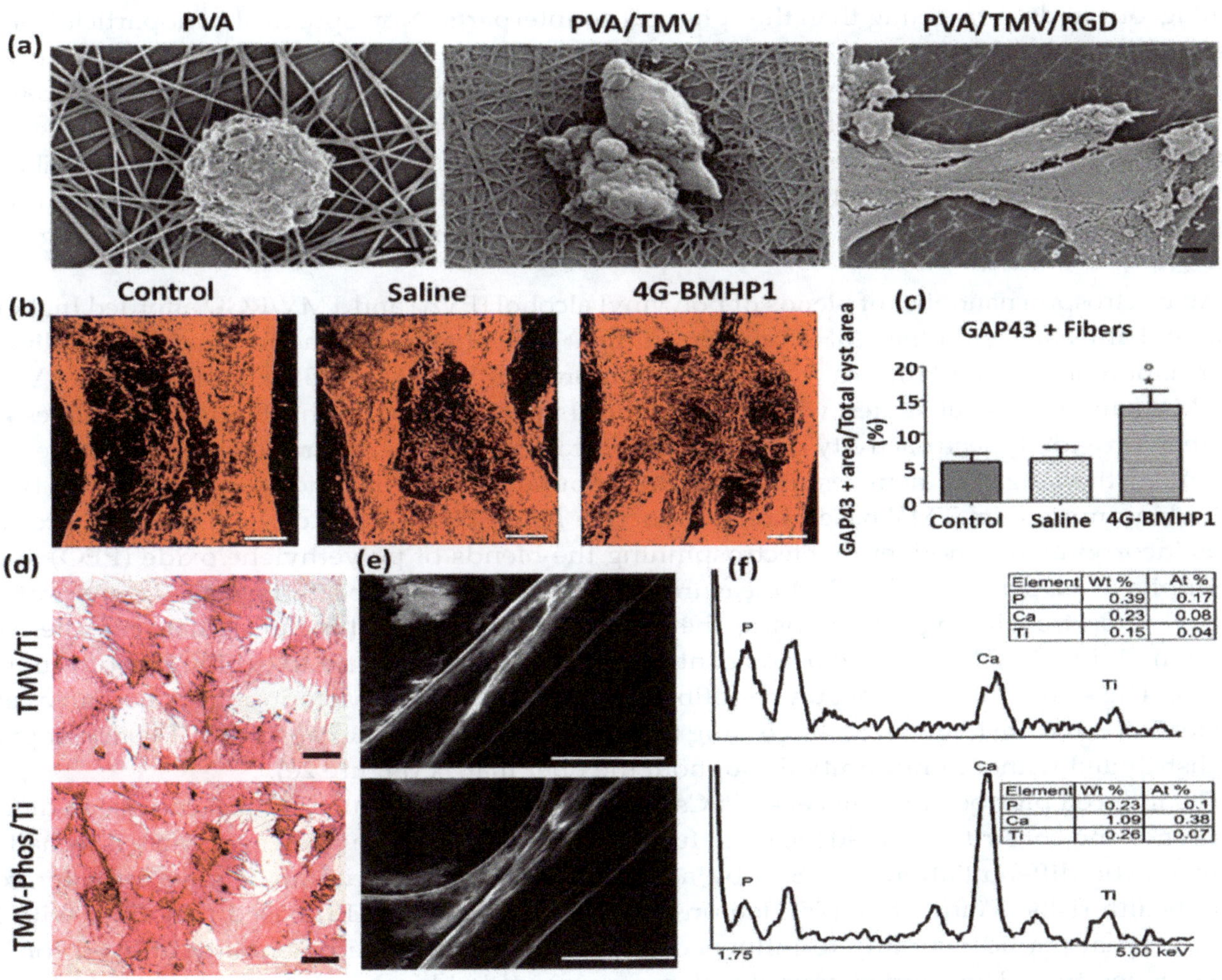

Figure 3. (**a**) Field emission scanning electron micrographs of Baby Hamster Kidney cells after 1 h incubation on electrospun nanofibrous substrates PVA, PVA-TMV, and PVA-TMV/RGD. Scale bar 2 µm. Reproduced with permission from [71]. Copyright John Wiley and Sons, 2014. (**b**) Immunofluorescence staining of longitudinal sections of total cyst area in spinal cord injury (SCI)-treated female Sprague-Dawley rats during the chronic phase. The synthesis of GAP-43 immunopositive fibers was significantly greater in bone marrow homing peptide-expressed phages (4G-BMHP1) treated group when compared to control groups (SCI control and saline). Scale bar 400 mm. The percentage of immunopositive fibers into the cyst area per total cyst area is represented in a bar graph (**c**) from six independent experiment results, and the values are reported with ± SEM. Significant factor * $p < 0.05$, 4G-BMHP1 vs. SCI control; 4G-BMHP1 vs. saline. Reproduced with permission from [72]. Copyright PLOS, 2011. (**d**) Bright-field optical microscope images of histochemical staining of alkaline phosphatase (ALPL) in bone marrow-derived stem cells (BMSCs) in TMV and phosphate grafted TMV (TMV-Phos)-coated Ti substrates after 14 days of culture. BMSCs have been shown to form neighboring well-spread cells under osteogenic conditions, which are stained highly positive for ALPL. Scale bar is 100 µm. (**e**,**f**) SEM micrographs and EDX analyses of TMV and TMV-Phos coating on Ti substrates. Scale bar for SEM is 100 µm. Reproduced with permission from [73]. Copyright Elsevier, 2010.

2.2. *Virus-Incorporated 2D Films and Nanofibers*

A combinatorial biomaterial consisting of PVX-based cyclic RGD, containing filament (RGD-PVX) and polyethylene glycol conjugated stealth filament (PEG-PVX), was developed to analyze biodistribution in mice xenograft models. The comparative studies demonstrated that PEG-PVX was preferentially accumulated into tumor cells, while RGD-PVX was trapped into the lung site in a large quantity. It has been reported that the filamentous and elongated nanoparticles are more advantageous in drug targeting than the spherical counterparts. Non-spherical nanoparticles present more ligands on their surfaces and show significant accumulation towards the vessel wall, improving the efficiency of tumor homing. Owing to the flexible nature of viral capsid, PVX-based nanoparticles could pass through restrictions in the complex biological environments and permeate into tissue cells without difficulty [74]. Wu et al. [56] successfully synthesized TMV-based electroactive nanofibers for neural tissue regeneration from the blend of polyaniline (PANI) and sodium polystyrene sulfonate (PSS). The morphology of the TMV/PANI/PSS nanofiber has been shown through TEM micrograph in Figure 2b.

An electrospun nanofiber of blends of polyvinyl alcohol (PVA) and TMV/RGD afforded higher cell density of baby hamster kidney (BHK) cells in culture. The enhanced cell adhesion and spreading and the formation of F-actin filaments were observed more on PVA/TMV/RGD nanofiber than PVA and PVA/TMV nanofibrous substrates, which were noticed in SEM micrographs (Figure 3a). The resulting nanofiber provided electroactivity and topographical cues to the neural cells and was reported to augment the length of neurites, increase the population of cells, and lead the cellular bipolar morphology more than TMV-based non-conductive nanofibers [71]. Korehei et al. [57] produced a virus incorporated nanofiber by electrospinning the blends of polyethylene oxide (PEO) and T4 bacteriophage suspension. The SEM measurement showed that T4 bacteriophages were protected from severe electrospinning conditions, as they were concentrated within the alginate capsule, as can be seen in Figure 2c. The alginate beads containing phages were found to exhibit smooth rounded surfaces. The size of the electrospun nanofiber of PEO/alginate/T4 bacteriophages had an average diameter 500 ± 100 nm (Figure 2d). According to TEM measurement, the capsules of T4 bacteriophages were distributed without uniformity throughout the fiber matrix (Figure 2e).

The induced pluripotent stem cells (iPSCs) are a promising cell source, which can rise to different cell lineages and construct a well-developed functional bone substitute. However, there is a challenge in osteoblastic differentiation of iPSCs by a conventional biomaterial, as it may form teratoma, raising health risks. Wang et al. [75] demonstrated that a phage (M13)-based nanofiber with four different signal peptides aiming to influence stem cell fate could be potentially utilized for bone tissue regeneration. The aligned nanofibrous matrix provided biochemical and biophysical cues to the cells promoting differentiation of iPSCs into osteoblasts. Among the signal peptides investigated by them were two adhesive-directing peptides RGD and RGD/PHSRN from fibronectin, and the remaining two included ALKRQGRTLYGFGG and KIPKASSVPTELSAISTLYL sequences, which are the growth regulating peptides from osteogenic growth factor and bone morphogenetic protein 2 (BMP2), respectively. The layer-by-layer technique produced a phage-assembled nanofiber assuming nanotopography of the ridge-groove structure, wherein the phage strands were parallel to each other but separated by grooves. Due to this specialized nanotopography of material, the occurrence of controlled osteoblastic differentiation was observed, even in the absence of osteogenic supplements. The research group reported that the phages displaying growth factor signal peptides could express a higher level of alkaline phosphatase (ALP) than the phages having adhesive signal peptides on the surface. The in vivo animal studies disclosed that iPSCs alone caused teratoma after one month of cells injection into nude mice, whereas the group of iPSC-derived osteoblasts did not. Cigognini and co-workers engineered an electrospun nanofibrous scaffold dispersing phage-displayed bone marrow homing peptide (BMHP (1) with sequence PFSSTKT and investigated its potential use in a chronically damaged spinal cord, which was caused by the degeneration of the central nervous system [72]. The clinical data showed that the biomimetic material enhanced nervous tissue regeneration, owing to porosity and nanostructure at the microscopic level, and improved the locomotor recovery of experimental rats. From Figure 3b,c, the histological analyses revealed that the scaffold affected increased cellular infiltration and axonal regeneration after eight weeks of experimental investigation in rats. They found a higher synthesis of growth-associated protein 43 (GAP-43) in engineered scaffold-treated animal when compared to saline and control groups with spinal cord defects. Our research group has explored electrospun nanofibrous matrices of PLGA containing self-assembled M13 bacteriophages along with additives RGD and graphene oxide to show enhanced differentiation of fibroblasts, smooth muscle cells, and myoblasts [76–80].

2.3. Virus-Incorporated 3D Hydrogel Scaffolds

Cell-laden-agarose hydrogel was prepared by dispersing genetically engineered rod-shaped PVX nanoparticles, which present functional RGD peptides and mineralization inducing peptides (MIP) on its surface, into agarose polymeric components [81]. Luckanagul et al. [59] prepared freeze-dried solid foam of a porous alginate hydrogel (PAH) comprising TMV. The incorporation of TMV nanoparticles resulted in large sized and well-defined spherical pores (100–500 μm) in TMV/PAH, analyzed by Field Emission Scanning Electron Microscope image (Figure 2g).

The PVX nanoparticles adopted a nano-filamentous structural network on coated surfaces. Exploiting the synergistic effect of both peptides, the PVX nanoparticles in hydrogel expressed significant cell adhesion as well as hydroxyapatite nucleation. Confirmed by SEM and immunostaining characterizations, it was further reported that the viral nanoparticles could be preserved over 14 days in hydrogel and the whole biomaterial could act as a promising bone substitute. Maturavongsadit et al. [82] developed an injectable TMV based hydrogel under physiological conditions to imitate a cartilage microenvironment. The hydrogel was prepared by cross-linking methacrylate hyaluronic acid polymers by cysteine inserted TMV mutants involving in situ Michael addition reaction. The hydrogel was reported to influence enhancement of cartilage tissue regeneration by promoting chondrogenesis via up-regulation of BMP-2. The interaction of TMV nanoparticles with the cells assisted the high-level expression of BMP-2, an effective inducer of differentiation of mesenchymal stem cells into chondrocytes.

Luckanagul et al. [58] investigated the performance of functional TMV-RGD-blended alginate hydrogel nanocomposites to treat in vivo cranial bone defects in Sprague-Dawley rats. The TMV-functionalized sponge-like hydrogel supported cell localization without triggering any systemic toxicity in the defect area, and hence was envisaged as an active bone replacement biomimetic material in the future direction of reconstructive orthopedic surgery. Shah et al. [83] studied an integrated co-assembled hydrogel system of peptide amphiphiles, in which M13 phage coat protein was modified to express a high density of binding peptide HSNGLPL to combine with transforming growth factor β1 (TGF-β1). The research group found an enhancement in articular cartilage tissue regeneration in a rabbit model with a full-thickness chondral defect because of the slower release of growth factor from the hydrogel, with approximately 60% of cumulative drug release at 72 h, which supported the viability and chondrogenic differentiation of mesenchymal stem cells in the defective site. The in vivo evaluation of the rabbit model showed that the hydrogel treated animal group had no apparent symptoms of chronic inflammatory responses after four weeks. All of the rabbits appeared with a full range of motion in their knees at the end of the investigation. Caprini et al. [84] isolated M13 phage-displayed peptide, KLPGWSG, which could adhere on the surface of murine neural stem cells. Subsequently, the research group designed a self-assembled KLPGWSG-based biomimetic hydrogel with tunable visco-elastic properties for the regeneration of the degenerated nervous system. It was discovered that the phage-based hydrogel favored cell adherence and differentiation in the range of 100–1000 Pa, suggesting that the elastic property of the matrix is a crucial factor in tissue regeneration.

2.4. Virus-Incorporated Organic-Inorganic Hybrid Nanocomposites

The interaction of organic and inorganic biocompatible materials in scaffolds bring about significant impacts in biomedical applications. Cementum, classified as a hard mineralized tissue, surrounds tooth root and has been a part of periodontal tissue that connects the tooth to the bone. When an infectious biofilm adheres to tooth root, triggering periodontal disease, the tooth loss is more enhanced. Gungormus et al. [85] demonstrated amelogenin-derived M13 phage-displayed peptide controlled hydroxyapatite biomineralization for dental tissue regeneration. It was reported that Amelogenin directed hydroxyapatite to form a protein matrix during the formation of enamel. Hence, the research group synthesized the cementomimetic material by applying an aqueous solution of the amelogenin-displayed peptide on the human demineralized root surface to form a layer, which was subsequently immersed into the solution of calcium and phosphate ions. Ramaraju et al. [86] isolated M13 phage-displayed peptides to design a dual functional apatite-coated film for effective bone tissue regeneration. They reported that one peptide sequence of the phage, VTKHLNQISQSY, had mineral (apatite) binding affinity with 25% hydrophobicity, whereas another peptide, DPIYALSWSGMA, had cell binding affinity with 50% hydrophobicity. Also, they discovered that the dual functional apatite-based biomaterial could stimulate the adhesion strength of human bone marrow stromal cells (hMSC) and subsequently increase cell proliferation and differentiation. Due to the mineral binding affinity, the film provided a platform for the adherence of osteogenic cells with osteoconductive and osteoinductive signals. Further, the biomimetic nanocomposite showed a greater extent of proliferation of hMSCs with an elevated level of Runx2 expression when compared to biomimetic apatite without functional peptides.

Wang et al. [58] prepared a 3D-printed biomimetic nanofiber with M13 phage-displayed RGD peptides residing in the pores of the scaffold to enhance bone tissue regeneration. The nanocomposite consisted of hydroxyapatite and tri-calcium phosphate showing an ordered pattern with interconnected micro and macro scale pores, which are shown in the TEM micrograph (Figure 2f). The research group implanted a MSC-seeded biomimetic scaffold into a rat radial bone defect and discovered that the order of regeneration was found as follows: scaffold filled with modified phages > scaffolds filled with wild-type phages > pure scaffold. He et al. [60] carried out a similar kind of research work, genetically modifying M13 phage to express oligonucleotide encoding E8 and inducing self-assembly followed by oriental mineralization to synthesize nanofibrous biomimetic materials under the influence of divalent calcium ions. The resulting mineralized phage bundle has been shown in TEM micrography (Figure 2h). Wang et al. [61] used Ca^{2+} ions to prompt self-assembly of fd phage-based anionic nanofibers and transform them into a bundle sheet (Figure 2i), which provided insights into biomineralization and fabrication of organic–inorganic hybrid nanocomposites. The divalent ion-triggered bundle not only acted as a biotemplate but also served as a Ca source to initiate the ordered nucleation and growth of crystalline hydroxyapatite in the biological fluid.

3. Other Formulations of Virus-Based Nanocomposites with Different Biomedical Applications

Apart from tissue regeneration, virus-based biomimetic nanocomposites have traced their steps in different biomedical applications, such as drug delivery, bioimaging, and biosensing. Wang et al. [87] studied f8/8 phage-based polymeric micelles from the self-assembly of polymeric PEG- diacyl lipid conjugates. These polymeric micelles were reported to have cell-targeting ability to release less water-soluble drugs with more specificity towards breast cancer Michigan Cancer Foundation-7 (MCF-7) cells. The non-toxic filamentous f88.4 bacteriophage viral nanoparticle, which was designed to display a single chain antibody, delivered the vectors to the different regions of the brain in albino, laboratory-bred nude mice (BALB/c), and hence was proposed for treating Alzheimer's disease with early diagnosis [88,89]. Wang et al. [90] studied a M13 phage-displayed peptide with the sequence HSQAAVP to target fibroblast growth factor 8b (FGF8b) to treat prostate cancer. The genetic level disturbances in homeostasis between prostate epithelial and stromal cells cause prostate cancer. The major isotherm of fibroblast growth factor 8 is FGF8b, which is associated with the stages of prostate cancer and has been a potential target for appropriate therapies. In this study, the research group revealed that the biomimetic material interrupted FGF8b binding to its receptors, and thereby prevented FGF8b-induced cell proliferation. Furthermore, they reported that the biomaterial had the potential to arrest the cell cycle at the phase G0/G1 by suppressing cyclin D1 and proliferating cell nuclear antigens (PCNA).

Carrico et al. [43] chemically modified the amino acid residues present on the surface of filamentous fd phage coat protein following a two-step transamination/oxime reaction for its potential use in characterizing breast cancer cells. The research group discovered that the chemical reaction selectively targets N-terminal groups but is not involved in transamination of lysine ε-amines. They conjugated PEG polymeric chains to the phage protein in order to reduce immunogenicity, decrease non-specific binding, and increase solubility in the aqueous environment. They observed that there were no significant differences in either absorption or emission properties after fluorophores were labeled with polymer conjugated phages. Fan et al. [91] isolated cyclic peptide CAGALCY from T7 phage nanoparticles in order to target the pial microvasculature of the brain and inhibit platelet adhesion. The presence of the bulky hydrophobic core, two cysteine residues at each end, and the tyrosine residue at the carboxy terminus are considered as remarkable features for selectively binding the brain microvasculature. When pharmacokinetic properties were assessed, the non-filamentous phage, T7, showed a fast clearance rate from the blood with a half-life of 12 min, whereas the filamentous phages M13 and fUSE5 had longer half-lives of 7 h and 9 h, respectively. To identify the specificity of the T7 phage-displayed peptide, they determined selectivity indices using plaque assay for various organs of mice, including lung, liver, brain, kidney, colon, small intestine, and large intestine. The characterization results exposed that T7 displayed peptide resided (accumulated) in the brain, with a selectivity index of 1000, whereas other organs possessed low specificity for the peptide, with selectivity indices less than 50.

Bean et al. [92] prepared a bacteriophage K (ΦK) by incorporating the virus into a photo cross-linked hyaluronic acid methacrylate (HAMA)-based hydrogel that resulted in a material with antimicrobial properties. The presence of two zinc finger genomes ($CX_2CX_{22}CX_2C$ and $CX_2CX_{23}CX_2C$) in the virus caused it to be virulent against a wide range of infective Staphylococci. The secretion of hyaluronidase enzyme-mediated *S. aureus* sensitizes HAMA and triggered degradation of the hydrogel, facilitating the release of ΦK at a sustained level to inhibit bacterial growth effectively. This stimuli-responsive hydrogel was shown to reduce pain, promote cell migration and tissue hydration in the wound site, and was suggested for the application of dermal tissue regeneration. Schmidt et al. [93] identified two different adenovirus phage-displayed peptides QTRFLLH and VPTQSSG to target neural precursor cells (NPC) in the hippocampal dentate gyrus of adult mice through adenovirus-mediated gene transfer. The peptides were found to be strongly internalized into NPCs when the investigated material was added to neurosphere culture containing clusters of neural stem cells.

Kelly KA et al. [94] isolated high-throughput fluorochrome-labeled M13 phage particles (Ph.D. C7C library) to rapidly identify ligands of biological interest in vivo using secreted protein acidic and rich in cysteine (SPARC) molecules and vascular cell adhesion molecules-1 (VCAM-1). The engineered phage particles led to higher sensitivity with an attachment of 800 fluorophores per phage. Wan et al. [37] developed an f8/8 phage-based biosensor exploiting magnetoelastic wireless detection system. The genetically modified phage-expressed peptide sequence EPRLSPHS on the surface of the target biological agent, *Bacillus anthracis* spore. The resonance frequency of the sensor decreased gradually depending on the binding agent on the surface. They reported that this affinity-based phage-displayed biosensor exhibited more longevity activity as a diagnostic probe to target numerous agents with more efficiency than antibody-based biosensors.

4. Conclusions and Perspectives

The potential application of virus-incorporated biomimetic nanocomposites in the form of self-assembled nanoparticles, nanofibers, hydrogels, and organic–inorganic hybrids in the field of tissue regeneration has been elucidated in this review. Though virus-based biomaterial has displayed many beneficial properties, there are some issues to be addressed. (1) Many research groups have expressed desired peptides on the surface of phage-based viral nanoparticles exploiting phage libraries. However, whether the number of peptides exhibited by each nanoparticle is the same is questionable. (2) Biodistribution of viral nanoparticles in different organs of animal tissues has been studied by some researchers. Still, a comprehensive study to describe bioavailability must be demonstrated. (3) It has been well documented that viral nanoparticles contribute to the enhancement in tissue regeneration. However, a systematic study is required to explain the phases of tissue regeneration, in which viral nanocomposites contribute more. (4) The viral nanocomposites in the form of polymeric micelles, vesicles, and dendrimers are less formulated and have not been explored enough for the application of tissue regeneration. The following are suggestions for the future of this field. (1) Sophisticated techniques and methodology to quantify the number of peptides expressed on each phage particle. (2) Pharmacokinetic and pharmacodynamics studies to determine the required dosage of viral nanoparticles in each organ type of tissue regeneration. (3) An extensive in vivo animal study to show the influence of viral-based nanocomposites in each phase of tissue regeneration. (4) Successful bioconjugation of viral nanoparticles with amphiphilic polymers or surfactants to design various oil-in-water-type emulsions. We hope that the researchers with interdisciplinary backgrounds will advance the field of tissue regeneration using viral-based biomimetic nanocomposites by considering the problems and the concerned suggestions.

Author Contributions: I.S.R. and D.-W.H. developed the idea and structure of the review article. I.S.R. and C.K. wrote the paper using material supplied by S.-J.S., Y.C.S., and M.S.K. S.-H.H. and J.-W.O. revised and improved the manuscript. J.-W.O. and D.-W.H. supervised the manuscript. All authors have given approval to the final version of the manuscript.

References

1. Shafiee, A.; Atala, A. Tissue engineering: Toward a new era of medicine. *Ann. Rev. Med.* **2017**, *68*, 29–40. [CrossRef] [PubMed]
2. Atala, A. Regenerative medicine strategies. *J. Pediatr. Surg.* **2012**, *47*, 17–28. [CrossRef] [PubMed]
3. Raja, I.S.; Fathima, N.N. Gelatin–cerium oxide nanocomposite for enhanced excisional wound healing. *ACS Appl. Biol. Mater.* **2018**, *1*, 487–495. [CrossRef]
4. Galliot, B.; Crescenzi, M.; Jacinto, A. Trends in tissue repair and regeneration. *Development* **2017**, *144*, 357–364. [CrossRef] [PubMed]
5. Dorati, R.; DeTrizio, A.; Modena, T.; Conti, B.; Benazzo, F.; Gastaldi, G.; Genta, I. Biodegradable scaffolds for bone regeneration combined with drug-delivery systems in osteomyelitis therapy. *Pharmaceuticals* **2017**, *10*, 96. [CrossRef] [PubMed]
6. Ahadian, S.; Khademhosseini, A. Smart scaffolds in tissue regeneration. *Regener. Biomater.* **2018**, *5*, 125–128. [CrossRef] [PubMed]
7. Lu, H.H.; El-Amin, S.F.; Scott, K.D.; Laurencin, C.T. Three-dimensional, bioactive, biodegradable, polymer-bioactive glass composite scaffolds with improved mechanical properties support collagen synthesis and mineralization of human osteoblast-like cells in vitro. *J. Biomed. Mater. Res. A* **2003**, *64*, 465–474. [CrossRef] [PubMed]
8. Shibuya, N.; Jupiter, D.C. Bone graft substitute: Allograft and xenograft. *Clin. Podiatr. Med. Surg.* **2015**, *32*, 21–34. [CrossRef] [PubMed]
9. Ferrara, J.L.M.; Levine, J.E.; Reddy, P.; Holler, E. Graft-versus-host disease. *Lancet* **2009**, *373*, 1550–1561. [CrossRef]
10. Higuchi, A.; Ling, Q.-D.; Chang, Y.; Hsu, S.-T.; Umezawa, A. Physical cues of biomaterials guide stem cell differentiation fate. *Chem. Rev.* **2013**, *113*, 3297–3328. [CrossRef]
11. Kular, J.K.; Basu, S.; Sharma, R.I. The extracellular matrix: Structure, composition, age-related differences, tools for analysis and applications for tissue engineering. *J. Tissue Eng.* **2014**, *5*, 2041731414557112. [CrossRef] [PubMed]
12. Taylor, D.A.; Sampaio, L.C.; Ferdous, Z.; Gobin, A.S.; Taite, L.J. Decellularized matrices in regenerative medicine. *Acta Biomater.* **2018**, *74*, 74–89. [CrossRef] [PubMed]
13. Zhao, P.; Gu, H.; Mi, H.; Rao, C.; Fu, J.; Turng, L.S. Fabrication of scaffolds in tissue engineering: A review. *Front. Mech. Eng.* **2018**, *13*, 107–119. [CrossRef]
14. Sultana, N. Natural-synthetic polymer blend composite scaffold for bone tissue engineering: Study of in vitro degradation and protein adsorption. *Appl. Mech. Mater.* **2014**, *554*, 42–46. [CrossRef]
15. Doulabi, A.H.; Mequanint, K.; Mohammadi, H. Blends and nanocomposite biomaterials for articular cartilage tissue engineering. *Materials* **2014**, *7*, 5327–5355. [CrossRef] [PubMed]
16. Gerhardt, L.C.; Boccaccini, A.R. Bioactive glass and glass-ceramic scaffolds for bone tissue engineering. *Materials* **2010**, *3*, 3867–3910. [CrossRef] [PubMed]
17. Zeng, Y.; Hoque, J.; Varghese, S. Biomaterial-assisted local and systemic delivery of bioactive agents for bone repair. *Acta Biomater.* **2019**, *93*, 152–168. [CrossRef]

18. Monteiro, N.; Martins, A.; Pires, R.; Faria, S.; Fonseca, N.A.; Moreira, J.N.; Reis, R.L.; Neves, N.M. Immobilization of bioactive factor-loaded liposomes on the surface of electrospun nanofibers targeting tissue engineering. *Biomater. Sci.* **2014**, *2*, 1195–1209. [CrossRef]
19. Raja, I.S.; Fathima, N.N. A gelatin based antioxidant enriched biomaterial by grafting and saturation: Towards sustained drug delivery from antioxidant matrix. *Colloids Surf. B Biointerfaces* **2015**, *128*, 537–543. [CrossRef]
20. Hasan, A.; Morshed, M.; Memic, A.; Hassan, S.; Webster, T.J.; Marei, H.E. Nanoparticles in tissue engineering: Applications, challenges and prospects. *Int. J. Nanomed.* **2018**, *13*, 5637–5655. [CrossRef]
21. Mieszawska, A.J.; Llamas, J.G.; Vaiana, C.A.; Kadakia, M.P.; Naik, R.R.; Kaplan, D.L. Clay enriched silk biomaterials for bone formation. *Acta Biomater.* **2011**, *7*, 3036–3041. [CrossRef] [PubMed]
22. Shin, Y.C.; Song, S.J.; Jeong, S.J.; Kim, B.; Kwon, I.K.; Hong, S.W.; Oh, J.W.; Han, D.W. Graphene-based nanocomposites as promising options for hard tissue regeneration. *Adv. Exp. Med. Biol.* **2018**, *1078*, 103–117. [PubMed]
23. Shin, Y.C.; Lee, J.H.; Kim, M.J.; Hong, S.W.; Kim, B.; Hyun, J.K.; Choi, Y.S.; Park, J.C.; Han, D.W. Stimulating effect of graphene oxide on myogenesis of C2C12 myoblasts on RGD peptide-decorated PLGA nanofiber matrices. *J. Biol. Eng.* **2015**, *9*, 22. [CrossRef] [PubMed]
24. Shin, Y.C.; Kim, J.; Kim, S.E.; Song, S.J.; Hong, S.W.; Oh, J.W.; Lee, J.; Park, J.C.; Hyon, S.H.; Han, D.W. RGD peptide and graphene oxide co-functionalized PLGA nanofiber scaffolds for vascular tissue engineering. *Regen. Biomater.* **2017**, *4*, 159–166. [CrossRef] [PubMed]
25. Saidykhan, L.; Abu Bakar, M.Z.; Rukayadi, Y.; Kura, A.U.; Latifah, S.Y. Development of nanoantibiotic delivery system using cockle shell-derived aragonite nanoparticles for treatment of osteomyelitis. *Int. J. Nanomed.* **2016**, *11*, 661–673. [CrossRef] [PubMed]
26. Kizilel, S.; Scavone, A.; Liu, X.; Nothias, J.M.; Ostrega, D.; Witkowski, P.; Millis, M. Encapsulation of pancreatic islets within nano-thin functional polyethylene glycol coatings for enhanced insulin secretion. *Tissue Eng. A* **2010**, *16*, 2217–2228. [CrossRef]
27. Yoon, H.; Lee, J.-S.; Yim, H.; Kim, G.; Chun, W. Development of cell-laden 3D scaffolds for efficient engineered skin substitutes by collagen gelation. *RSC Adv.* **2016**, *6*, 21439–21447. [CrossRef]
28. Mohan, K.; Weiss, G.A. Chemically modifying viruses for diverse applications. *ACS Chem. Biol.* **2016**, *11*, 1167–1179. [CrossRef]
29. Manoukian, O.S.; Matta, R.; Letendre, J.; Collins, P.; Mazzocca, A.D.; Kumbar, S.G. Electrospun nanofiber scaffolds and their hydrogel composites for the engineering and regeneration of soft tissues. *Methods Mol. Biol.* **2017**, *1570*, 261–278.
30. Shin, Y.C.; Lee, J.H.; Kim, M.J.; Park, J.H.; Kim, S.E.; Kim, J.S.; Oh, J.W.; Han, D.W. Biomimetic hybrid nanofiber sheets composed of RGD peptide-decorated PLGA as cell-adhesive substrates. *J. Funct. Biomater.* **2015**, *6*, 367–378. [CrossRef]
31. Steinmetz, N.F. Viral nanoparticles as platforms for next-generation therapeutics and imaging devices. *Nanomedicine* **2010**, *6*, 634–641. [CrossRef] [PubMed]
32. Howard-Varona, C.; Hargreaves, K.R.; Abedon, S.T.; Sullivan, M.B. Lysogeny in nature: Mechanisms, impact and ecology of temperate phages. *ISME J.* **2017**, *11*, 1511. [CrossRef] [PubMed]
33. Sunderland, K.S.; Yang, M.; Mao, C. Phage-enabled nanomedicine: From probes to therapeutics in precision medicine. *Angew. Chem. Int. Ed.* **2017**, *56*, 1964–1992. [CrossRef] [PubMed]
34. Kale, A.; Bao, Y.; Zhou, Z.; Prevelige, P.E.; Gupta, A. Directed self-assembly of CdS quantum dots on bacteriophage P22 coat protein templates. *Nanotechnology* **2013**, *24*, 045603. [CrossRef] [PubMed]
35. Mohan, K.; Weiss, G.A. Dual genetically encoded phage-displayed ligands. *Anal. Biochem.* **2014**, *453*, 1–3. [CrossRef] [PubMed]
36. Miller-Ensminger, T.; Garretto, A.; Brenner, J.; Thomas-White, K.; Zambom, A.; Wolfe, A.J.; Putonti, C. Bacteriophages of the urinary microbiome. *J. Bacteriol.* **2018**, *200*, e00738–17. [CrossRef] [PubMed]
37. Cao, B.; Li, Y.; Yang, T.; Bao, Q.; Yang, M.; Mao, C. Bacteriophage-based biomaterials for tissue regeneration. *Adv. Drug Deliv. Rev.* **2018**. [CrossRef]
38. Wan, J.; Shu, H.; Huang, S.; Fiebor, B.; Chen, I.; Petrenko, V.A.; Chin, B.A. Phage-based magnetoelastic wireless biosensors for detecting bacillus Anthracis spores. *IEEE Sens. J.* **2007**, *7*, 470–477. [CrossRef]
39. Smith, G.P. Filamentous fusion phage: Novel expression vectors that display cloned antigens on the virion surface. *Science* **1985**, *228*, 1315–1317. [CrossRef]

40. Williams, K.A.; Glibowicka, M.; Li, Z.; Li, H.; Khan, A.R.; Chen, Y.M.; Wang, J.; Marvin, D.A.; Deber, C.M. Packing of coat protein amphipathic and transmembrane helices in filamentous bacteriophage M13: Role of small residues in protein oligomerization. *J. Mol. Biol.* **1995**, *252*, 6–14. [CrossRef]
41. Cross, T.A.; Tsang, P.; Opella, S.J. Comparison of protein and deoxyribonucleic acid backbone structures in fd and Pf1 bacteriophages. *Biochemistry* **1983**, *22*, 721–726. [CrossRef] [PubMed]
42. Xu, J.; Dayan, N.; Goldbourt, A.; Xiang, Y. Cryo-electron microscopy structure of the filamentous bacteriophage IKe. *Proc. Natl. Acad. Sci. USA* **2019**, *116*, 5493–5498. [CrossRef] [PubMed]
43. Wang, Y.; Ju, Z.; Cao, B.; Gao, X.; Zhu, Y.; Qiu, P.; Xu, H.; Pan, P.; Bao, H.; Wang, L.; et al. Ultrasensitive rapid detection of human serum antibody biomarkers by biomarker-capturing viral nanofibers. *ACS Nano* **2015**, *9*, 4475–4483. [CrossRef] [PubMed]
44. Carrico, Z.M.; Farkas, M.E.; Zhou, Y.; Hsiao, S.C.; Marks, J.D.; Chokhawala, H.; Clark, D.S.; Francis, M.B. N-Terminal labeling of filamentous phage to create cancer marker imaging agents. *ACS Nano* **2012**, *6*, 6675–6680. [CrossRef] [PubMed]
45. Tian, F.; Tsao, M.-L.; Schultz, P.G. A phage display system with unnatural amino acids. *J. Am. Chem. Soc.* **2004**, *126*, 15962–15963. [CrossRef]
46. Li, K.; Chen, Y.; Li, S.; Nguyen, H.G.; Niu, Z.; You, S.; Mello, C.M.; Lu, X.; Wang, Q. Chemical modification of M13 bacteriophage and its application in cancer cell imaging. *Bioconjug. Chem.* **2010**, *21*, 1369–1377. [CrossRef]
47. Sandman, K.E.; Benner, J.S.; Noren, C.J. Phage display of selenopeptides. *J. Am. Chem. Soc.* **2000**, *122*, 960–961. [CrossRef]
48. Jaworski, J.W.; Raorane, D.; Huh, J.H.; Majumdar, A.; Lee, S.-W. Evolutionary screening of biomimetic coatings for selective detection of explosives. *Langmuir* **2008**, *24*, 4938–4943. [CrossRef]
49. Cho, W.; Fowler, J.D.; Furst, E.M. Targeted binding of the M13 bacteriophage to thiamethoxam organic crystals. *Langmuir* **2012**, *28*, 6013–6020. [CrossRef]
50. Li, Y.; Cao, B.; Yang, M.; Zhu, Y.; Suh, J.; Mao, C. Identification of novel short $BaTiO_3$-binding/nucleating peptides for phage-templated in situ synthesis of $BaTiO_3$ polycrystalline nanowires at room temperature. *ACS Appl. Mater. Interfaces* **2016**, *8*, 30714–30721. [CrossRef]
51. Sanghvi, A.B.; Miller, K.P.; Belcher, A.M.; Schmidt, C.E. Biomaterials functionalization using a novel peptide that selectively binds to a conducting polymer. *Nat. Mater.* **2005**, *4*, 496–502. [CrossRef] [PubMed]
52. McGuire, M.J.; Li, S.; Brown, K.C. Biopanning of phage displayed peptide libraries for the isolation of cell-specific ligands. *Methods Mol. Biol.* **2009**, *504*, 291–321. [PubMed]
53. Cao, B.; Mao, C. Identification of microtubule-binding domains on microtubule-associated proteins by major coat phage display technique. *Biomacromolecules* **2009**, *10*, 555–564. [CrossRef] [PubMed]
54. Lee, K.L.; Uhde-Holzem, K.; Fischer, R.; Commandeur, U.; Steinmetz, N.F. Genetic engineering and chemical conjugation of potato virus X. *Methods Mol. Biol.* **2014**, *1108*, 3–21. [PubMed]
55. Li, L.; Xu, C.; Zhang, W.; Secundo, F.; Li, C.; Zhang, Z.-P.; Zhang, X.-E.; Li, F. Cargo-compatible encapsulation in virus-based nanoparticles. *Nano Lett.* **2019**, *19*, 2700–2706. [CrossRef] [PubMed]
56. Wu, Y.; Feng, S.; Zan, X.; Lin, Y.; Wang, Q. Aligned electroactive TMV nanofibers as enabling scaffold for neural tissue engineering. *Biomacromolecules* **2015**, *16*, 3466–3472. [CrossRef] [PubMed]
57. Korehei, R.; Kadla, J. Incorporation of T4 bacteriophage in electrospun fibres. *J. Appl. Microbiol.* **2013**, *114*, 1425–1434. [CrossRef] [PubMed]
58. Wang, J.; Yang, M.; Zhu, Y.; Wang, L.; Tomsia, A.P.; Mao, C. Phage nanofibers induce vascularized osteogenesis in 3D printed bone scaffolds. *Adv. Mater.* **2014**, *26*, 4961–4966. [CrossRef] [PubMed]
59. Luckanagul, J.; Lee, L.A.; Nguyen, Q.L.; Sitasuwan, P.; Yang, X.; Shazly, T.; Wang, Q. Porous alginate hydrogel functionalized with virus as three-dimensional scaffolds for bone differentiation. *Biomacromolecules* **2012**, *13*, 3949–3958. [CrossRef]
60. He, T.; Abbineni, G.; Cao, B.; Mao, C. Nanofibrous bio-inorganic hybrid structures formed through self-assembly and oriented mineralization of genetically engineered phage nanofibers. *Small* **2010**, *6*, 2230–2235. [CrossRef]
61. Wang, F.; Cao, B.; Mao, C. Bacteriophage bundles with prealigned Ca^{2+} initiate the oriented nucleation and growth of hydroxylapatite. *Chem. Mater.* **2010**, *22*, 3630–3636. [CrossRef] [PubMed]
62. Sitasuwan, P.; Lee, L.A.; Li, K.; Nguyen, H.G.; Wang, Q. RGD-conjugated rod-like viral nanoparticles on 2D scaffold improve bone differentiation of mesenchymal stem cells. *Front. Chem.* **2014**, *2*, 31. [CrossRef] [PubMed]

63. Luckanagul, J.A.; Metavarayuth, K.; Feng, S.; Maneesaay, P.; Clark, A.Y.; Yang, X.; García, A.J.; Wang, Q. Tobacco mosaic virus functionalized alginate hydrogel scaffolds for bone regeneration in rats with cranial defect. *ACS Biomater. Sci. Eng.* **2016**, *2*, 606–615. [CrossRef]
64. Pi, Y.; Zhang, X.; Shi, J.; Zhu, J.; Chen, W.; Zhang, C.; Gao, W.; Zhou, C.; Ao, Y. Targeted delivery of non-viral vectors to cartilage in vivo using a chondrocyte-homing peptide identified by phage display. *Biomaterials* **2011**, *32*, 6324–6332. [CrossRef] [PubMed]
65. Jarvinen, T.A.; Ruoslahti, E. Molecular changes in the vasculature of injured tissues. *Am. J. Pathol.* **2007**, *171*, 702–711. [CrossRef] [PubMed]
66. Hsu, T.; Mitragotri, S. Delivery of siRNA and other macromolecules into skin and cells using a peptide enhancer. *Proc. Natl. Acad. Sci. USA* **2011**, *108*, 15816–15821. [CrossRef] [PubMed]
67. Samoylova, T.I.; Smith, B.F. Elucidation of muscle-binding peptides by phage display screening. *Muscle Nerve* **1999**, *22*, 460–466. [CrossRef]
68. Sun, Z.; Dai, X.; Li, Y.; Jiang, S.; Lou, G.; Cao, Q.; Hu, R.; Huang, Y.; Su, Z.; Chen, M.; et al. A novel Nogo-66 receptor antagonist peptide promotes neurite regeneration in vitro. *Mol. Cell. Neurosci.* **2016**, *71*, 80–91. [CrossRef]
69. Collett, S.; Torresi, J.; Earnest-Silveira, L.; Christiansen, D.; Elbourne, A.; Ramsland, P.A. Probing and pressing surfaces of hepatitis C virus-like particles. *J. Colloid Interface Sci.* **2019**, *545*, 259–268. [CrossRef]
70. Engel, J.L.; Ardehali, R. Sendai virus based direct cardiac reprogramming: What lies ahead? *Stem Cell Investig.* **2018**, *5*, 37. [CrossRef]
71. Zhao, X.; Lin, Y.; Wang, Q. Virus-based scaffolds for tissue engineering applications. *Wiley Interdiscip. Rev. Nanomed. Nanobiotechnol.* **2015**, *7*, 534–547. [CrossRef] [PubMed]
72. Cigognini, D.; Satta, A.; Colleoni, B.; Silva, D.; Donega, M.; Antonini, S.; Gelain, F. Evaluation of early and late effects into the acute spinal cord injury of an injectable functionalized self-assembling scaffold. *PLoS ONE* **2011**, *6*, e19782. [CrossRef] [PubMed]
73. Kaur, G.; Wang, C.; Sun, J.; Wang, Q. The synergistic effects of multivalent ligand display and nanotopography on osteogenic differentiation of rat bone marrow stem cells. *Biomaterials* **2010**, *31*, 5813–5824. [CrossRef] [PubMed]
74. Shukla, S.; DiFranco, N.A.; Wen, A.M.; Commandeur, U.; Steinmetz, N.F. To target or not to target: Active Vs. passive tumor homing of filamentous nanoparticles based on Potato virus X. *Cell. Mol. Bioeng.* **2015**, *8*, 433–444. [CrossRef] [PubMed]
75. Wang, J.; Wang, L.; Yang, M.; Zhu, Y.; Tomsia, A.; Mao, C. Untangling the effects of peptide sequences and nanotopographies in a biomimetic niche for directed differentiation of iPSCs by assemblies of genetically engineered viral nanofibers. *Nano Lett.* **2014**, *14*, 6850–6856. [CrossRef] [PubMed]
76. Han, J.; Devaraj, V.; Kim, C.; Kim, W.G.; Han, D.W.; Hong, S.W.; Kang, Y.C.; Oh, J.W. Fabrication of self-assembled nanoporous structures from a self-templating M13 bacteriophage. *ACS Appl. Nano Mater.* **2018**, *1*, 2851–2857. [CrossRef]
77. Shin, Y.C.; Lee, J.H.; Jin, L.; Kim, M.J.; Oh, J.W.; Kim, T.W.; Han, D.W. Cell-adhesive RGD peptide-displaying M13 bacteriophage/PLGA nanofiber matrices for growth of fibroblasts. *Biomater. Res.* **2014**, *18*, 14. [CrossRef]
78. Shin, Y.C.; Lee, J.H.; Jin, O.S.; Lee, E.J.; Jin, L.H.; Kim, C.S.; Hong, S.W.; Han, D.W.; Kim, C.; Oh, J.-W. RGD peptide-displaying M13 bacteriophage/PLGA nanofibers as cell-adhesive matrices for smooth muscle cells. *J. Korean Phys. Soc.* **2015**, *66*, 12–16. [CrossRef]
79. Shin, Y.C.; Lee, J.H.; Jin, L.; Kim, M.J.; Kim, C.; Hong, S.W.; Oh, J.W.; Han, D.W. Cell-adhesive matrices composed of RGD peptide-displaying M13 bacteriophage/poly(lactic-co-glycolic acid) nanofibers beneficial to myoblast differentiation. *J. Nanosci. Nanotechnol.* **2015**, *15*, 7907–7912. [CrossRef]
80. Shin, Y.C.; Kim, C.; Song, S.J.; Jun, S.; Kim, C.S.; Hong, S.W.; Hyon, S.H.; Han, D.W.; Oh, J.W. Ternary aligned nanofibers of RGD peptide-displaying M13 bacteriophage/PLGA/graphene oxide for facilitated myogenesis. *Nanotheranostics* **2018**, *2*, 144–156. [CrossRef]
81. Lauria, I.; Dickmeis, C.; Roder, J.; Beckers, M.; Rutten, S.; Lin, Y.Y.; Commandeur, U.; Fischer, H. Engineered Potato virus X nanoparticles support hydroxyapatite nucleation for improved bone tissue replacement. *Acta Biomater.* **2017**, *62*, 317–327. [CrossRef] [PubMed]
82. Maturavongsadit, P.; Luckanagul, J.A.; Metavarayuth, K.; Zhao, X.; Chen, L.; Lin, Y.; Wang, Q. Promotion of in vitro chondrogenesis of mesenchymal stem cells using in situ hyaluronic hydrogel functionalized with rod-like viral nanoparticles. *Biomacromolecules* **2016**, *17*, 1930–1938. [CrossRef] [PubMed]

83. Shah, R.N.; Shah, N.A.; Del Rosario Lim, M.M.; Hsieh, C.; Nuber, G.; Stupp, S.I. Supramolecular design of self-assembling nanofibers for cartilage regeneration. *Proc. Natl. Acad. Sci. USA* **2010**, *107*, 3293–3298. [CrossRef] [PubMed]
84. Caprini, A.; Silva, D.; Zanoni, I.; Cunha, C.; Volonte, C.; Vescovi, A.; Gelain, F. A novel bioactive peptide: Assessing its activity over murine neural stem cells and its potential for neural tissue engineering. *New Biotechnol.* **2013**, *30*, 552–562. [CrossRef] [PubMed]
85. Gungormus, M.; Oren, E.E.; Horst, J.A.; Fong, H.; Hnilova, M.; Somerman, M.J.; Snead, M.L.; Samudrala, R.; Tamerler, C.; Sarikaya, M. Cementomimetics-constructing a cementum-like biomineralized microlayer via amelogenin-derived peptides. *Int. J. Oral Sci.* **2012**, *4*, 69–77. [CrossRef] [PubMed]
86. Ramaraju, H.; Miller, S.J.; Kohn, D.H. Dual-functioning peptides discovered by phage display increase the magnitude and specificity of BMSC attachment to mineralized biomaterials. *Biomaterials* **2017**, *134*, 1–12. [CrossRef]
87. Wang, T.; Petrenko, V.A.; Torchilin, V.P. Optimization of landscape phage fusion protein-modified polymeric PEG-PE micelles for improved breast cancer cell targeting. *J. Nanomed. Nanotechnol.* **2012**, *4*, 008.
88. Frenkel, D.; Solomon, B. Filamentous phage as vector-mediated antibody delivery to the brain. *Proc. Natl. Acad. Sci. USA* **2002**, *99*, 5675–5679. [CrossRef]
89. Enshell-Seijffers, D.; Smelyanski, L.; Gershoni, J.M. The rational design of a 'type 88' genetically stable peptide display vector in the filamentous bacteriophage fd. *Nucleic Acids Res.* **2001**, *29*, e50. [CrossRef]
90. Wang, W.; Chen, X.; Li, T.; Li, Y.; Wang, R.; He, D.; Luo, W.; Li, X.; Wu, X. Screening a phage display library for a novel FGF8b-binding peptide with anti-tumor effect on prostate cancer. *Exp. Cell Res.* **2013**, *319*, 1156–1164. [CrossRef]
91. Fan, X.; Venegas, R.; Fey, R.; van der Heyde, H.; Bernard, M.A.; Lazarides, E.; Woods, C.M. An in vivo approach to structure activity relationship analysis of peptide ligands. *Pharm. Res.* **2007**, *24*, 868–879. [CrossRef] [PubMed]
92. Bean, J.E.; Alves, D.R.; Laabei, M.; Esteban, P.P.; Thet, N.T.; Enright, M.C.; Jenkins, A.T.A. Triggered release of bacteriophage K from agarose/hyaluronan hydrogel matrixes by Staphylococcus aureus virulence factors. *Chem. Mater.* **2014**, *26*, 7201–7208. [CrossRef]
93. Schmidt, A.; Haas, S.J.; Hildebrandt, S.; Scheibe, J.; Eckhoff, B.; Racek, T.; Kempermann, G.; Wree, A.; Putzer, B.M. Selective targeting of adenoviral vectors to neural precursor cells in the hippocampus of adult mice: New prospects for in situ gene therapy. *Stem Cells* **2007**, *25*, 2910–2918. [CrossRef] [PubMed]
94. Kelly, K.A.; Waterman, P.; Weissleder, R. In vivo imaging of molecularly targeted phage. *Neoplasia* **2006**, *8*, 1011–1018. [CrossRef] [PubMed]

12

Functional RNA Structures in the 3′UTR of Tick-Borne, Insect-Specific and No-Known-Vector Flaviviruses

Roman Ochsenreiter [1], Ivo L. Hofacker [1,2] and Michael T. Wolfinger [1,2,*]

1 Department of Theoretical Chemistry, University of Vienna, Währingerstraße 17, 1090 Vienna, Austria; romanoch@tbi.univie.ac.at (R.O.); ivo@tbi.univie.ac.at (I.L.H.)
2 Research Group BCB, Faculty of Computer Science, University of Vienna, Währingerstraße 29, 1090 Vienna, Austria
* Correspondence: michael.wolfinger@univie.ac.at

Abstract: Untranslated regions (UTRs) of flaviviruses contain a large number of RNA structural elements involved in mediating the viral life cycle, including cyclisation, replication, and encapsidation. Here we report on a comparative genomics approach to characterize evolutionarily conserved RNAs in the 3′UTR of tick-borne, insect-specific and no-known-vector flaviviruses in silico. Our data support the wide distribution of previously experimentally characterized exoribonuclease resistant RNAs (xrRNAs) within tick-borne and no-known-vector flaviviruses and provide evidence for the existence of a cascade of duplicated RNA structures within insect-specific flaviviruses. On a broader scale, our findings indicate that viral 3′UTRs represent a flexible scaffold for evolution to come up with novel xrRNAs.

Keywords: flavivirus; non-coding RNA; secondary structure

1. Introduction

Flaviviruses are small, single-stranded positive-sense RNA viruses that are typically transmitted between arthropod vectors and vertebrate hosts. They are endemic in tropic and sub-tropic regions and represent a global health threat, although humans are considered dead end hosts in many cases.

The genus *Flavivirus* within the *Flaviviridae* family comprises more than 70 species, which are organized into four groups, each with a specific host association: Mosquito-borne flaviviruses (MBFVs) and tick-borne flaviviruses (TBFVs) spread between vertebrate (mammals and birds) and invertebrate (mosquitoes and ticks) hosts, whereas insect-specific flaviviruses (ISFVs) replicate specifically in mosquitoes and no-known-vector flaviviruses (NKVs) have only been found in rodents and bats, respectively. This natural host-range-based classification is in good agreement with sequence-based phylogenetic clustering, mainly because all flaviviruses share a common genome organization [1]. Conversely, epidemiology, disease association [2] and transmission cycles [3] are fundamentally different among different flavivirus groups.

Emerging and re-emerging MBFVs such as Dengue virus (DENV), Japanese encephalitis virus (JEV), West Nile virus (WNV), Yellow fever virus (YFV) or Zika virus (ZIKV) are the causative agents of large-scale outbreaks that result in millions of human and veterinary infections every year [4]. Likewise, tick-borne encephalitis virus (TBEV), Powassan virus (POWV) and other members of the tick-borne serocomplex are neuropathogenic agents that cause a large number of infections every year, resulting in a massive incidence increase since the 1970ies [5]. Consequently, much research effort has gone into studying MBFV and TBFV biology, biochemistry and phylogeny [6]. The two remaining groups, ISFVs and NKVs, however, have received limited attention in the research community, mainly because they are generally not associated with human or veterinary disease and therefore are still underrepresented in the literature. The phylogenetic relationship among the four ecological flavivirus groups is shown in Figure 1. Table A1 lists all viral species considered in the present study.

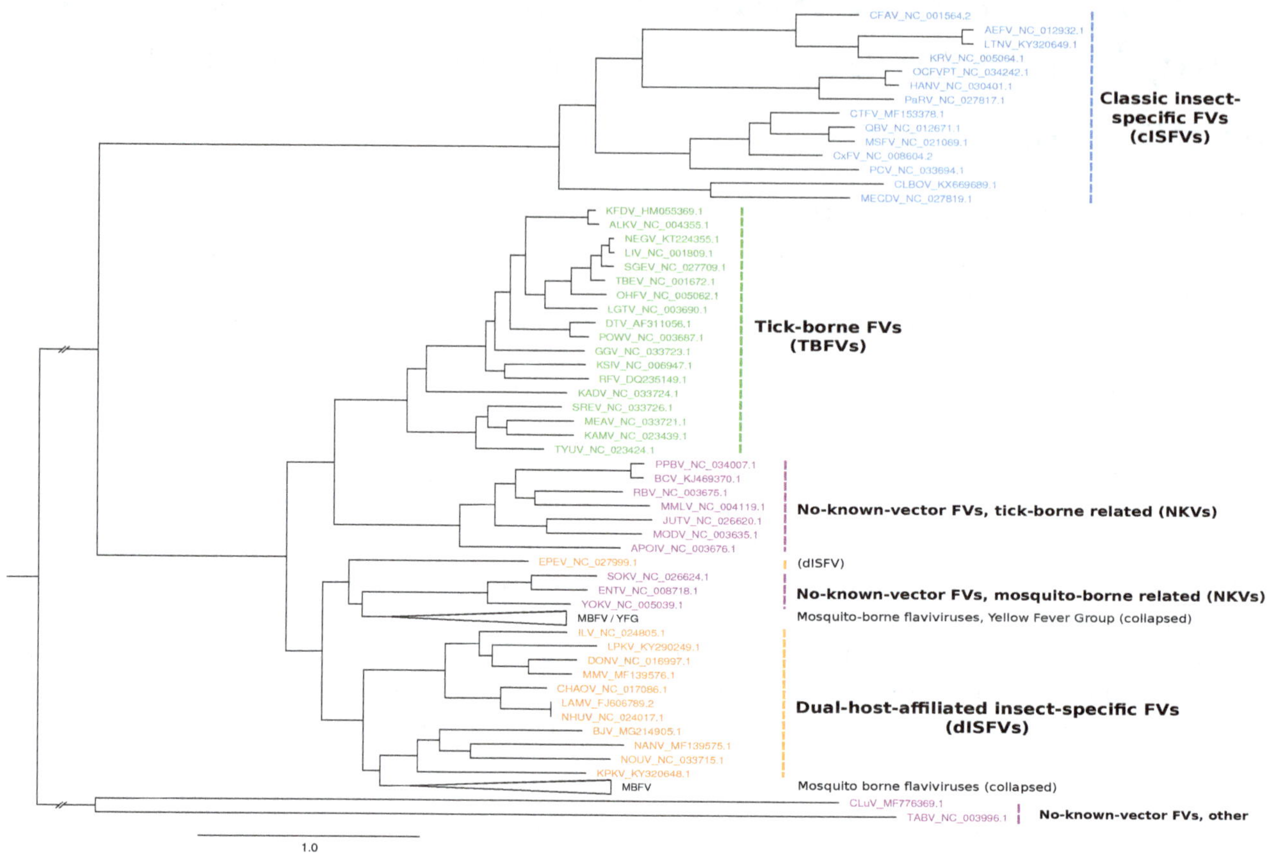

Figure 1. Maximum-likelihood phylogenetic tree of the genus *Flavivirus*, highlighting the major groups ISFVs (blue), dISFVs(orange), TBFVs (green), and NKVs (magenta). The MBFV Yellow Fever virus group (YFG) and the main MBFV branch were not covered in this study and are both collapsed. The tree has been computed from a `MAFFT` alignment of complete polyprotein amino acid sequences with `iq-tree`. Figure rendered with `FigTree`.

TBFVs form a monophyletic group consisting of a single serocomplex, although pathology and clinical manifestations vary among different viruses. They comprise more than a dozen of recognized species and separate into three groups: Mammalian tick-borne flaviviruses (M-TBFV), seabird tick-borne flaviviruses (S-TBFV) and the Kadam virus group. See [7] for a comprehensive review.

ISFVs naturally infect hematophagous *Diptera* and are typically divided into two groups [8]: Classical insect-specific flaviviruses (cISFVs) naturally infect mosquitoes and excursively replicate in mosquito cells in vitro. They form a phylogenetically distinct clade among known flaviviruses, appearing at the root of the MBFV, TBFV and NKV branches. The cISFV group separates into two clades, one associated with *Aedes* spp. mosquitoes and the other associated with *Culex* spp. mosquitoes, respectively [9]. They lack the ability to infect vertebrates and to replicate in vertebrate cell lines and have not been in the research spotlight until very recently. The second group is comprised of arbovirus-related or dual-host affiliated insect-specific flaviviruses (dISFVs), which represent a non-monophyletic group which is phylogenetically and antigenically related to mosquito/vertebrate flaviviruses, although they do not appear to infect vertebrate cells [10]. Insect-specific viruses play a crucial role in the mosquito microbiome and have been shown to modulate the replication of other arboviruses [11]. In this line, they are currently considered as biological control agents and vaccine platforms [12].

NKVs represent an ecologically and phylogenetically diverse set of viruses which have been isolated exclusively from vertebrates (mainly bats and rodents), without evidence for transmission by arthropod vectors. They form a non-monophyletic group among flaviviruses and are typically divided into bat- (B-NKV) and rodent-associated (R-NKV) groups, see Table A1. B-NKVs can be further separated into Entebbe virus group, which is phylogenetically closer to MBFVs, and Rio Bravo virus group, which is a sister clade to TBFVs. Species in the R-NKV group form the Modoc virus group, which is phylogenetically close to the B-NKV Rio Bravo group [13]. While NKVs are poorly characterized they represent a valuable resource to study evolutionary traits related to host-switch capacity mediated by conserved genomic elements.

Conserved RNA Structures Mediate Pathogenesis

Conserved RNA structures in the untranslated regions (UTRs) of RNA viruses are of particular interest because they mediate the viral life cycle by promoting or enhancing replication, as proposed for elements in both 5′UTRs [14] and 3′UTRs [15–18]. Mosquito/vertebrate viruses must operate efficiently in vectors and hosts, phylogenetically distinct organisms with different cellular machineries. This requires a high level of flexibility of viral regulatory elements to evade various antiviral response strategies while assuring proper replication conditions required for maintaining a stable quasispecies population. To achieve this resilience in host adaptation, RNA duplication strategies have been proposed as an evolutionary trait for MBFVs [19]. Tandem RNA structures within DENV 3′UTR are under different selective pressures in alternating hosts, suggesting the idea that duplicated RNA structures differentially evolved to accommodate specific functions in the two hosts [20]. Likewise, there is evidence for evolutionary pressure on maintaining the primary sequence of parts of duplicated RNA elements, as recently shown for flaviviral dumbbell (DB) elements in the context of finding a biophysical model for explaining a possible route for ZIKV-induced neurotropism [21].

Viral RNA genomes are different from procaryotic and eucaryotic mRNA. In addition to coding for and regulating the viral machinery, viral genomic RNA (gRNA) exhibits functional regions that act upon different stages of the viral life cycle. The 10-12kB flaviviral gRNA is capped, but non-polyadenylated and encodes a single open reading frame (ORF). Upon translation, a polyprotein is produced, which is then cleaved by viral and cellular enzymes into structural and non-structural proteins [22]. The ORF is flanked by highly structured untranslated regions (UTRs), which contain evolutionary conserved RNA elements that are crucially related to regulation of the viral life cycle, thereby inducing processes such as genome circularization, viral replication and packaging [23–25].

Upon flavivirus infection, accumulation of both gRNA as well as viral long non-coding RNAs (lncRNAs) is observed. These lncRNAs, which have been referred to as subgenomic flaviviral RNAs (sfRNAs) [26] are stable decay intermediates produced by exploiting the host's mRNA degradation machinery [27] and are associated with viral replication, pathogenesis and cytopathicity [28,29]. The production of sfRNA is induced by partial degradation of viral gRNA by the 5′-3′ exoribonuclease Xrn1, an enzyme associated with the cell's RNA turnover machinery [30,31]. Mechanistically, sfRNAs are generated by stalling Xrn1 at conserved structural elements in the viral 3′UTR, termed xrRNA (exoribonuclease-resistant RNA elements). These structures efficiently stall Xrn1 from progressing through from the 5′ direction, thus protecting the downstream RNA from degradation, while pass-through in the 3′-5′ direction, as required for viral RNA-dependent RNA-polymerase is still possible [32]. In particular, different types of stem-loop (SL) and dumbbell (DB) elements found in many MBFVs and TBFVs have been related to quantitative protection of downstream virus RNA against exoribonuclease degradation [33].

Xrn1 stalling results in dysregulation of cellular function with the aim of promoting viral infections. In this regard, functions of sfRNA in modulating cellular mRNA decay and RNAi pathways [34] as well as modulating anti-viral interferon response [35,36] have been reported.

The genomic architecture of flaviviruses has been extensively studied to understand the molecular principles required for sfRNA production. Chemical and enzymatic probing methods [37], together with x-ray crystallography revealed the 3′UTR structure of the MBFVs WNV [38], YFV [39], DENV [40], Murray Valley encephalitis virus (MVEV) [41], ZIKV [42] and recently different species of the TBFV and NKV groups [33], highlighting the possibility that exoribonuclease resistance might be a pervasive mechanism of the viral world. Interestingly, several conserved RNA structural elements in viral 3′UTRs have been predicted in our group [43–47], some of which have later been attributed to xrRNA functionality [26]. To further expand the set of potential xrRNAs, we report here on a comparative genomics survey aimed at characterization of evolutionary conserved RNA structures in flavivirus 3′UTRs, focusing on TBFVs and the hitherto understudied groups of ISFVs and NKVs. A detailed study on the evolutionary traits of conserved RNAs in MBFV 3′UTRs will be published elsewhere.

2. Materials and Methods

Viral genome data for the present study were obtained from the public National Center for Biotechnology Information (NCBI) `refseq` (https://www.ncbi.nlm.nih.gov/refseq/) and `genbank` (https://www.ncbi.nlm.nih.gov/genbank/) databases on 28 May 2018. We downloaded all complete viral genomes under taxonomy ID 11051 (genus *Flavivirus*) and filtered for TBFV, ISFV and NKV species listed in Table A1. Whenever `refseq` annotation was not available for a species, we selected the longest complete genome from the `genbank` set as representative sequence. In total, the data set is comprised of 86 ISFV, 275 TBFV and 27 NKV isolates, respectively. The number of isolates with available 3′UTR sequence data per species varies between 1 and 167.

2.1. *Phylogeny Reconstruction*

The polyprotein/coding sequence (CDS) regions of most flaviviruses can be aligned consistently, however, UTRs typically show large variance both in length and sequence composition, rendering them ill-suited for phylogeny reconstruction. A phylogeny of all members of the genus *Flavivirus* (Figure 1) was therefore reconstructed via a multiple sequence alignment (MSA) of the nucleotide sequences of the CDS regions only. The MSA was computed with `MAFFT`[48] and subsequent maximum-likelihood tree reconstruction was performed using `iq-tree` [49] using the GTR+F+R7 substitution model.

2.2. *Structural Homology Search with Covariance Models*

The present study is centered around structural homology of RNA elements among phylogenetically narrow subgroups. A straightforward approach to finding novel homologous RNA structures is to search RNA sequence databases with Covariance Models (CMs), i.e., statistical models of RNA structure that extend classic Hidden-Markov-Models (HMMs) to simultaneously represent sequence and secondary structure. CMs, as implemented in the `infernal` package [50] allow for rapid screening of large RNA sequence databases to find even weakly conserved sequence-only or structurally homologous RNAs. We have recently applied this approach to identify novel telomerase RNAs in *Saccharomycetes* [51].

Here, structural multiple sequence alignments of the viral 3′UTR sequences were generated with `locARNA` [52] and CMs were built for known or experimentally verified xrRNAs [33]. All 3′UTR sequences were then screened and novel candidate sequences were added to perform iterative refinement until convergence. Weak sequence conservation of putative xrRNA elements resulted in initially fragile results, indicating that `infernal` default parameters are typically not optimal. Adjusting parameters, in particular disabling both heuristic filtering and local end detection, however, allowed our CMs to find homologs with strongly conserved secondary structures in presence of large sequence deviation from the original sequence the CM was built from. Likewise, `cmsearch` E-values turned out unsuitable for assessing hit quality in case of major sequence divergence. We therefore employed a cutoff approach, requiring a hit to form at least 75% of all base pairs listed in the CM in order to be considered significant.

2.3. *De novo Discovery of Conserved RNA Elements*

Beside characterization of RNAs with homology to known structurally conserved elements, we aimed at identifying novel elements, considering both thermodynamic stability and sequence covariation as evolutionary traits. In this line, `locARNA`-generated structural alignments of full UTR sequences were cut manually into blocks corresponding to conserved secondary structures. Alternatively, we employed `RNALalifold` from the `ViennaRNA` package [53] to compute locally stable secondary structures for aligned UTR sequences. A CM was built for each structure and searched against all flavivirus 3′UTRs, keeping only CMs that scored well multiple times per UTR. The rationale here is that the occurrence of multiple copies hints towards a possible functional role of a structural element, given that the ability of two or more independently evolving sequences to form a common structure is unlikely.

The above approach is implemented as a set of custom Perl and Python scripts for semi-automatic characterization and annotation of conserved RNAs in viral UTR sequences. Internally, these scripts build on the `ViennaRNA` scripting language interface for thermodynamics calculations, the `ViennaNGS` [54] suite for extraction of genomic loci, the `RNAaliSplit` package [55] for splitting alignments into subparts with common consensus structures (i.e., common structures formed by all individual sequences), `R2R` [56] for visualization, and the `ETE3` framework [57] for tree annotation and visualization.

3. Results

Several flaviviruses have previously been studied in great detail, yielding a varied landscape of repeated RNA sequence and structure elements within the 3′UTRs of these viruses, which are likely to have evolved from numerous duplications [19,58,59]. Many of these studies relied on single sequence predictions, which resulted in a good understanding of both structure and genomic position of conserved elements in individual species. A unified picture of homologous RNAs within the 3′UTRs of flaviviruses, however, has not been available.

The comparative approach applied in the present study outperforms single sequence predictions by considering consensus structures formed by all sequences. This allows us not only to confirm previously described RNA structures but also to elucidate hitherto unrecognized tandem repeats in many species. In this line our results can help in understanding the complex evolution of flavivirus 3′UTRs.

3.1. *Construction of Seed Alignments*

Based on recent experimental evidence for the existence of xrRNAs in TBFVs, ISFVs and NKVs, and previously characterized conserved RNA elements in flaviviral 3′UTRs, we built seed alignments for initial CMs, which were then refined iteratively, i.e., subjected to multiple rounds of screening and incorporation of best hits into the CM. Likewise, candidate structures from `RNALalifold` calculations were used as seeds for identification of conserved RNA structures. Figure 2 shows an overview of refined consensus structures for each ecologic group of flaviviruses analyzed here.

The four ecologic groups of flaviviruses show a varied 3′UTR architecture, however, the terminal 3′ stem-loop structure (3′SL, also referred to as 3′ long stable hairpin, 3′LSH) has been shown to be associated with panhandle-formation during virus replication and is therefore present in the terminal region of all flaviviruses [14]. The element is listed in Rfam as RF00185 (Flavivirus 3′UTR cis-acting replication element, Flavi_CRE) and we could use it to consistently identify terminal regions within 3′UTRs. Absence of this element from a UTR sequence is indicative of incomplete or truncated sequence data. The underlying sequences generally form a stable stem-loop structure upon structural alignment and single sequence folding. We built individual 3′SL seed alignments and CMs for each ecologic group, termed T.3SL, N.3SL and I.3SL, respectively.

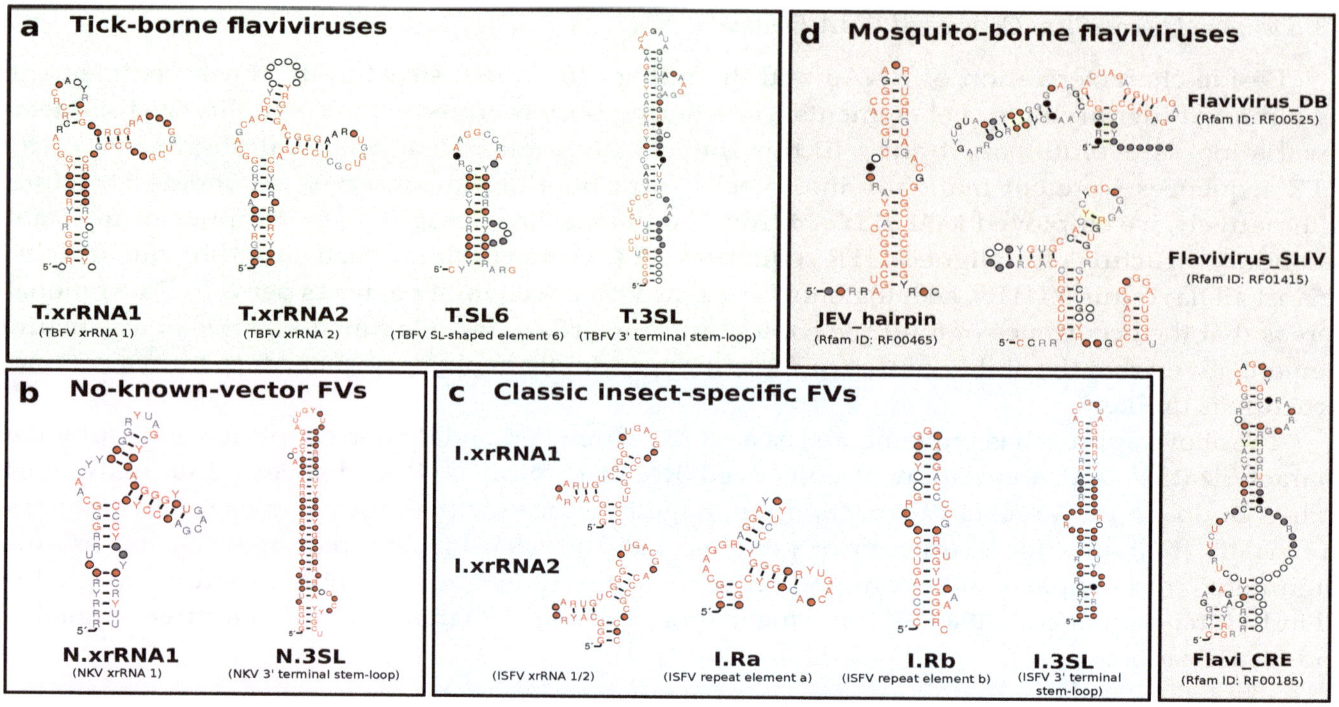

Figure 2. Overview of consensus structures of all CMs used for the annotation of flavivirus 3′UTRs. TBFV, ISFV, and NKV elements were refined from published experimental data (T.xrRNA1/2, I.xrRNA1/2, N.xrRNA) or identified computationally (T.SL6, I.Ra, I.Rb as well as all 3′-terminal stem-loop structures). MBFV elements were obtained from Rfam. Throughout this paper, all CMs are referred to by the name written in bold. References to xrRNA-like structures refer to the generalized xrRNA CM (Section 3.6).

3.2. Tick-Borne Flaviviruses

MacFadden et al. [33] suggested two different exoribonuclease-resistant structures in TBFV 3′UTRs. We used the proposed sequences from TBEV, POWV, Karshi virus (KSIV), Langat virus (LGTV), Louping ill virus (LIV), Omsk hemorrhagic fever virus (OHFV) and Alkhumra hemorrhagic fever virus (ALKV) as templates for a set of initial structural alignments and CMs. These models were then employed to search for high confidence hits within all TBFV 3′UTRs to construct seed alignments of the two exoribonuclease resistant structures in TBFVs, termed T.xrRNA1 and T.xrRNA2. These models allowed us to construct highly specific CMs for both TBFV xrRNAs, which were subsequently used to annotate xrRNA instances in already studied and previously unstudied TBFV species (Figure 3a).

The full structural alignment of the 3′UTRs of selected tick-borne species moreover suggests a short stem-loop element in several species, which is characterized by high sequence heterogeneity but heavily conserved structure supported by multiple covariations. Evidence for this element, termed stem-loop 6 (SL6), has been reported earlier for at least TBEV, LGTV and OHFV [58,59]. We kept this nomenclature and identified the exact position in each TBFV 3′UTR (Figure 3d).

Our data further shows that both TBFV xrRNA CMs (Figure 2a), as well as NKV xrRNA CMs (Figure 2b and Section 3.5) consistently yield plausible hits with a high degree of structure conservation immediately upstream of the strongly conserved terminal stem-loop element. Existence of a Y-shaped element (termed Y1) and putative similarity to NKVs has been proposed earlier based on single sequence structure predictions [58]. Structural `locARNA` alignment and subsequent `RNAalifold` consensus structure prediction indicates strong secondary structure conservation with frequent structure-conserving sequence covariations. Taken together, this suggests good evidence that respective regions in TBFVs harbor a putatively structured and functional xrRNA-like RNA (Y1, Figure 3d).

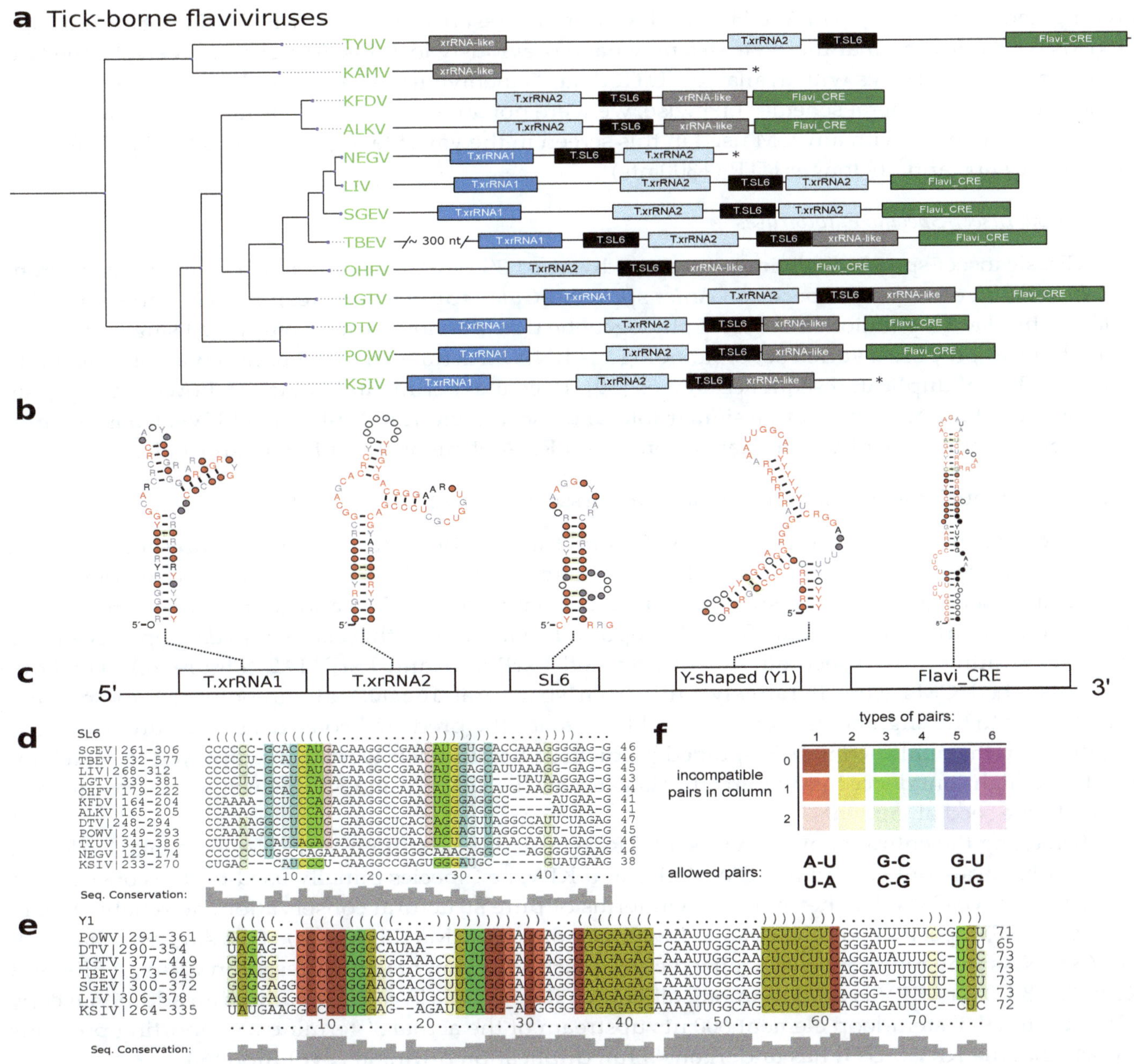

Figure 3. (**a**) Annotated 3′UTRs of TBFVs. The phylogenetic tree on the left has been computed from complete coding sequence nucleotide alignments and corresponds to the TBFV subtree in Figure 1. For each species with available 3′UTR sequence a sketch of the 3′UTR is drawn to scale next to the leaves of the tree. Colored boxes represent conserved RNA structural elements. Identifiers within the boxes indicate the CM which was used to infer homology at this position. Asterisks indicate incomplete 3′UTR sequences. Species without available 3′UTR are not shown. (**b**) Consensus structure plots of CM hits as calculated by *mlocarna*. (**c**) Schematic depiction of the common structural architecture of TBFV 3′UTRs. (**d**,**e**) Structural alignments of elements SL6 and Y1. (**f**) `RNAalifold` coloring scheme for paired columns in alignments. Colors indicate the number of basepair combinations found in pair of columns. Fainter colors indicate that some sequences cannot form a base pair.

Despite the differences in length and sequence composition, the 3′UTRs of most species in the TBFV group share a common architecture. Similar to MBFV SL-elements [19], two copies of xrRNAs are found in almost every species of this ecologic group, generally succeeded by one instance of SL6 and Y1. Likewise, the terminal 3′ stem loop is conserved in all TBFVs and can be reliably annotated by both our CM, T.3SL, and the Rfam Flavi_CRE model, which is used in Figure 3. Among all investigated species, only ALKV, OHFV and Kyasanur forest disease virus (KFDV) do not have a copy of xrRNA1, indicating that these viruses may have previously lost this element. Conversely, the two seabird-associated TBFVs with available 3′UTR data, Tyuleniy virus (TYUV) and Kama virus (KAMV) do not fit into this general scheme. Likewise, we were not able to annotate additional homologous or conserved structures with any CM used in this screen in the variable region of the 3′UTR of TBEV [5], despite the substantially longer UTR (+300 nts).

3.3. *Classic Insect-Specific Flaviviruses*

Classic insect-specific flaviviruses present diverged 3′UTR architectures, which likely result from the association of different species to *Aedes* spp. and *Culex* spp. vectors, respectively, which are also reflected by clade separation in the ISFV phylogenetic tree. Previous studies employed single sequence predictions to propose a varied set of homologous RNA structures in combination with an unusually large number of duplicated sequence signals [59]. Recent experimental evidence, however, suggests the presence of xrRNAs that have a similar fold to those known from MBFVs in cISFVs. Consequently, we set out to independently characterize conserved RNA elements for different subclades.

3.3.1. Exoribonuclease-Resistant RNAs in *Aedes*-Associated cISFVs

MacFadden et al. [33] used SHAPE structure probing to report the presence of two exoribonuclease-resistant RNAs in Cell fusing agent virus (CFAV), and provided evidence for a duplicated set of homologous structures in Aedes flavivirus (AEFV) and Kamiti river virus (KRV). We constructed initial alignments from the reported sequences in this clade in *Aedes* spp. associated viruses, resulting in two seed alignments, termed I.xrRNA1 and I.xrRNA2 (Figure 2c). For both elements, seed CMs were iteratively built from structural `locARNA` alignments. Minor manual adjustments to the alignments were required here, since the predicted consensus structures diverged slightly from the published SHAPE-guided prediction. Both models were then employed to search for additional high confidence hits within other isolates of CFAV, AEFV and KRV which were subsequently added to the seed alignments.

Screening the entire set of flavivirus 3′UTRs revealed that both ISFV xrRNA elements, I.xrRNA1 and I.xrRNA2, are only found in CFAV, AEFV and KRV, i.e., species initially used for the construction of the respective CMs. Furthermore, also in terms of pure structural conservation, no reliable hits in any other ISFV species could be obtained with any of these CMs (Figure 4a). This suggests that both ISFV xrRNAs may represent a specialized class of xrRNA elements only present in CFAV, AEFV and KRV. The 3′UTR of KRV is unique among all known flaviviruses because it harbors an additional copy of the terminal 3′ stem-loop element 600 nts upstream of the actual 3′-terminus, supporting previous reports that the KRV 3′UTR has undergone a full duplication during its evolution [60].

3.3.2. Conserved Structures in *Culex*-Associated cISFVs

The second distinct clade of cISFVs includes Culex flavivirus (CxFV), Quang Binh virus (QBV), Mosquito flavivirus (MSFV), Palm Creek virus (PCV), Culex theileri flavivirus (CTFV) as well as a few other species with only partial genome sequence availability [9] and is associated with *Culex* spp. vectors. An interesting observation in this clade is that no other CM from any of the four ecologic flavivirus groups shows a hit, not even with remote sequence or structure conservation. We therefore set out to produce a a high quality structural alignment of the complete 3′UTRs of CxFV, QBV and MSFV. Consensus structure folding of the full alignment revealed each species to harbor 3–4 repeats of two highly conserved elements supported by multiple co-varying base pairs (Figure 4b,c). We termed these "Repeat element a/b", respectively (Ra and Rb). Both elements, while strongly conserving their folds, show highly variable loop regions as well as weak sequence conservation in the case of the Ra element. Structure conservation and occurrence in multiple copies, as typically seen with other exoribonuclease-stalling elements, hints towards possible functional importance. These results complement earlier reports of sequence repeats in the 3′UTR of CxFV [61] with the identification of a quadruplicated pair of conserved structures.

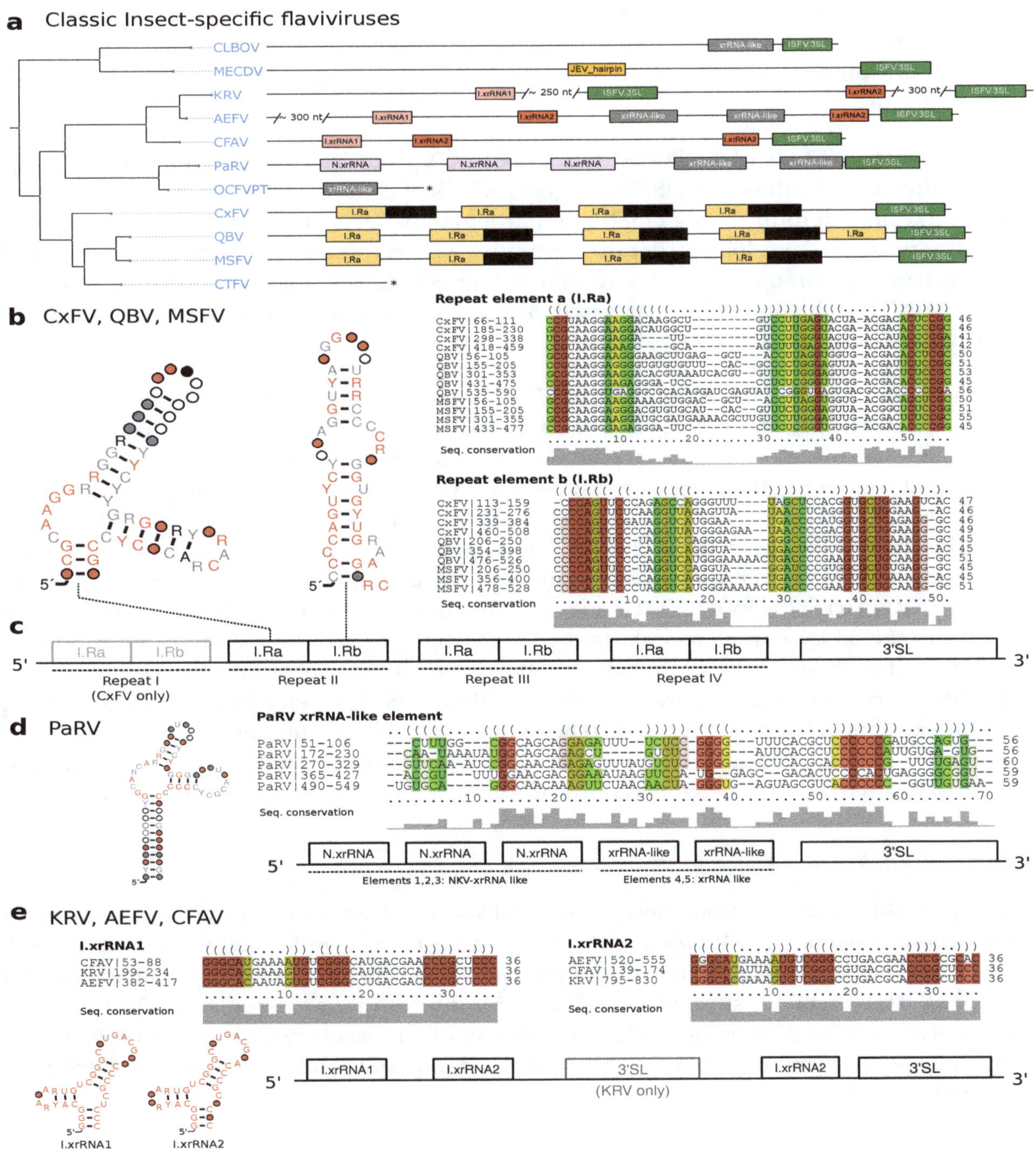

Figure 4. (**a**) Annotated Tree of cISFV 3′UTRs. Asterisks denote incomplete 3′UTR sequences. Species without available 3′UTR are not shown. (**b**) Consensus secondary structure plots and structural alignments of CM hits of Repeat a/b elements in CxFV, QBV, and MSFV. (**c**) Schematic of the common architecture of CxFV, QBV, and MSFV. Element I-IV refers to the respective repeat of elements. (**d**) Consensus structure plot and structural alignment of all CM hits of xrRNA-like elements in PaRV. (**e**) Proposed 3′UTR architecture of KRV, AEFV, and CFAV with consensus structure plots and structural alignments of I.xrRNA1 and I.xrRNA2.

3.3.3. Diverged 3′UTR Architecture in Many cISFVs

Interestingly, a screen of all available CMs in Parramatta River virus (PaRV) revealed five xrRNA-like elements (Figure 4a), with elements 1–3 sharing sequence and structure properties with NKV xrRNAs (Section 3.5), while elements 4 and 5 only conserve N.xrRNAs structure. All five hits can be structurally aligned into a consistent consensus structure (Figure 4d), despite the overall weak sequence consensus.

Conversely, the 3′UTRs of Calbertado virus (CLBOV) and Mercadeo virus (MECDV) appear structurally different from the other cISFVs. A general lack of characteristic CM hits lets these species appear more like an outgroup among cISFVs. In particular, we could only find significant hits for the omnipresent terminal 3′stem-loop structure, a putative xrRNA-like element in CLBOV and a single instance of a structure homologous to the Rfam model RF00465 (Japanese encephalitis virus hairpin structure) in MECDV. Still, limited availability of 3′UTR sequence data renders the characterization of conserved elements and interpretation difficult here.

Our data suggests that the 3′UTRs of cISFVs, in contrast to TBFVs (Section 3.2 and dISFVs (Section 3.4), do not appear to have a consistent architectural organization. In agreement with the cISFV phylogenetic subtree (Figure 1) we constitute three diverged groups with common 3′UTR organization that conform to their respective sub-clades: (i) CFAV-AEFV-KRV, each with two instances of xrRNAs, (ii) CxFV-QBV-MSFV with 3–4 copies of I.Ra/I.Rb elements and (iii) PaRV with 4–5 copies of xrRNA-like structures. Although no full 3′UTR sequences are available for the phylogenetically closest relatives of PaRV, HANV and OCFVPT, an xrRNA-like element in the small available fragment (syntenic to PaRV UTR) of OCFVPT 3′UTR suggests that both viruses might be organized in a similar manner, as supported by earlier reports that these viruses should be classified within the same species [1]. For CLBOV and MECDV, no clear pattern of conserved elements can be identified with our CMs. Both viruses either employ an entirely different class of elements or might not require capability for exoribonuclease stalling at all. The only element shared universally among all cISFVs is the 3′-terminal stem-loop, although cISFVs seem to diverge from other flaviviruses here, indicated by the inability of Rfam model RF00185 (Flavivirus CRE) to reliably annotate any cISFV 3′UTR.

3.4. Dual-Host Affiliated Insect-Specific Flaviviruses

Isolated almost exclusively from mosquitoes, dISFVs do not seem to infect vertebrate cells, despite their phylogenetic proximity to MBFVs (Figure 1). This association is reflected by good hits of the Rfam covariance models RF00525 (Flavivirus DB element) and RF00465 (Japanese encephalitis virus hairpin structure) in all dISFV isolates studied here (Figure 5). Interestingly, we could not find evidence for any sequences or structures homologous to tick-borne or other insect-specific flaviviruses. In this line, our data is in good agreement with the phylogenetic location of these viruses, which share ancestral roots with MBFVs [10].

An unusual species within this group is Ecuador Paraiso Escondido virus (EPEV), which has been isolated from New World sandflies and has been classified as insect-specific virus. EPEV phylogenetically appears at the root of the Entebbe bat virus group (ENTVG), a clade comprised of the three NKVs Entebbe bat virus (ENTV), Sokoluk virus (SOKV) and Yokose virus (YOKV). While all of these viruses contain homologs of conserved stem-loop (SL) and dumbbell (DB) elements found in MBFVs, ENTVG species may have lost their vector dependence [1].

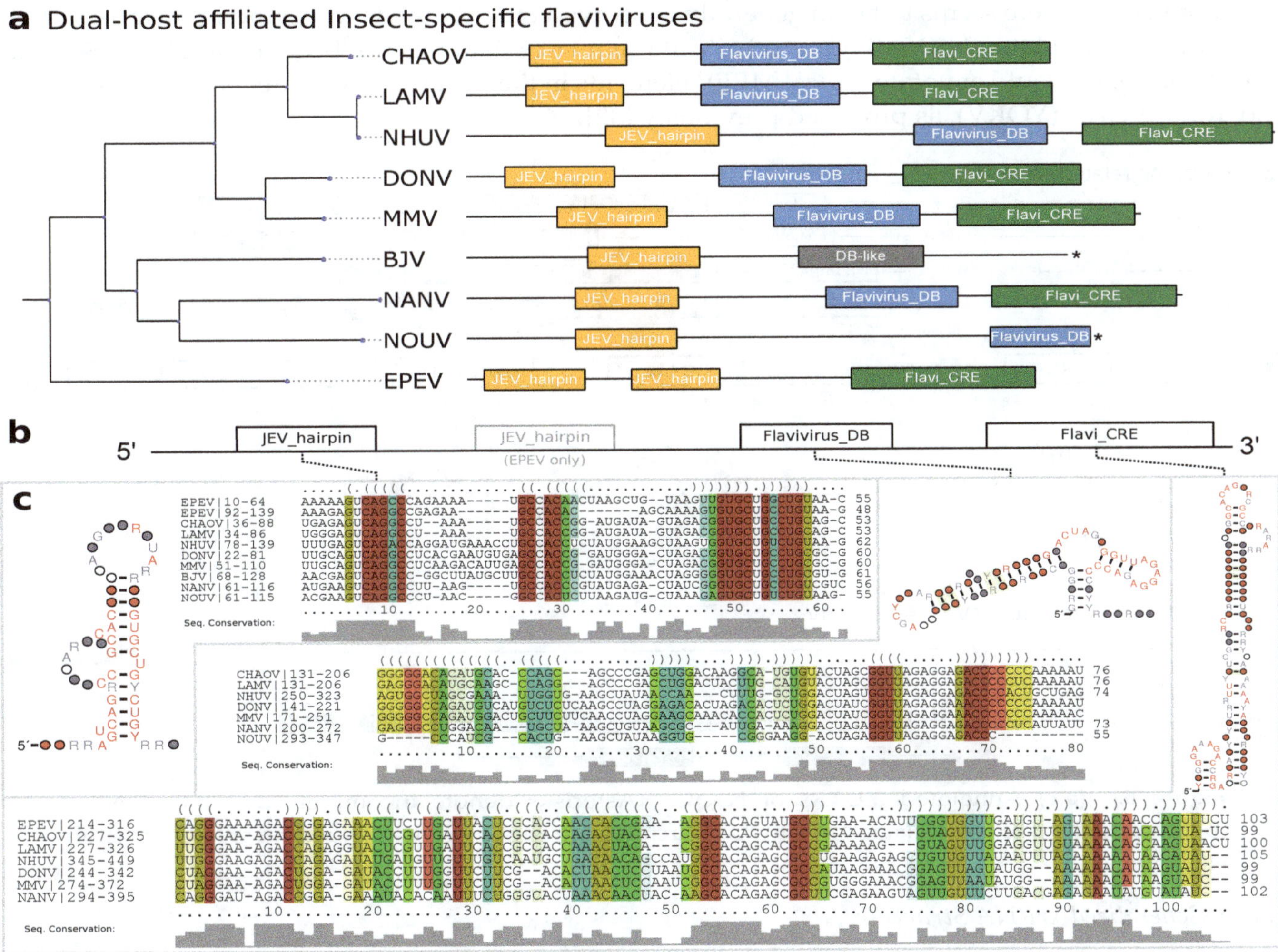

Figure 5. (**a**) Annotated Tree of dISFV 3′UTRs. Asterisks denote incomplete 3′UTR sequences. Species without available 3′UTR are not shown. (**b**) Schematic Architecture of the dISFV 3′UTR. (**c**) Structural alignments and consensus structure plots of dISFV elements.

3.5. No-Known-Vector Flaviviruses

Rather than forming a monophyletic group, the no-known-vector flaviviruses can be separated into two distinct lineages, which are closely related to either TBFVs or MBFVs, respectively (Figure 1). Two additional NKVs, Tamana bat virus (TABV) and Cyclopterus lumpus virus (CLuV), are phylogenetically distant and serve as an outgroup to all flaviviruses. In analogy to the procedure outlined above for TBFVs (Section 3.2) and ISFVs (Section 3.3), we built a CM for experimentally verified xrRNAs in tick-borne related NKVs, termed N.xrRNA.

We found multiple hits of this CM at various loci within the 3′UTRs of tick-borne related NKVs, indicating that these species, in contrast to TBFVs, do not conserve a common 3′UTR architecture (Figure 6a). Surprisingly, we could identify several high-quality hits of the Rfam model RF00525 (Flavivirus DB element), an element typically found in MBFVs, in Rio Bravo virus (RBV), Montana myotis leukoencephalitis virus (MMLV) and Modoc virus (MODV). This is in so far remarkable as there is no evidence for conservation of this element in TBFVs, which phylogenetically cluster with this clade of NKVs. This element might have been introduced by an ancestral recombination event. Alternatively, conservation of an MBFV element in NKVs might be indicative of an association with an unknown vector, in agreement with the hypothesis that vector specificity is mediated by characteristic 3′UTR elements [19].

Conversely, there seems to be no generally conserved 3′UTR architecture among members of the mosquito-borne related NKVs (Figure 6c). While sequence data has not been available for Sokoluk virus (SOKV), we could annotate typical MBFV elements in the next relatives Entebbe bat virus (ENTV) and Yokose virus (YOKV), as proposed previously [32].

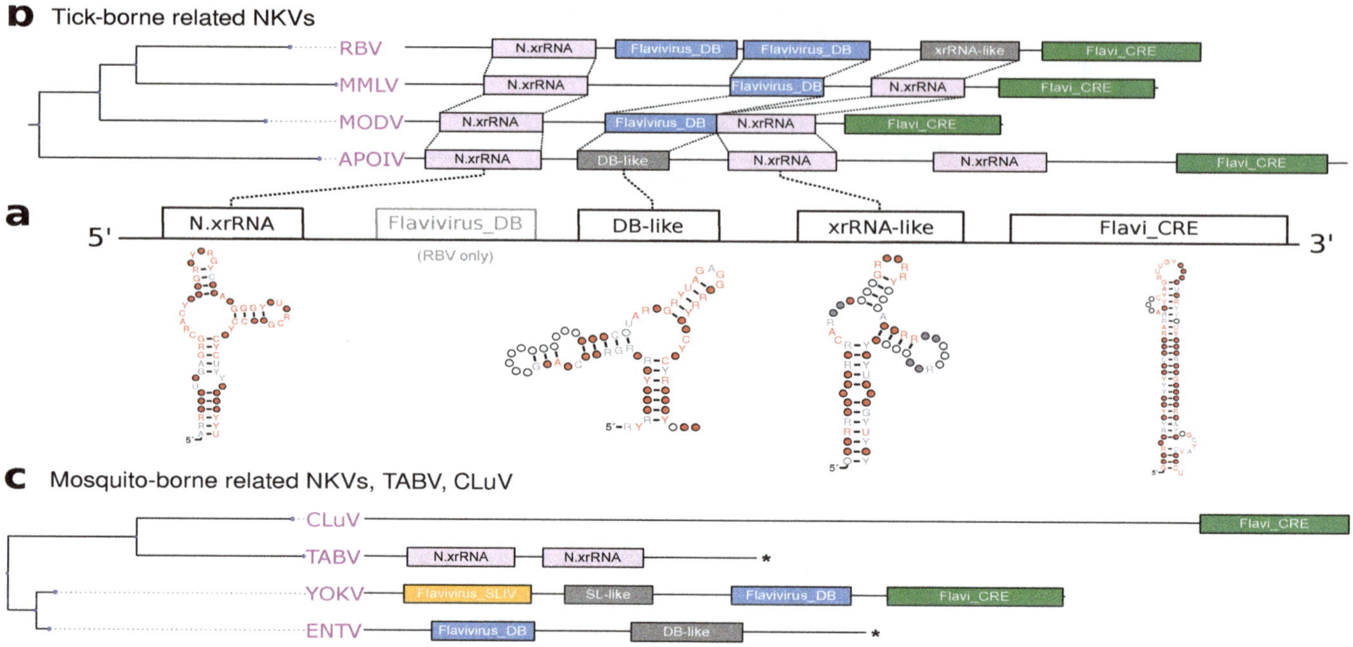

Figure 6. (**a**,**c**) Annotated 3′UTRs of NKVs. Asterisks denote incomplete 3′UTR sequences. (**b**) Schematic of TBFV-associated NKV-FV UTR architecture with consensus structures of NKV structure elements.

3.6. A Generalized xrRNA Structure

Earlier work suggested that xrRNAs from TBFVs and tick-borne related NKVs fall into a more general structural class of xrRNAs [33]. Following this line of reasoning, we investigated whether all high confidence hits obtained with our TBFV and NKV CMs could be assembled into one coherent CM that conserves the xrRNA-typical fold. A further advantage of a generalized CM would be higher sensitivity, allowing for identification of common features and eventually lead to annotation of previously unannotated xrRNAs.

Structural alignment and consensus structure prediction revealed all high confidence hits to fold into a common secondary structure (Figure 7a,b). While most of the consensus structure is characterized by low sequence conservation, stem 3 (S3) and loop 1 (L1) show medium to high degree of sequence conservation. The length of all stems is well conserved, although both major loop regions L2 and L3 show large fluctuations, with the length of L3 being de facto constant and L2 showing a high degree of flexibility.

We further investigated whether any high confidence hits from I.xrRNA1/2 in cISFVs could be aligned to the generalized xrRNA model. Although both cISFV xrRNAs (Figure 2c) bear some similarity to the generalized model, in particular to S3 and L3, we were not able to build a common alignment or consensus structure. Despite seemingly similar shape, our data also suggests that cISFV xrRNAs form a separate xrRNA subclass, unrelated to MBFV xrRNAs. In particular, we could not obtain hits of Rfam CMs (which can be seen as representatives of MBFV elements) in cISFVs, nor could we confirm any hits of cISFV-specific elements in MBFVs.

In addition to learning xrRNA features, a more generalized CM enabled us to detect xrRNA-like structures (indicated as such in all annotation plots) that could not be found previously.

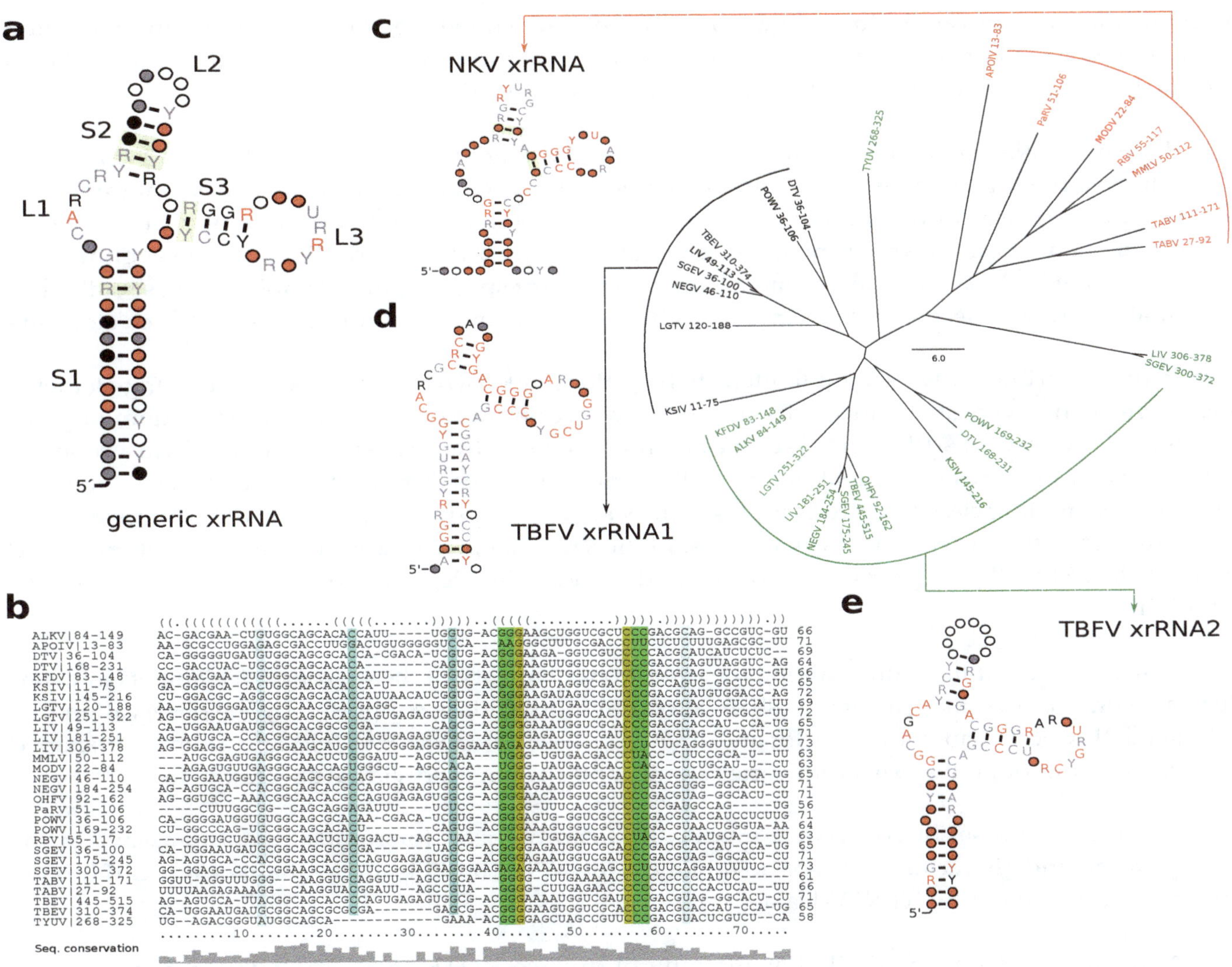

Figure 7. Generalized structure of all high confidence (cmsearch evalue $< 10^{-5}$) hits of T.xrRNA1, T.xrRNA2 and N.xrRNA. (**a**) Consensus structure prediction and (**b**) structural alignment of all high confidence hits. (**c–e**) Neighbor-joining tree of all high confidence hits of N.xrRNA (**c**), T.xrRNA1 (**d**), and T.xrRNA2 (**e**). Leaves are grouped and colored by the CM used for annotation, coordinates correspond to the position in the respective 3′UTR. For each group a separate structural alignment was computed, the consensus structures are shown.

4. Discussion

Mediated gRNA decay in the form of exoribonuclease resistance seems to be a pervasive strategy employed by viruses to circumvent host immune responses. Evidence of sfRNA production following incomplete Xrn1 degradation has not only been observed in different members of the *Flavivirus* genus [62], but also in other species of the *Flaviviridae* family, however, with major differences in xrRNA structure and sfRNA characteristics. While MBFV produce a 300-500nt sfRNA that corresponds to degradation products of the gRNA 3′UTR, hepaciviruses and pestiviruses produce a long subgenomic RNA whose 5′ end is located within the first 130nt of the viral gRNA [63].

Moreover, recent studies have identified xrRNA functionality in several phylogenetically distant RNA viruses, such as animal-infecting, segmented viruses of the *Bunyaviridae* and *Arenavividae* [64] families, as well as plant-infecting viruses of the *Tombusuviridae* and *Luteoviridae* families [65,66]. The interesting question whether exoribonucleases other than Xrn1 would be blocked as well has recently been answered. MacFadden et al. [33] could show that both RNAse J1 and Dxo1 are stalled by MBFV xrRNAs, thereby demonstrating the general nature of this structure-induced blocking mechanism. These novel findings, together with previous knowledge of Xrn1 stalling in segmented plant viruses [67,68] provide evidence for a convergent evolution scenario where xrRNAs depend on a specific folded RNA structure and form a distinct class of functional RNAs.

Repeated RNA elements appear to be a hallmark of flavivirus 3′UTR architecture. While there seems to be a plethora of conserved structure classes, our data emphasizes the consistent trend that these elements typically do not occur as single copies. Rather, duplicate or even multiple occurrences of these elements hint towards functional relevance. This is further underlined by the evolutionary conservation of both patterns and elements among different species. Since, exoribonuclease stalling is presumably never perfect, it makes sense that viruses might employ multiple copies of such elements.

In this contribution we set out to identify homologs of known exoribonuclease stalling elements and novel conserved structures. To this end, we computationally characterized homologs of experimentally verified xrRNA in tick-borne and no-known vector viruses that seem to form a coherent class of RNA structures with capability to stall exoribonucleases (T.xrRNA1/2, N.xrRNA). Likewise, we identified another class of xrRNAs in classic insect-specific flaviviruses (I.xrRNA1/2) which appears to be only distantly related to the former class. In the same line, we predicted a set of novel conserved elements in cISFVs that appear in quadruples and do not coincide with other insect-specific elements (I.Ra,I.Rb).

While we did not focus on studying the evolutionary history of these groups of elements in detail, our data suggests that many elements share ancestral roots. This is supported by the observation that at least the tick-borne, no-known-vector and *Aedes* spp. associated xrRNAs fold into a similar Y-shaped substructure, although the exact fold varies significantly among individual species.

We compiled a set of covariance models that can be used for rapid screening assays in the identification and characterization of novel flaviviruses. All models are available from GitHub via https://github.com/mtw/ITNFV-Data.

A major problem is the limited availability of diverse 3′UTR sequence data for many viruses analyzed here, particularly within cISFVs. Many novel ISFVs have previously been discovered, but 3′UTR sequences were only available for a subset of them. Future studies are required to shed more light on the evolutionary history of 3′UTR evolution in flaviviruses.

Author Contributions: M.T.W. conceived the study. R.O. and M.T.W. analyzed the data, characterized conserved RNA elements and performed phylogenetic studies. All authors contributed to writing the paper and approved of the submitted version.

Appendix A. Data Set Analyzed in This Study

Table A1. Viral genomes considered in this study. Flaviviruses are categorized into the groups tick-borne flaviviruses (TBFV), insect-specific flaviviruses (ISFV) and no-known-vector flaviviruses (NKV). The length of the 3′UTR is listed for each isolate. † Representative accession number from the `refseq` database. Whenever a `refseq` genome was not available, the isolate with the longest 3′UTR was selected as representative species. [a] Mammalian TBFVs. [b] Seabird TBFVs. [c] Classic ISFVs. [d] Dual-host affiliated ISFVs. [e] Rodent-associated NKVs. [f] Bat-associated NKVs. [N/A] 3′UTR partial or not available in the `refseq` data set.

Group	Accession Number †	Acronym	Scientific Name	3′UTR Length (nt)	Isolates
ISFV [c]	NC_012932.1	AEFV	Aedes flavivirus	942	3
ISFV [c]	NC_001564.2	CFAV	Cell fusing agent virus	553	3
ISFV [c]	KX669689.1	CLBOV	Calbertado virus	546	8
ISFV [c]	MF153378.1	CTFV	Culex theileri flavivirus	112	1

Table A1. *Cont.*

Group	Accession Number †	Acronym	Scientific Name	3′UTR Length (nt)	Isolates
ISFV [c]	NC_008604.2	CxFV	Culex flavivirus	654	13
ISFV [c]	NC_030401.1	HANV	Hanko virus	*N/A*	*N/A*
ISFV [c]	NC_005064.1	KRV	Kamiti River virus	1208	3
ISFV [c]	NC_027819.1	MECDV	Mercadeo virus	638	3
ISFV [c]	NC_021069.1	MSFV	Mosquito flavivirus	674	6
ISFV [c]	NC_034242.1	OCFVPT	Ochlerotatus caspius flavivirus	148	2
ISFV [c]	NC_027817.1	PaRV	Parramatta River virus	629	2
ISFV [c]	NC_033694.1	PCV	Palm Creek virus	*N/A*	*N/A*
ISFV [c]	NC_012671.1	QBV	Quang Binh virus	673	2
ISFV [d]	MG214905.1	BJV	Barkedji virus	335	1
ISFV [d]	NC_017086.1	CHAOV	Chaoyang virus	326	2
ISFV [d]	NC_016997.1	DONV	Donggang virus	343	2
ISFV [d]	NC_027999.1	EPEV	Paraiso Escondido virus	316	2
ISFV [d]	NC_024805.1	ILV	Ilomantsi virus	*N/A*	*N/A*
ISFV [d]	KY320648.1	KPKV	Kampung Karu virus	*N/A*	*N/A*
ISFV [d]	FJ606789.2	LAMV	Lammi virus	326	1
ISFV [d]	KY290249.1	LPKV	Long Pine Key virus	*N/A*	*N/A*
ISFV [d]	KY320649.1	LTNV	La Tina virus	*N/A*	*N/A*
ISFV [d]	MF139576.1	MMV	Marisma mosquito virus	376	1
ISFV [d]	MF139575.1	NANV	Nanay virus	399	1
ISFV [d]	NC_024017.1	NHUV	Nhumirim virus	451	1
ISFV [d]	NC_033715.1	NOUV	Nounane virus	347	3
NKV [e]	NC_003676.1	APOIV	Apoi virus	576	1
NKV	KJ469370.1	BCV	Batu Cave virus	*N/A*	*N/A*
NKV	MF776369.1	CLuV	Cyclopterus lumpus virus	601	1
NKV [f]	NC_008718.1	ENTV	Entebbe bat virus	308	3
NKV [e]	NC_026620.1	JUTV	Jutiapa virus	*N/A*	*N/A*
NKV [f]	NC_004119.1	MMLV	Montana myotis leukoencephalitis virus	460	1
NKV [e]	NC_003635.1	MODV	Modoc virus	366	1
NKV [f]	NC_034007.1	PPBV	Phnom Penh bat virus	*N/A*	*N/A*
NKV [f]	NC_003675.1	RBV	Rio Bravo virus	486	2
NKV [f]	NC_026624.1	SOKV	Sokoluk virus	*N/A*	*N/A*
NKV	NC_003996.1	TABV	Tamana bat virus	241	1
NKV [f]	NC_005039.1	YOKV	Yokose virus	429	1
TBFV [a]	NC_004355.1	ALKV	Alkhumra hemorrhagic fever virus	393	21
TBFV [a]	AF311056.1	DTV	Deer tick virus	459	1
TBFV [a]	NC_033723.1	GGV	Gadgets Gully virus	*N/A*	*N/A*
TBFV	NC_033724.1	KADV	Kadam virus	*N/A*	*N/A*
TBFV [b]	NC_023439.1	KAMV	Kama virus	282	2
TBFV [a]	HM055369.1	KFDV	Kyasanur forest disease virus	392	6
TBFV [a]	NC_006947.1	KSIV	Karshi virus	381	3
TBFV [a]	NC_003690.1	LGTV	Langat virus	568	5
TBFV [a]	NC_001809.1	LIV	Louping ill virus	500	5
TBFV [b]	NC_033721.1	MEAV	Meaban virus	*N/A*	*N/A*
TBFV [a]	KT224355.1	NEGV	Negishi virus	266	1
TBFV [a]	NC_005062.1	OHFV	Omsk hemorrhagic fever virus	410	4
TBFV [a]	NC_003687.1	POWV	Powassan virus	480	23
TBFV	DQ235149.1	RFV	Royal Farm virus	*N/A*	*N/A*
TBFV [a]	NC_027709.1	SGEV	Spanish goat encephalitis virus	493	2
TBFV [b]	NC_033726.1	SREV	Saumarez Reef virus	*N/A*	*N/A*
TBFV [a]	NC_001672.1	TBEV	Tick-borne encephalitis virus	764	167
TBFV [b]	NC_023424.1	TYUV	Tyuleniy virus	591	3

References

1. Kuno, G.; Chang, G.J.J.; Tsuchiya, K.R.; Karabatsos, N.; Cropp, C.B. Phylogeny of the Genus *Flavivirus*. *J. Virol.* **1998**, *72*, 73–83.
2. Gaunt, M.W.; Sall, A.A.; de Lamballerie, X.; Falconar, A.K.; Dzhivanian, T.I.; Gould, E.A. Phylogenetic relationships of flaviviruses correlate with their epidemiology, disease association and biogeography. *J. Gen. Virol.* **2001**, *82*, 1867–1876. [CrossRef]

3. Vasilakis, N.; Weaver, S.C. Flavivirus transmission focusing on Zika. *Curr. Opin. Virol.* **2017**, *22*, 30–35. [CrossRef]
4. Barrows, N.J.; Campos, R.K.; Liao, K.C.; Prasanth, K.R.; Soto-Acosta, R.; Yeh, S.C.; Schott-Lerner, G.; Pompon, J.; Sessions, O.M.; Bradrick, S.S.; et al. Biochemistry and Molecular Biology of Flaviviruses. *Chem. Rev.* **2018**, *118*, 4448–4482. [CrossRef]
5. Kellman, E.; Offerdahl, D.; Melik, W.; Bloom, M. Viral Determinants of Virulence in Tick-Borne Flaviviruses. *Viruses* **2018**, *10*, 329. [CrossRef] [PubMed]
6. Braack, L.; de Almeida, A.P.G.; Cornel, A.J.; Swanepoel, R.; De Jager, C. Mosquito-borne arboviruses of African origin: Review of key viruses and vectors. *Parasite Vector* **2018**, *11*, 29. [CrossRef] [PubMed]
7. Gritsun, T.; Lashkevich, V.; Gould, E. Tick-borne encephalitis. *Antivir. Res.* **2003**, *57*, 129–146. [CrossRef]
8. Guzman, H.; Contreras-Gutierrez, M.A.; da Rosa, A.P.T.; Nunes, M.R.; Cardoso, J.F.; Popov, V.L.; Young, K.I.; Savit, C.; Wood, T.G.; Widen, S.G.; et al. Characterization of three new insect-specific flaviviruses: Their relationship to the mosquito-borne flavivirus pathogens. *Am. J. Trop. Med. Hyg.* **2018**, *98*, 410–419. [CrossRef] [PubMed]
9. Blitvich, B.; Firth, A. Insect-Specific Flaviviruses: A Systematic Review of Their Discovery, Host Range, Mode of Transmission, Superinfection Exclusion Potential and Genomic Organization. *Viruses* **2015**, *7*, 1927–1959. [CrossRef] [PubMed]
10. Alkan, C.; Zapata, S.; Bichaud, L.; Moureau, G.; Lemey, P.; Firth, A.E.; Gritsun, T.S.; Gould, E.A.; de Lamballerie, X.; Depaquit, J.; et al. Ecuador Paraiso Escondido virus, a new flavivirus isolated from New World sandflies in Ecuador is the first representative of a novel clade in the genus *Flavivirus*. *J. Virol.* **2015**, *89*, 11773–11785. [CrossRef] [PubMed]
11. Halbach, R.; Junglen, S.; van Rij, R.P. Mosquito-specific and mosquito-borne viruses: Evolution, infection, and host defense. *Curr. Opin. Insect. Sci.* **2017**, *22*, 16–27. [CrossRef] [PubMed]
12. Bolling, B.; Weaver, S.; Tesh, R.; Vasilakis, N. Insect-Specific Virus Discovery: Significance for the Arbovirus Community. *Viruses* **2015**, *7*, 4911–4928. [CrossRef]
13. Blitvich, B.; Firth, A. A Review of Flaviviruses that Have No Known Arthropod Vector. *Viruses* **2017**, *9*, 154. [CrossRef] [PubMed]
14. Filomatori, C.V.; Lodeiro, M.F.; Alvarez, D.E.; Samsa, M.M.; Pietrasanta, L.; Gamarnik, A.V. A 5′ RNA element promotes dengue virus RNA synthesis on a circular genome. *Gene Dev.* **2006**, *20*, 2238–2249. [CrossRef] [PubMed]
15. Alvarez, D.E.; Ezcurra, A.L.D.L.; Fucito, S.; Gamarnik, A.V. Role of RNA structures present at the 3′ UTR of dengue virus on translation, RNA synthesis, and viral replication. *Virology* **2005**, *339*, 200–212. [CrossRef] [PubMed]
16. Manzano, M.; Reichert, E.D.; Polo, S.; Falgout, B.; Kasprzak, W.; Shapiro, B.A.; Padmanabhan, R. Identification of Cis-Acting Elements in the 3′-Untranslated Region of the Dengue Virus Type 2 RNA That Modulate Translation and Replication. *J Biol. Chem.* **2011**, *286*, 22521–22534. [CrossRef]
17. Brinton, M.A.; Basu, M. Functions of the 3′ and 5′ genome RNA regions of members of the genus Flavivirus. *Virus Res.* **2015**, *206*, 108–119. [CrossRef]
18. Ng, W.; Soto-Acosta, R.; Bradrick, S.; Garcia-Blanco, M.; Ooi, E. The 5′ and 3′ untranslated regions of the flaviviral genome. *Viruses* **2017**, *9*, 137. [CrossRef] [PubMed]
19. Villordo, S.M.; Carballeda, J.M.; Filomatori, C.V.; Gamarnik, A.V. RNA Structure Duplications and Flavivirus Host Adaptation. *Trends Microbiol.* **2016**, *24*, 270–283. [CrossRef] [PubMed]
20. de Borba, L.; Villordo, S.M.; Marsico, F.L.; Carballeda, J.M.; Filomatori, C.V.; Gebhard, L.G.; Pallarés, H.M.; Lequime, S.; Lambrechts, L.; Vargas, I.S.; et al. RNA Structure Duplication in the Dengue Virus 3′ UTR: Redundancy or Host Specificity? *mBio* **2019**, *10*, e02506–18. [CrossRef]
21. de Bernardi Schneider, A.; Wolfinger, M.T. Musashi binding elements in Zika and related Flavivirus 3′UTRs: A comparative study in silico. *bioRxiv* **2018**. [CrossRef]
22. Rice, C.M.; Lenches, E.M.; Eddy, S.R.; Shin, S.; Sheets, R.; Strauss, J. Nucleotide sequence of yellow fever virus: Implications for flavivirus gene expression and evolution. *Science* **1985**, *229*, 726–733. [CrossRef] [PubMed]
23. Hahn, C.S.; Hahn, Y.S.; Rice, C.M.; Lee, E.; Dalgarno, L.; Strauss, E.G.; Strauss, J.H. Conserved elements in the 3′ untranslated region of flavivirus RNAs and potential cyclization sequences. *J. Mol. Biol.* **1987**, *198*, 33–41. [CrossRef]

24. Villordo, S.M.; Alvarez, D.E.; Gamarnik, A.V. A balance between circular and linear forms of the dengue virus genome is crucial for viral replication. *RNA* **2010**, *16*, 2325–2335. [CrossRef] [PubMed]
25. de Borba, L.; Villordo, S.M.; Iglesias, N.G.; Filomatori, C.V.; Gebhard, L.G.; Gamarnik, A.V. Overlapping local and long-range RNA-RNA interactions modulate dengue virus genome cyclization and replication. *J. Virol.* **2015**, *89*, 3430–3437. [CrossRef] [PubMed]
26. Pijlman, G.P.; Funk, A.; Kondratieva, N.; Leung, J.; Torres, S.; Van der Aa, L.; Liu, W.J.; Palmenberg, A.C.; Shi, P.Y.; Hall, R.A.; et al. A highly structured, nuclease-resistant, noncoding RNA produced by flaviviruses is required for pathogenicity. *Cell Host Microbe* **2008**, *4*, 579–591. [CrossRef] [PubMed]
27. Akiyama, B.M.; Eiler, D.; Kieft, J.S. Structured RNAs that evade or confound exonucleases: Function follows form. *Curr. Opin. Struct. Biol.* **2016**, *36*, 40–47. [CrossRef] [PubMed]
28. Moon, S.L.; Anderson, J.R.; Kumagai, Y.; Wilusz, C.J.; Akira, S.; Khromykh, A.A.; Wilusz, J. A noncoding RNA produced by arthropod-borne flaviviruses inhibits the cellular exoribonuclease XRN1 and alters host mRNA stability. *RNA* **2012**, *18*, 2029–2040. [CrossRef]
29. Roby, J.A.; Pijlman, G.P.; Wilusz, J.; Khromykh, A.A. Noncoding subgenomic flavivirus RNA: Multiple functions in West Nile virus pathogenesis and modulation of host responses. *Viruses* **2014**, *6*, 404–427. [CrossRef]
30. Jones, C.I.; Zabolotskaya, M.V.; Newbury, S.F. The 5′-> 3′ exoribonuclease Xrn1/Pacman and its functions in cellular processes and development. *Wiley Interdiscip. Rev. RNA* **2012**, *3*, 455–468. [CrossRef]
31. Antic, S.; Wolfinger, M.T.; Skucha, A.; Hosiner, S.; Dorner, S. General and miRNA-mediated mRNA degradation occurs on ribosome complexes in Drosophila cells. *Mol. Cell Biol.* **2015**, doi:10.1128/MCB.01346-14. [CrossRef]
32. Kieft, J.S.; Rabe, J.L.; Chapman, E.G. New hypotheses derived from the structure of a flaviviral Xrn1-resistant RNA: Conservation, folding, and host adaptation. *RNA Biol.* **2015**, *12*, 1169–1177. [CrossRef]
33. MacFadden, A.; O'Donoghue, Z.; Silva, P.A.; Chapman, E.G.; Olsthoorn, R.C.; Sterken, M.G.; Pijlman, G.P.; Bredenbeek, P.J.; Kieft, J.S. Mechanism and structural diversity of exoribonuclease-resistant RNA structures in flaviviral RNAs. *Nat. Commun.* **2018**, *9*, 119. [CrossRef]
34. Hussain, M.; Torres, S.; Schnettler, E.; Funk, A.; Grundhoff, A.; Pijlman, G.P.; Khromykh, A.A.; Asgari, S. West Nile virus encodes a microRNA-like small RNA in the 3′ untranslated region which up-regulates GATA4 mRNA and facilitates virus replication in mosquito cells. *Nucleic Acids Res.* **2012**, *40*, 2210–2223. [CrossRef] [PubMed]
35. Schuessler, A.; Funk, A.; Lazear, H.M.; Cooper, D.A.; Torres, S.; Daffis, S.; Jha, B.K.; Kumagai, Y.; Takeuchi, O.; Hertzog, P.; et al. West Nile virus noncoding subgenomic RNA contributes to viral evasion of the type I interferon-mediated antiviral response. *J. Virol.* **2012**, *86*, 5708–5718. [CrossRef] [PubMed]
36. Bidet, K.; Dadlani, D.; Garcia-Blanco, M.A. G3BP1, G3BP2 and CAPRIN1 are required for translation of interferon stimulated mRNAs and are targeted by a dengue virus non-coding RNA. *PLoS Pathog.* **2014**, *10*, e1004242. [CrossRef] [PubMed]
37. Lorenz, R.; Wolfinger, M.T.; Tanzer, A.; Hofacker, I.L. Predicting RNA Structures from Sequence and Probing Data. *Methods* **2016**, *103*, 86–98. [CrossRef] [PubMed]
38. Funk, A.; Truong, K.; Nagasaki, T.; Torres, S.; Floden, N.; Melian, E.B.; Edmonds, J.; Dong, H.; Shi, P.Y.; Khromykh, A.A. RNA structures required for production of subgenomic flavivirus RNA. *J. Virol.* **2010**, *84*, 11407–11417. [CrossRef]
39. Silva, P.A.; Pereira, C.F.; Dalebout, T.J.; Spaan, W.J.; Bredenbeek, P.J. An RNA pseudoknot is required for production of yellow fever virus subgenomic RNA by the host nuclease XRN1. *J. Virol.* **2010**, *84*, 11395–11406. [CrossRef] [PubMed]
40. Sztuba-Solinska, J.; Teramoto, T.; Rausch, J.W.; Shapiro, B.A.; Padmanabhan, R.; Le Grice, S.F. Structural complexity of Dengue virus untranslated regions: cis-acting RNA motifs and pseudoknot interactions modulating functionality of the viral genome. *Nucleic Acids Res.* **2013**, *41*, 5075–5089. [CrossRef] [PubMed]
41. Chapman, E.G.; Costantino, D.A.; Rabe, J.L.; Moon, S.L.; Wilusz, J.; Nix, J.C.; Kieft, J.S. The structural basis of pathogenic subgenomic flavivirus RNA (sfRNA) production. *Science* **2014**, *344*, 307–310. [CrossRef] [PubMed]
42. Akiyama, B.M.; Laurence, H.M.; Massey, A.R.; Costantino, D.A.; Xie, X.; Yang, Y.; Shi, P.Y.; Nix, J.C.; Beckham, J.D.; Kieft, J.S. Zika virus produces noncoding RNAs using a multi-pseudoknot structure that confounds a cellular exonuclease. *Science* **2016**, *354*, 1148–1152. [CrossRef]

43. Rauscher, S.; Flamm, C.; Mandl, C.W.; Heinz, F.X.; Stadler, P.F. Secondary structure of the 3′-noncoding region of flavivirus genomes: comparative analysis of base pairing probabilities. *RNA* **1997**, *3*, 779–791. [PubMed]
44. Hofacker, I.L.; Fekete, M.; Flamm, C.; Huynen, M.A.; Rauscher, S.; Stolorz, P.E.; Stadler, P.F. Automatic detection of conserved RNA structure elements in complete RNA virus genomes. *Nucleic Acid Res.* **1998**, *26*, 3825–3836. [CrossRef] [PubMed]
45. Witwer, C.; Rauscher, S.; Hofacker, I.L.; Stadler, P.F. Conserved RNA secondary structures in Picornaviridae genomes. *Nucleic Acid Res.* **2001**, *29*, 5079–5089. [CrossRef]
46. Hofacker, I.L.; Stadler, P.F.; Stocsits, R.R. Conserved RNA secondary structures in viral genomes: A survey. *Bioinformatics* **2004**, *20*, 1495–1499. [CrossRef]
47. Thurner, C.; Witwer, C.; Hofacker, I.L.; Stadler, P.F. Conserved RNA secondary structures in Flaviviridae genomes. *J. Gen. Virol.* **2004**, *85*, 1113–1124. [CrossRef]
48. Katoh, K.; Standley, D.M. MAFFT multiple sequence alignment software version 7: Improvements in performance and usability. *Mol. Biol. Evol.* **2013**, *30*, 772–780. [CrossRef]
49. Nguyen, L.T.; Schmidt, H.A.; von Haeseler, A.; Minh, B.Q. IQ-TREE: A fast and effective stochastic algorithm for estimating maximum-likelihood phylogenies. *Mol. Biol. Evol.* **2014**, *32*, 268–274. [CrossRef] [PubMed]
50. Nawrocki, E.P.; Eddy, S.R. Infernal 1.1: 100-fold Faster RNA Homology Searches. *Bioinformatics* **2013**, *29*, 2933–2935. [CrossRef]
51. Waldl, M.; Thiel, B.C.; Ochsenreiter, R.; Holzenleiter, A.; de Araujo Oliveira, J.V.; Walter, M.E.M.; Wolfinger, M.T.; Stadler, P.F. TERribly Difficult: Searching for Telomerase RNAs in Saccharomycetes. *Genes* **2018**, *9*, 372. [CrossRef] [PubMed]
52. Will, S.; Reiche, K.; Hofacker, I.L.; Stadler, P.F.; Backofen, R. Inferring noncoding RNA families and classes by means of genome-scale structure-based clustering. *PLoS Comp. Biol.* **2007**, *3*, e65. [CrossRef] [PubMed]
53. Lorenz, R.; Bernhart, S.H.; Zu Siederdissen, C.H.; Tafer, H.; Flamm, C.; Stadler, P.F.; Hofacker, I.L. ViennaRNA Package 2.0. *Algorithm Mol. Biol.* **2011**, *6*, 26. [CrossRef] [PubMed]
54. Wolfinger, M.T.; Fallmann, J.; Eggenhofer, F.; Amman, F. ViennaNGS: A toolbox for building efficient next-generation sequencing analysis pipelines. *F1000Research* **2015**, *4*. [CrossRef]
55. Wolfinger, M.T. Bio::RNA::RNAaliSplit 0.09. 2019. Available online: https://github.com/mtw/Bio-RNA-RNAaliSplit (accessed on 23 March 2019).
56. Weinberg, Z.; Breaker, R.R. R2R-software to speed the depiction of aesthetic consensus RNA secondary structures. *BMC Bioinform.* **2011**, *12*, 3. [CrossRef]
57. Huerta-Cepas, J.; Serra, F.; Bork, P. ETE 3: Reconstruction, analysis, and visualization of phylogenomic data. *Mol. Biol. Evol.* **2016**, *33*, 1635–1638. [CrossRef] [PubMed]
58. Gritsun, T.; Gould, E. Origin and Evolution of 3′UTR of Flaviviruses: Long Direct Repeats as A Basis for the Formation of Secondary Structures and Their Significance for Virus Transmission. *Adv. Virus Res.* **2006**, *69*, 203–248.
59. Gritsun, D.J.; Jones, I.M.; Gould, E.A.; Gritsun, T.S. Molecular Archaeology of Flaviviridae Untranslated Regions: Duplicated RNA Structures in the Replication Enhancer of Flaviviruses and Pestiviruses Emerged via Convergent Evolution. *PLoS ONE* **2014**, *9*, e92056. [CrossRef]
60. Gritsun, T.; Gould, E. The 3′ untranslated regions of Kamiti River virus and Cell fusing agent virus originated by self-duplication. *J. Gen. Virol.* **2006**, *87*, 2615–2619. [CrossRef]
61. Hoshino, K.; Isawa, H.; Tsuda, Y.; Yano, K.; Sasaki, T.; Yuda, M.; Takasaki, T.; Kobayashi, M.; Sawabe, K. Genetic characterization of a new insect flavivirus isolated from *Culex pipiens* mosquito in Japan. *Virology* **2007**, *359*, 405–414. [CrossRef]
62. Clarke, B.; Roby, J.; Slonchak, A.; Khromykh, A. Functional non-coding RNAs derived from the flavivirus 3′ untranslated region. *Virus Res.* **2015**, *206*, 53–61. [CrossRef]
63. Moon, S.L.; Blackinton, J.G.; Anderson, J.R.; Dozier, M.K.; Dodd, B.J.; Keene, J.D.; Wilusz, C.J.; Bradrick, S.S.; Wilusz, J. X1 Stalling in the 5′UTR of Hepatitis C Virus and Bovine Viral Diarrhea Virus Is Associated with Dysregulated Host mRNA Stability. *PLoS Pathog.* **2015**, *11*, e1004708. [CrossRef]
64. Charley, P.A.; Wilusz, C.J.; Wilusz, J. Identification of Phlebovirus and Arenavirus RNA sequences that stall and repress the exoribonuclease XRN1. *J. Biol. Chem.* **2018**, *293*, 285–295. [CrossRef] [PubMed]

65. Steckelberg, A.L.; Akiyama, B.M.; Costantino, D.A.; Sit, T.L.; Nix, J.C.; Kieft, J.S. A folded viral noncoding RNA blocks host cell exoribonucleases through a conformationally dynamic RNA structure. *Proc. Natl. Acad. Sci. USA* **2018**, *115*, 6404–6409. [CrossRef] [PubMed]
66. Steckelberg, A.L.; Vicens, Q.; Kieft, J.S. Exoribonuclease-Resistant RNAs Exist within both Coding and Noncoding Subgenomic RNAs. *mBio* **2018**, *9*, e02461-18. [CrossRef] [PubMed]
67. Iwakawa, H.o.; Mizumoto, H.; Nagano, H.; Imoto, Y.; Takigawa, K.; Sarawaneeyaruk, S.; Kaido, M.; Mise, K.; Okuno, T. A viral noncoding RNA generated by cis-element-mediated protection against 5′- 3′ RNA decay represses both cap-independent and cap-dependent translation. *J. Virol.* **2008**, *82*, 10162–10174. [CrossRef] [PubMed]
68. Flobinus, A.; Hleibieh, K.; Klein, E.; Ratti, C.; Bouzoubaa, S.; Gilmer, D. A viral noncoding RNA complements a weakened viral RNA silencing suppressor and promotes efficient systemic host infection. *Viruses* **2016**, *8*, 272. [CrossRef] [PubMed]

13

The Utility of Data Transformation for Alignment, De Novo Assembly and Classification of Short Read Virus Sequences

Avraam Tapinos [1,*], Bede Constantinides [1,2], My V. T. Phan [3], Samaneh Kouchaki [1,4], Matthew Cotten [3,5,6] and David L. Robertson [1,5]

1 School of Biological Sciences, The University of Manchester, Manchester M13 9PT, UK; bede.constantinides@manchester.ac.uk (B.C.); samaneh.kouchaki@eng.ox.ac.uk (S.K.); david.l.robertson@glasgow.ac.uk (D.L.R.)
2 Modernising Medical Microbiology Consortium, Nuffield Department of Clinical Medicine, John Radcliffe Hospital, University of Oxford, Oxford OX3 9DU, UK
3 Department of Viroscience, Erasmus Medical Centre, Doctor Molewaterplein 40, 3015 GD Rotterdam, The Netherlands; v.t.m.phan@erasmusmc.nl (M.V.T.P.); mlcotten13@gmail.com (M.C.)
4 Department of Engineering Science, Institute of Biomedical Engineering, University of Oxford, Oxford OX3 7DQ, UK
5 MRC-University of Glasgow Centre for Virus Research, Glasgow G61 1QH, UK
6 MRC/UVRI & LSHTM Uganda Research Unit Entebbe, P.O. Box 49 Entebbe, Uganda
* Correspondence: avraam.tapinos@manchester.ac.uk

Abstract: Advances in DNA sequencing technology are facilitating genomic analyses of unprecedented scope and scale, widening the gap between our abilities to generate and fully exploit biological sequence data. Comparable analytical challenges are encountered in other data-intensive fields involving sequential data, such as signal processing, in which dimensionality reduction (i.e., compression) methods are routinely used to lessen the computational burden of analyses. In this work, we explored the application of dimensionality reduction methods to numerically represent high-throughput sequence data for three important biological applications of virus sequence data: reference-based mapping, short sequence classification and de novo assembly. Leveraging highly compressed sequence transformations to accelerate sequence comparison, our approach yielded comparable accuracy to existing approaches, further demonstrating its suitability for sequences originating from diverse virus populations. We assessed the application of our methodology using both synthetic and real viral pathogen sequences. Our results show that the use of highly compressed sequence approximations can provide accurate results, with analytical performance retained and even enhanced through appropriate dimensionality reduction of sequence data.

Keywords: alignment; assembly; taxonomic classification; time series; data transformation; DWT; DFT; PAA; data compression; compressive genomics

---

1. Introduction

Next-generation sequencing (NGS) enables massively parallel determination of nucleotide order within genetic material, making it possible to rapidly sequence the genomes of individuals, populations and metagenomic samples [1–5]. However, the sequences generated by these instruments are almost always considerably shorter in length than the genomic regions studied. Genomic analyses often begin with the process of sequence assembly, where sequence fragments (reads) are reconstructed into the larger sequences from which they originated. Computational methods play a vital role in the assembly of short reads, and a variety of assemblers and related tools have been developed in tandem with emerging sequencing platforms [6]. All subsequent analyses and investigations depend upon the quality, accuracy and speed of this crucial sequence assembly process.

There are many computational methods to generate consensus sequences representing the genomes of species in a sample. Such approaches include seed-and-extend alignment methods using suffix array derivatives, such as the Burrows-Wheeler Transform (BWT) for aligning short reads informed by a known reference sequence [7,8], graph-based methods employing Overlap Layout Consensus (OLC) [9,10] and de Bruijn graphs of *k*-mers [11–13] for reference-free de novo sequence assembly. However, for sequencing projects to characterise genetic variation within populations (deep sequencing), metagenomics and pathogen discovery, the effectiveness of the aforementioned approaches varies considerably [14].

Samples with mixed viral infections, especially those comprising divergent variants, present a number of analytical and computational problems. The use of a reference sequence, even the use of a data specific generated sequence, can lead to valuable read information being discarded during the alignment process [15]. On the other hand, while de novo approaches require little a priori knowledge of target sequence composition, the methods are computationally intensive, and their performance scales poorly with datasets of increasing size [9]. Aggressive heuristics must be employed, to traverse graphs and deal with mismatches, reduce the running time of de novo assemblers, which, in turn, can compromise assembly quality. Indexing structures such as the BWT and its relatives are widely used to reduce the burden of pairwise sequence comparison, for both reference-based mapping and de novo assembly. However, they cannot process mismatches within reads, necessitating the use of computationally expensive heuristics to establish relationships between divergent sequences. Increasing sequence length further affects the performance of these approaches [16].

A major challenge in working with NGS data from metagenomic studies is the high levels of diversity present, particularly for the virus genetic material. Also, the number of sequences generated challenge many computational systems for a feasible working solution in terms of time and the computational resources typically available in biological laboratories. For biologists working on outbreak responses or pathogen discovery, both the accuracy of the assembly results and the speed of sequence analyses (e.g., assembly, alignment and pathogen classification) are crucial for crisis response and management. The ability to run analyses in the field on portable computer systems without internet connectivity is also important. Here, we explore the utility of data transform methods to extract major features from viral NGS sequence data and use the features to analyse data in a lower dimensional space.

Similar analytical challenges involving high dimensional sequential data are encountered in other data-intensive fields, such as signal and image processing, and time series analysis, where data transforms and approximation techniques are used for data dimensionality reduction. Data transform/approximation techniques include the discrete Fourier transform (DFT) [17], the discrete wavelet transform (DWT) [18,19] and piece-wise aggregate approximation (PAA) [20,21]. The DFT or DWT are used to transform data to their frequency domains, allowing feature extraction [22], and PAA is used as a data approximation approach. In data-intensive fields, data transformations/approximations are commonly used as dimensionality reduction approaches for obtaining fast approximate solutions for a given problem. Due to the ordered nature of genetic data, many of these transformation approaches can be applied to sequences of nucleotides [23] or amino acids [24]. An example of a successful implementation of a Fourier transform in computational biology is the multiple sequence alignment based on fast Fourier transform alignment algorithm MAFFT [25] where the physiochemical properties of amino acids are used to represent sequences for fast matching of homologous sequence regions for alignment. Since most transformation approaches are suitable only for numerical sequences, the strings of letters representing genetic sequences must be mapped into numerical space using a numerical sequence representation method [26].

In addition to the DFT, the DWT and PAA, suitable methods for measuring the pairwise similarity of sequential data or transformations include the Lp-norms [27], dynamic time warping (DTW) [28], longest common subsequence (LCS) [29], and alignment approaches, such as the Needleman-Wunsch and

Smith-Waterman algorithms. Euclidean distance is arguably the most widely used Lp-norm method for sequential data comparison but can only be used on sequences of the same length. Furthermore, Lp-norm methods do not accommodate shifts in the x-axis (time or position) and are thus limited in their ability to identify similar features within offset data. Elastic similarity/dissimilarity methods, such as LCS, unbounded DTW and various alignment algorithms, permit comparison of data with different dimensions and tolerate shifts in the x-axis. These properties of elastic similarity methods can be very useful in the analysis of speech signals, for example, but can be computationally expensive [30]. Several approaches have been proposed to permit fast searching with DTW, including the introduction of global constraints (wrapping path) or the use of lower bounding techniques, such as LB_keogh [28].

While pairwise comparison methods may be used for clustering, classification and similarity searches, they are very time consuming for large datasets ($O(n^2)$ time complexity). Indexing structures, such as the R^*-tree, KD-tree, VP-tree and MVP-tree have significantly lower time complexity ($O(n \log(n))$) for similarity search [31] and are more appropriate for efficient analysis of large datasets. The R^*-tree [32,33] and KD-tree [34] indexing structures are very accurate for low dimensional datasets. However, their performance deteriorates significantly in high dimensional space [31], a phenomenon known as the 'curse of dimensionality' [35,36]. Metric trees, such as the VP-tree [37] and MVP-tree [38], are less prone to this limitation. Metric space indexing structures make use of geometric properties for partitioning data and work efficiently on both low and high dimensional data [39]. The curse of dimensionality can be further mitigated using data approximations, such as the DFT, the DWT and the PAA, to partition a dataset in an approximated space without loss of generality [21].

Here, we investigate the performance of three established dimensionality reduction techniques on three common analysis tasks involving viral short read sequence data: classification, reference-based mapping/alignment and de novo assembly. We benchmarked the accuracy of our proposed methodology against existing tools, and demonstrate the applicability of time series and signal processing data mining techniques for the analysis of viral NGS data.

2. Materials and Methods

2.1. Symbolic to Numeric Sequence Representations

Various numeric sequence representation methods can be used for symbolising a nucleotide sequence to a numerical space (see 51). Depending on the chosen numerical representation, each nucleotide is associated with a specific numerical value or vector. The specific values are assigned to the position of each nucleotide indicating the presence of a nucleotide at each sequence position (Equation 1). R_i is the indicator for a specific nucleotide in the ith position of the sequence S with a length of n nucleotides. Values $v_1 \ldots v_5$ correspond to the numerical value or numerical vector associated with each nucleotide.

$$R = \begin{cases} v_1 \; if \; i = A \\ v_2 \; if \; i = T \\ v_3 \; if \; i = C \quad \forall i \in S \\ v_4 \; if \; i = G \\ v_5 \; otherwise \end{cases} \tag{1}$$

Methods, such as the electron-ion interaction pseudopotentials (EIIP) [40] and the atomic representation approach [41], aim to mimic the biochemical properties of nucleic acids but introduce some mathematical bias that does not exist in reality [26]. Other methods, like the Voss indicator [42] and the Tetrahedron approach, do not introduce internucleotide mathematical bias, meaning the pairwise distances between each non-identical transformed nucleotide are the same (for example, the distance between A and T is equal to the distances between A and C as well as A and G). Furthermore, the cumulative sum of a numerical representation R can be used to indicate the trajectory of a sequence in nucleotide space. Table 1 indicates the values used for different representation methods [26].

Table 1. Numerical nucleotide sequence representation methods.

Method	Numerical Representation
Integer number	$A = 1,\ C = -1,\ G = 2,\ T = -2,\ N = 0$
Real number	$A = -1.5,\ C = 0.5, G = -0.5,\ T = 1.5,\ N = 0.0$
EIIP	$A = 0.1260,\ C = 0.1340,\ G = 0.0806,\ T = 0.1335,\ N = 0$
Atomic	$A = 70,\ C = 58,\ G = 78,\ T = 66,\ N = 0$
Pair	$A\ or\ T = 1,\ C\ or\ G = -1,\ N = 0$
Complex number	$A = 1 + 1i,\ C = -1 + 1i,\ G = -1 - 1i,\ T = 1 - 1i,\ N = 0 + 0i$
DNA Walk	$A = [1,0],\ C = [0,1],\ G = [0,-1],\ T = [-1,0],\ N = [0,0]$
Tetrahedron	$A = [0,0,1],\ C = \left[-\frac{\sqrt{2}}{3}, -\frac{\sqrt{6}}{3}, \frac{1}{3}\right],$ $G = \left[-\frac{\sqrt{2}}{3},\ -\frac{\sqrt{6}}{3},\ -\frac{1}{3}\right],\ T = \left[2\times\frac{\sqrt{2}}{3},\ 0, -\frac{1}{3}\right],\ N = [0,0,0]$
Voss indicator	$A = [0,0,1,0],\ C = [1,0,0,0],\ G = [0,1,0,0],\ T = [0,0,0,1],\ N = [0,0,0,0]$

2.2. *Sequence Transformation*

Effective methods for transforming/approximating sequential data should: (i) accurately transform/approximate data without loss of useful information, (ii) have low computational overheads, (iii) facilitate rapid comparison of data and (iv) provide lower bounding—where the distance between data representations is always less than or equal to that of the original data—precluding false negative results [43]. The lower bounding property guarantees that if two data points are nearby in their original space, they will remain so in their transformed/approximate space. We employ the DFT and the DWT transformation methods and the PAA approximation method as they satisfy the above requirements, and these are widely used for analysing discrete signals [44] and can be used to transform/approximate nucleotide sequence numerical representations to different levels of resolution, permitting reduced dimensionality sequence analysis.

Figure 1A illustrates an example of the DFT and DWT transformations and PAA approximation of a short nucleotide sequence. The DFT and the fast Fourier transform (FFT) convert data from their original domain into the frequency domain. In principle, the DFT decomposes a numerically represented nucleotide sequence with n positions (dimensions) into a series of n frequency components ordered by their frequency. A subset of the resulting Fourier frequencies are used to approximate the original sequence in a lower dimensional space [17], and the tradeoff between analytical speed and accuracy can be varied according to the number of frequencies considered [45].

The DWT transforms data into the time-frequency domain, capturing both frequency and temporal location information [18,46,47], in contrast to DFT, which only provides frequency information. DWT is a set of averaging and differencing functions that may be used recursively to represent sequential data at different resolutions, and each resolution can be used as an approximation of the original data. Figure 1B depicts DWT transformations of a short nucleotide sequence.

In PAA, a numerical sequence is divided into n equally sized windows, the mean values of which together form a compressed sequence representation [20,21]. The selection of n determines the resolution of the compressed or approximate representation. While PAA is faster and easier to implement than the DFT and the DWT, unlike these two methods, PAA is irreversible, meaning that the original sequence cannot be recovered from the approximation. Figure 1C depicts an example of the PAA transformations of a short nucleotide sequence.

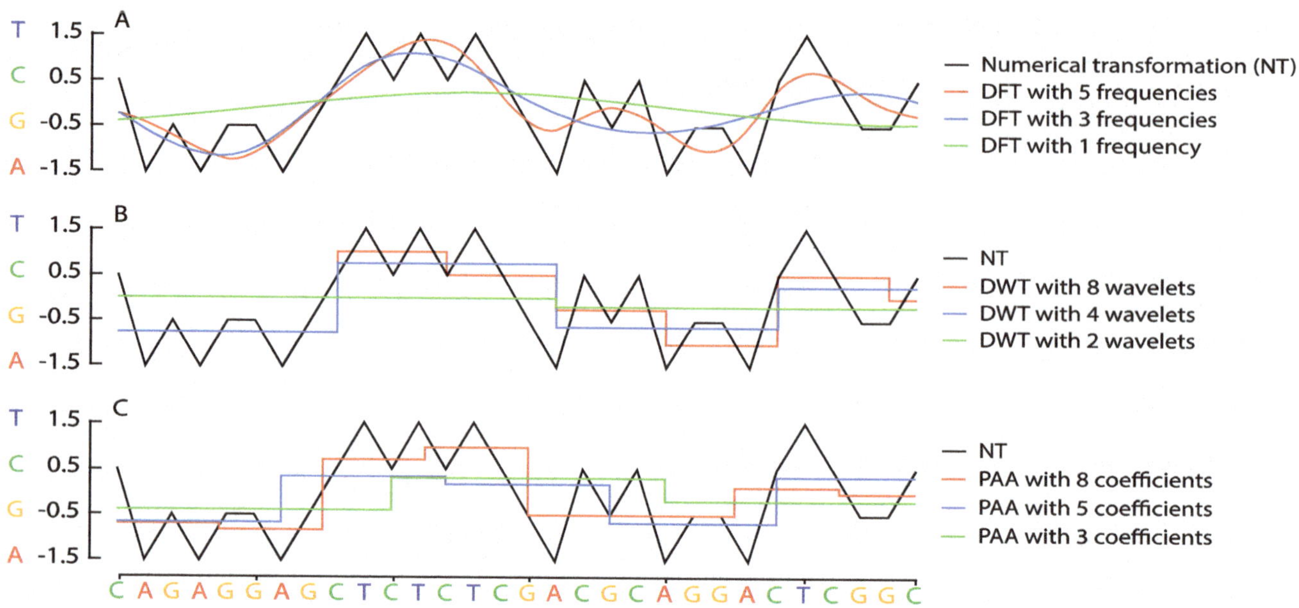

Figure 1. A numerically represented DNA sequence transformed at various levels of spatial resolution using the discrete Fourier transform (DFT) of the whole sequence (**A**), the Haar discrete wavelet transform (DWT) (**B**) and piecewise aggregate approximation (PAA) (**C**). A 30 nucleotide sequence (x-axis) is represented as a numerical sequence (black lines) using the real number representation method (y-axis where T = 1.5, C = 0.5, G = −0.5 and A = −1.5) for DFT approximations of the sequence with 5 (red), 3 (blue) and 1 (green) Fourier frequencies (**A**); DWT approximations of the same sequence with 8 level wavelets (red), 4 level wavelets (blue) and 2 level wavelets (green) (**B**); PAA approximations of the same sequence with 8 (red), 5 (blue) and 3 (green) coefficients (**C**).

2.3. Similarity Search Approaches for Sequential Data

Here, we adopt the Euclidian distance and *VP*-tree index to perform a fast *k*-nearest neighbour (*k*-NN) similarity search for aligning the reads to a reference genome.

In a *VP*-tree indexing structure, data is segregated using the distance between data points, thus implementing data partitioning in a metric space. A data point to use as a vantage point is selected (either randomly or by applying some heuristic to find and use the furthest point in the dataset [37]), and the rest of the data points are partitioned into two nodes based on their distance to that point. Data found to be closer to the vantage point than a given threshold (the median distance between all the data points and the vantage point) are assigned to the same node, and the rest of the data points to a different node. This function is repeated recursively in order to complete the partitioning process. The resulting indexing structure can then be used for fast identification of a *k*-nearest neighbour (*k*-NN) search. A *k*-NN-search returns the data points that are closest to a query q. Initially, the distance between the query q and the vantage point in the top node is calculated. If the distance between q and the vantage point satisfies a set of given conditions (the distance is smaller or larger than a given threshold – this threshold being the median distance between the vantage point and other data points within the node), a decision is made to visit either one or both of the child nodes. This process is repeated until the entire tree has been traversed. The k data points—in this case, reads—found closest to our query are the k-nearest neighbours to the query q.

2.4. Proposed Short Reads Processing Methodology

Our methodology for taxonomic classification, reference-based mapping and de novo assembly of short reads used time series and digital signal processing data transformation techniques. Figure 2 illustrates the fundamental concept of our approach. The short reads and reference genomes are mapped to a numerical space using an appropriate method from Table 1. Subsequently, lower dimensional approximations were generated for all data using the appropriate data transformation method, such as DFT, DWT and PAA. A *VP*-tree was constructed to allow fast data comparison. Depending on the application, the *VP*-tree was constructed either by using *k*-mer transformations obtained from the reference genomes or by using the short reads' transformations. Consequently, the best matches for our short reads' transformations were identified using a *k*-NN search approach on the *VP*-tree. As a final step, the results obtained from the *k*-NN search were re-evaluated in the original space to remove potential false positive results.

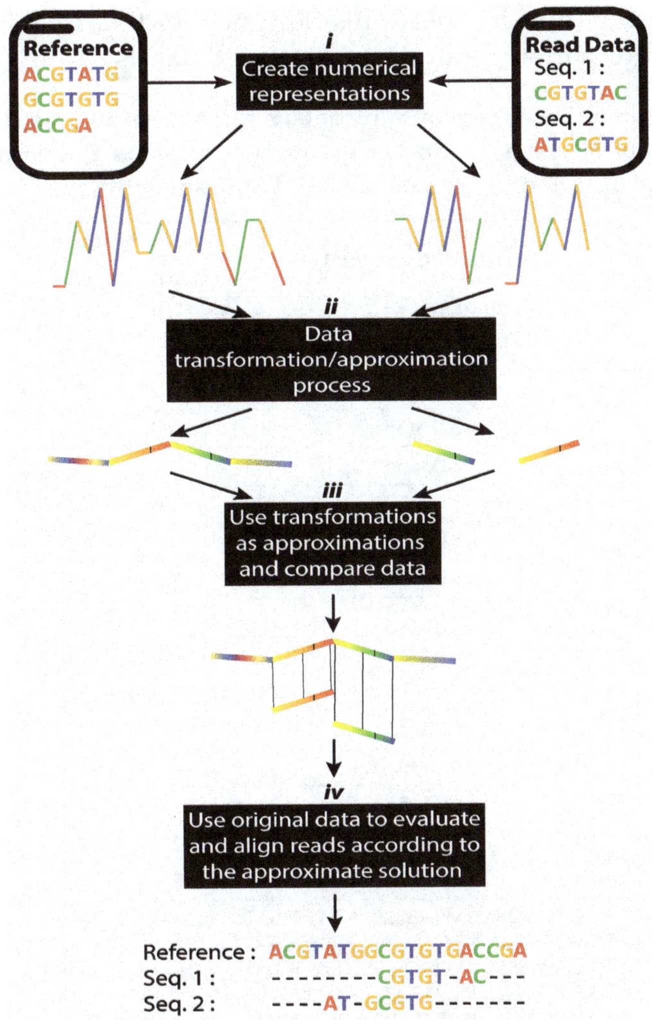

Figure 2. Overview of our proposed methodology using time series transformation/approximation methods: (*i*) Creation of numerical representations of input sequences. (*ii*) Application of an appropriate signal decomposition method to transform sequences into their feature space. (*iii*) Use of approximated transformations to perform rapid data analysis in lower dimensional space. (*iv*) Validation of inferences against original, full-resolution input sequences. In the case of reference-based alignment and taxonomic classification, approximated read transformations were compared with a reference sequence. In our de novo implementation, pairwise comparisons were performed between all of the approximated read transformations.

2.5. Data

The implementations of our proposed methodologies were assessed with both simulated and real virus datasets. The simulated datasets were generated using CuReSim [48] and WGSIM (https://github.com/lh3/wgsim). Simulated data included information, such as the reference genome used, the alignment position and alignment direction, for each read, enabling rigorous evaluation of the proposed techniques. We used two simulators to examine our approach in a variety of use cases. CuReSim is highly customisable, allowing the user to control the type of variation (insertion, deletion and substitution) to simulate. WGSIM can simulate genomes with uniform insertion, deletion and substitution variation.

CuReSim was used to generate 16 HIV-1 HXB2 simulated datasets with different levels and types of variation. WGSIM was used to generate 4 mixed virus datasets with different levels of variation. Each simulation contained 200,000 reads generated using 5 Norovirus, 5 Ebola virus and 5 Respiratory syncytial virus (RSV) genomes, with various types and extents of simulated variation. HXB2 and simulated mixed virus datasets and corresponding reference genomes used to simulate them are deposited on GitHub (https://github.com/Avramis/Supporting-data/tree/master/Simulated%20Data). Table 2 contains detailed information about the simulated datasets.

Table 2. Simulated read data. Each row contains details for each simulated dataset (i.e., virus family, virus, GenBank ID, variation type, variation level, number of reads and simulator used to generate data). Abbreviations: Ins, insertions; Del, deletions and Sub, substitutions.

Family	Virus	GenBank Genome ID	Variation Type (%)			Reads	Simulator
			Ins	Del	Sub		
HIV	HXB2	K03455	0.0	0.0	0.0	2133	CuReSim
HIV	HXB2	K03455	0.0	0.0	1.0	2133	CuReSim
HIV	HXB2	K03455	0.0	0.0	2.0	2133	CuReSim
HIV	HXB2	K03455	0.0	0.0	3.0	2133	CuReSim
HIV	HXB2	K03455	0.0	0.0	4.0	2133	CuReSim
HIV	HXB2	K03455	0.0	0.0	5.0	2133	CuReSim
HIV	HXB2	K03455	0.5	0.5	0.0	2133	CuReSim
HIV	HXB2	K03455	1.0	1.0	0.0	2133	CuReSim
HIV	HXB2	K03455	1.5	1.5	0.0	2133	CuReSim
HIV	HXB2	K03455	2.0	2.0	0.0	2133	CuReSim
HIV	HXB2	K03455	2.5	2.5	0.0	2133	CuReSim
HIV	HXB2	K03455	0.5	0.5	1.0	2133	CuReSim
HIV	HXB2	K03455	1.0	1.0	2.0	2133	CuReSim
HIV	HXB2	K03455	1.5	1.5	3.0	2133	CuReSim
HIV	HXB2	K03455	2.0	2.0	4.0	2133	CuReSim
HIV	HXB2	K03455	2.5	2.5	5.0	2133	CuReSim
Mixed Viruses: Caliciviridae, Filoviridae, Pneumoviridae	Norovirus, Ebola virus, RSV	KM198529, KM198528, KM198511, KM198500, KM198486, KU296608, KU296553, KU296549, KU296528, KU296416, KP317952, KP317946, KP317934, KP317923, KP317922	0.0	0.0	0.0	200,000	WGSIM
Mixed Viruses: Caliciviridae, Filoviridae, Pneumoviridae	Norovirus, Ebola virus, RSV	KM198529, KM198528, KM198511, KM198500, KM198486, KU296608, KU296553, KU296549, KU296528, KU296416, KP317952, KP317946, KP317934, KP317923, KP317922	1.0	1.0	1.0	200,000	WGSIM
Mixed Viruses, Caliciviridae, Filoviridae, Pneumoviridae	Norovirus, Ebola virus, RSV	KM198529, KM198528, KM198511, KM198500, KM198486, KU296608, KU296553, KU296549, KU296528, KU296416, KP317952, KP317946, KP317934, KP317923, KP317922	3.33	3.33	3.33	100,000	WGSIM
Mixed Viruses, Caliciviridae, Filoviridae, Pneumoviridae	Norovirus, Ebola virus, RSV	KM198529, KM198528, KM198511, KM198500, KM198486, KU296608, KU296553, KU296549, KU296528, KU296416, KP317952, KP317946, KP317934, KP317923, KP317922	6.66	6.66	6.66	200,000	WGSIM

Furthermore, 15 publicly available real virus datasets were used for the evaluation of our methodology. The real datasets comprise 5 Norovirus, 5 Ebola virus and 5 human respiratory syncytial virus (RSV) short read datasets. Norovirus NGS datasets (ERR225628, ERR225629, ERR225631, ERR225632, ERR225633) were generated from diarrhoeal patients in Vietnam [49]. Group A rotavirus datasets were obtained from human and pig samples from Vietnam [50]. Human coronavirus NL63 datasets were obtained from Kenya [51]. The Ebola virus datasets (SRR3107337, SRR3107338, SRR3107340, SRR3107342, SRR3107343) were retrieved from the bioproject PRJNA309162, generated during the outbreaks in West Africa in 2013–2016 [52]. The human respiratory syncytial virus (RSV) datasets (ERR303259, ERR303260, ERR303261, ERR303262, ERR303263) [53] were generated from humans in Kenya. All 15 datasets are publicly available. The accession numbers of Sequence Read Archive (SRA) and European Nucleotide Archive (ENA) can be found in Table 3.

Table 3. Real short reads data. Rows contain information for each real reads' dataset (i.e., virus family, virus, genome strain GenBank ID, SRA project ID, number of reads and technology used to sequence data). SRA: Sequence Read Archive; ENA: European Nucleotide Archive.

Family	Virus	Amplicon/Random Primer	GenBank Genome ID	ENA/SRA_ID	Reads	Sequencing Technology
Caliciviridae	Norovirus	Amplicon	KM198486	ERR225628	2126502	Illumina MiSeq
Caliciviridae	Norovirus	Amplicon	KM198500	ERR225629	3037674	Illumina MiSeq
Caliciviridae	Norovirus	Amplicon	KM198511	ERR225631	3285078	Illumina MiSeq
Caliciviridae	Norovirus	Amplicon	KM198528	ERR225632	4361884	Illumina MiSeq
Caliciviridae	Norovirus	Amplicon	KM198529	ERR225633	5187234	Illumina MiSeq
Filoviridae	Ebola virus	Amplicon	KU296608	SRR3107337	522968	Ion Torrent PGM
Filoviridae	Ebola virus	Amplicon	KU296549	SRR3107338	771031	Ion Torrent PGM
Filoviridae	Ebola virus	Amplicon	KU296416	SRR3107340	186657	Ion Torrent PGM
Filoviridae	Ebola virus	Amplicon	KU296553	SRR3107342	478346	Ion Torrent PGM
Filoviridae	Ebola virus	Amplicon	KU296528	SRR3107343	42410	Ion Torrent PGM
Pneumoviridae	RSV	Amplicon	KP317934	ERR303259	7275032	Illumina MiSeq
Pneumoviridae	RSV	Amplicon	KP317922	ERR303260	9278070	Illumina MiSeq
Pneumoviridae	RSV	Amplicon	KP317946	ERR303261	11111114	Illumina MiSeq
Pneumoviridae	RSV	Amplicon	KP317923	ERR303262	13293226	Illumina MiSeq
Pneumoviridae	RSV	Amplicon	KP317952	ERR303263	15237848	Illumina MiSeq

The HIV-1 HXB2 genome (K03455) was used as a reference index to align and/or run the taxonomic classification analysis for the HXB2 simulated dataset. The Norovirus genome KM198509, the Ebola virus genome KM034562 and the RSV genome KP317934 were used as a reference index to align and/or run the taxonomic classification analysis for the mixed virus datasets. The Norovirus genome KM198509 was used to run the taxonomic classification analysis on the real Norovirus datasets, the Ebola virus genome KM034562 was used to run the taxonomic classification analysis on the real Ebola datasets and the RSV genome KP317934 was used to perform the taxonomic classification analysis on the real RSV datasets. All reference genomes used in this study are available from the NCBI (https://www.ncbi.nlm.nih.gov/genome), and accession numbers can be found in Table 4.

Table 4. Reference genomes used during classification and reference-based alignment.

Family	Virus	GenBank ID:	Length (nt)
Retroviridae	Human immunodeficiency virus 1 (HXB2)	K03455	9179
Caliciviridae	Norovirus	KM198509.1	7425
Filoviridae	Zaire ebolavirus	KM034562.1	18957
Pneumoviridae	Human orthopneumovirus (Respiratory Syncytial Virus)	KP317934.1	15233

2.6. Classification and Alignment Evaluation

The accuracy of a classification and an alignment tool can be quantified in terms of the *F*-measure [48], a balanced measure of precision and recall, with precision = true positive/true positive + false positive, recall = true positive/true positive + false negative and the *F*-measure = 2 × (precision *x* recall)/(precision + recall) [48]. In the case of simulated data, information concerning the position of the read on the reference and alignment direction can be used to establish the correctness of alignment, and thereby provide a more informative *F*-measure score. Unclassified reads are considered a false negative result. Any reported match to the correct region of the genome in the correct direction is considered a true positive result. However, if the alignment position or direction information is unavailable, the *F*-measure can be calculated from the number of hits reported for a read, or the absence of a hit. Again, unclassified reads are considered false negative results, and classified reads are considered true positive results. In the case of mixed genome data, the *F*-measure score can be calculated by taking into consideration the number of hits that are reported for a read, as well as if a read is assigned to a reference genome from the same family. If a read is assigned to a genome from a different virus family, it is considered a false positive result, while unclassified reads are considered a false negative result.

3. Results

3.1. Classification by Numbers (CBN)

For the taxonomic classification analysis, a classification tool was implemented in C++ (https://github.com/Avramis/ClassificationByNumbers). The implementation was developed to evaluate our methodology but was not optimised for speed. Users might specify parameters, such as the representation method, transformation method, search stringency and the *k*-mer length. A *VP*-tree indexing structure classified reads using a given set of genomic references. *VP*-tree construction began with the extraction of all unique *k*-mers, of a user-specified length *k*, from the set of supplied reference genomes. Each unique *k*-mer was represented in numerical sequence and then transformed into a lower dimensional space. The transformed data were then used to generate the *VP*-tree indexing structure. Subsequently, each short read from a query set was converted into numerical space, transformed to a lower dimensional space and evaluated against the *VP*-tree. The approximate solution arising from this was then evaluated using the original data to identify false positive matches. The CBN algorithm generated two output files. The first output was a text file providing detailed information on all of the classification matches generated for each read, including the reference name, the direction in which the query read was aligned to the reference, the start and end position of the query on the reference, the alignment score, the CIGAR string describing how the read aligns with the reference and the actual alignment of the query read on the reference genome. The second tabular output file provided a brief overview of the alignment. Each line contained the name of the read, the number of classifications generated for that particular read, the highest classification score obtained, the name of the reference, which provided the highest classification score, the alignment direction and starting position on the reference.

The CBN tool was evaluated against NCBI-BLAST 2.8.1 BLASTn [54] and Kaiju 1.6.3 [55] classifier tools. BLASTn performs the analysis in nucleotide space, whereas Kaiju translates nucleotide sequences from every possible reading frame and performs the analysis in protein sequence space. Figures 3–5 illustrate the results of the classification evaluation process. Both BLASTn and Kaiju were evaluated using their default parameters. CBN was evaluated using *k*-mers of 100, 150, 200, 250 and 300 for the HXB2 simulated reads and 50, 100 and 150 for the mixed virus and real datasets. For the DFT and PAA methods, we evaluated the use of transformation/approximations with 2, 4, 6, 8, 10 and 12 Fourier frequencies or PAA coefficients, respectively. For the DWT variant, we tested the cases of 2, 4, 8, 16 and 32 wavelets.

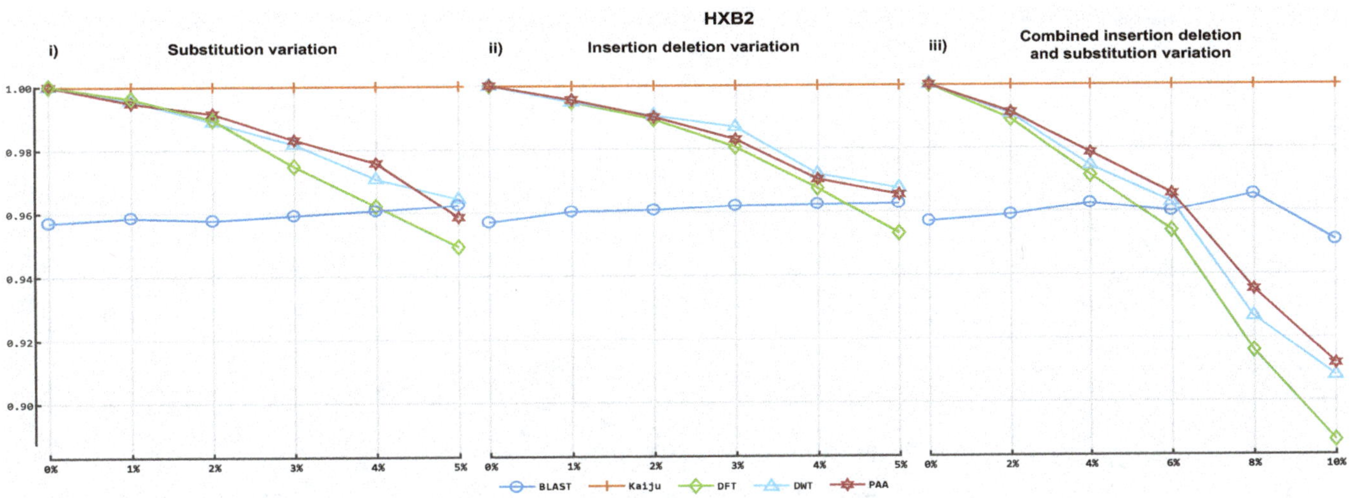

Figure 3. Accuracy of our prototype classification implementation and two established tools on HIV-1 HXB2 simulated datasets. All plots illustrate the *F*-measures obtained on the 16 different HIV datasets. The y-axis indicates the *F*-measure score, and the x-axis depicts the reads data files. Plot 3-i depicts the *F*-measures obtained for each classifier on the simulations with 0% to 5% of substitution variation rate. Plot 3-ii illustrates the *F*-measures obtained for each classifier on the simulations with 0% to 5% uniform insertion/deletion variation, and plot 3-iii illustrates the *F*-measures obtained for each tool on simulations of uniform 0% to 10% insertion/deletion and substitution variation.

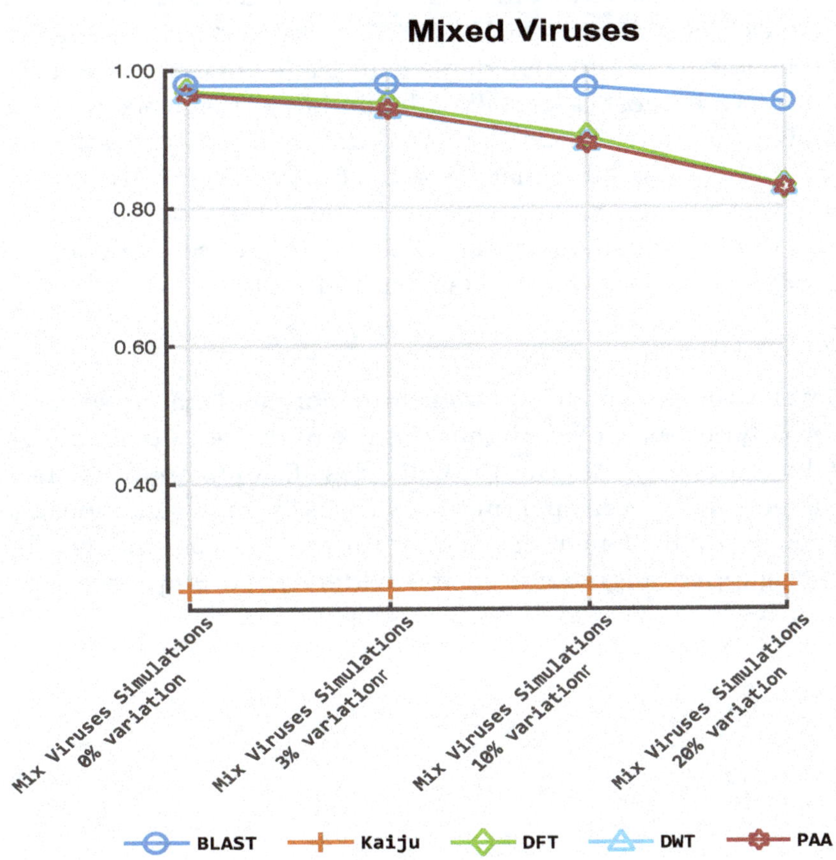

Figure 4. Accuracy of our prototype classification implementation and two established tools on mixed viruses simulated datasets. The *y*-axis indicates the *F*-measure score, and the *x*-axis depicts the reads data files. The plot depicts the *F*-measures obtained for each classifier on the mixed virus simulations. DFT: discrete Fourier transform; DWT: discrete wavelet transform; PAA: piece-wise aggregate approximation.

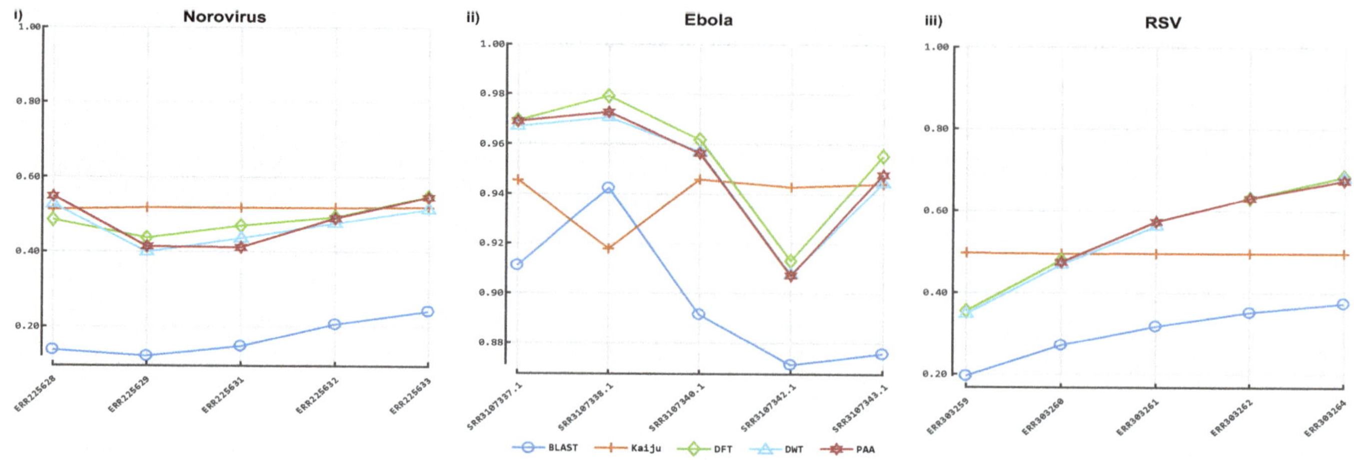

Figure 5. Accuracy of our prototype classification implementation and two established tools on real sequences. The *y*-axis indicates the *F*-measure score, and the *x*-axis depicts the reads data files. Plot 5-i depicts the *F*-measures obtained for each classifier on the Norovirus sequences data. Plot 5-ii illustrates the *F*-measures obtained for each classifier on the Ebola sequence data. Plot 5-iii illustrates the *F*-measures obtained for each tool on Respiratory syncytial virus (RSV) sequence data. DFT: discrete Fourier transform; DWT: discrete wavelet transform; PAA: piece-wise aggregate approximation.

Figure 3 shows the results obtained from the classification process on HIV- 1 HXB2 data. Figure 4 illustrates the results of the mixed virus datasets. Figure 5 illustrates the results obtained from the real data. For taxonomic classification of HIV-1 HXB2 simulated reads, where the short reads were classified against the genome used to generate them, Kaiju reported the highest accuracy scores. CBN outperformed BLASTn in most cases, falling behind in terms of accuracy only on datasets with high variation rates. For the mixed virus simulated datasets, where reads were classified against species strains related to those used to generate reads, BLASTn correctly assigned the most species, followed closely by CBN and finally Kaiju. In the evaluation of the tools on the real data, where reads were classified using a publicly available species-specific reference sequence, CBN generated more accurate results than other tools, followed by Kaiju and BLASTn.

3.2. Alignment by Numbers (ALBN)

To test the applicability of sequential data transformations and feature selection for read alignment, we implemented a prototype *k*-NN read aligner (Figure 6) in C++ (available at https://github.com/Avramis/Alignment_by_numbers). As with the CBN classification analysis, the ALBN code was not optimised for speed. Users might specify parameters, such as the representation method, transformation method, search stringency and the *k*-mer length used for seeding alignments. The algorithm's output was used to construct gapped alignments in the widely used Sequence Alignment/Map (SAM) file format.

```
1) Represent short reads and reference genome as numerical sequences.
2) Select a k-mer length.
3) Create transformations of each reference sequence k-mer, build VP-tree, and create
transformations of the initial k-mer of each short read.
4) Identify candidate alignments using data transformations.
for each read i
    candidate_alignments[i] = VPtree.k-NNSearch(query i)
end
5) Align approximate results with original data using the Smith-Waterman (SW)
algorithm:
for each read i
    best_score = null
   best_aln = []
    for each k neighbour in candidate_alignments[i]
        if SW_score(k neighbour, read i)
            best_score = SW_score(k neighbour, read i)
            best_aln = SW_aln(k neighbour, read i)
        end
    end
end
6) Output alignment in Sequence Alignment/Map (SAM) format.
```

Figure 6. Pseudocode for the alignment procedure.

The ALBN tool was evaluated against a set of well-established, widely used, state-of-the-art tools, such as Bowtie2 (version 2.3.1) [56], BWA-MEM (version 0.7.16) [7], GraphMap (version 0.5.2) [57] and Segmehl (version 0.3.4) [58]. Existing state-of-the-art tools were evaluated with default settings. ALBN was evaluated using *k*-mer lengths of 100, 150, 200, 250 and 300 for the HXB2 simulated reads and 50, 100 and 150 for the mixed virus and real datasets. For the DFT and PAA variants, we evaluated the use of transformation/approximations with 2, 4, 6, 8, 10 and 12 frequencies and PAA coefficients accordingly. For the DWT variant, we tested the cases of 2, 4, 8, 16 and 32 wavelets.

Each aligner's accuracy was quantified in terms of the *F*-measure [48]. CuReSim provides information, such as the simulated read's origin on the reference genome and its alignment direction, enabling evaluation of each aligner's output and calculation of alignment accuracy in terms of the *F*-measure. For mixed virus datasets, tool performance was evaluated in terms of ability to match and align reads to the appropriate virus reference genome. For the real data, *F*-measures were calculated according to the number of reads aligned to the given genome or otherwise.

Figures 7–9 illustrate the *F*-measures obtained by evaluating alignments from each aligner. Figure 7 illustrates alignment performance for each of the 16 datasets simulated using the K03455 HIV-1 HXB2 reference genome. Figure 8 illustrates the alignment performance for virus reads simulated with Norovirus genome KM198509.1, Ebola genome KM034562.1 and the RSV genome KP317934.1. Figure 9i–iii illustrate alignment performance (*F*-measure) for alignments of real Norovirus, Ebola virus and RSV sequences against the same reference genomes as those used for simulation.

ALBN provided accurate results in all scenarios tested. Regarding the HIV-1 HXB2 data, where short reads were aligned to the genome used to generate them, ALBN provided the most accurate results in all 16 cases, followed by Bowtie2. This was also the case for the mixed virus datasets, where reads were aligned to reference strains related to those used to generate the dataset. In both cases, GraphMap and BWA-MEM were third and fourth in terms of accuracy, respectively. ALBN also generated the most accurate alignment results using real data, where reads were aligned to species-specific reference genomes.

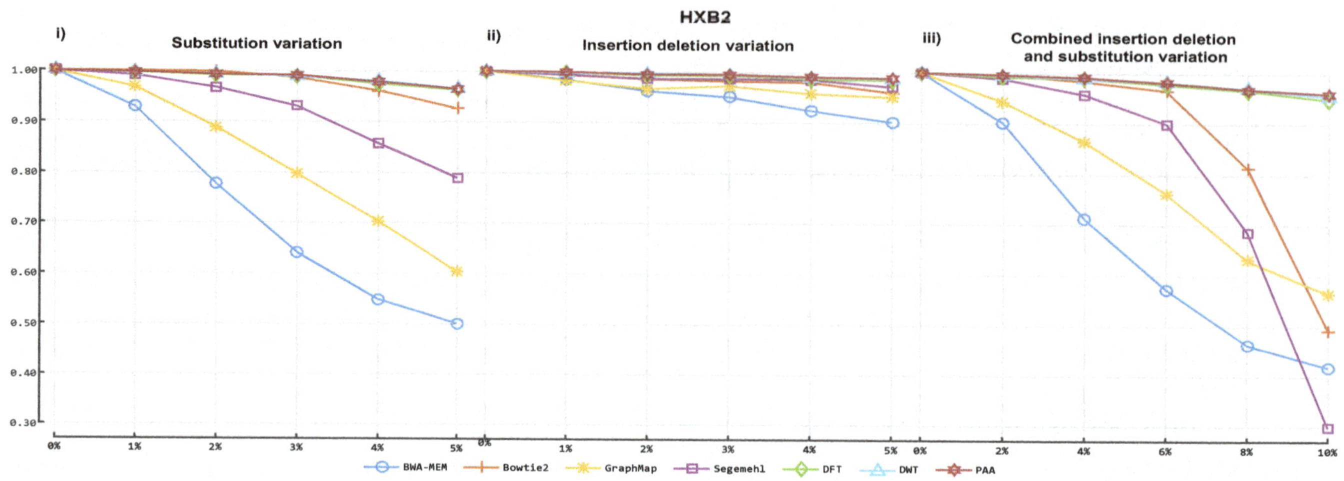

Figure 7. Accuracy of our prototype reference alignment implementation and four established tools on HIV-1 HXB2 simulated datasets. This Figure illustrates the *F*-measures obtained on the 16 different HIV datasets. Plot 6-(**i**) depicts the *F*-measures obtained for each aligner on the simulations with 0% to 5% of substitution variation rate. Plot 6-(**ii**) illustrates the *F*-measures obtained for each aligner on the simulations with 0% to 5% uniform insertion/deletion variation, and plot 6-(**iii**) illustrates the *F*-measures obtained for each tool on simulations of uniform 0% to 10% insertion/deletion and substitution variation. DFT: discrete Fourier transform; DWT: discrete wavelet transform; PAA: piece-wise aggregate approximation.

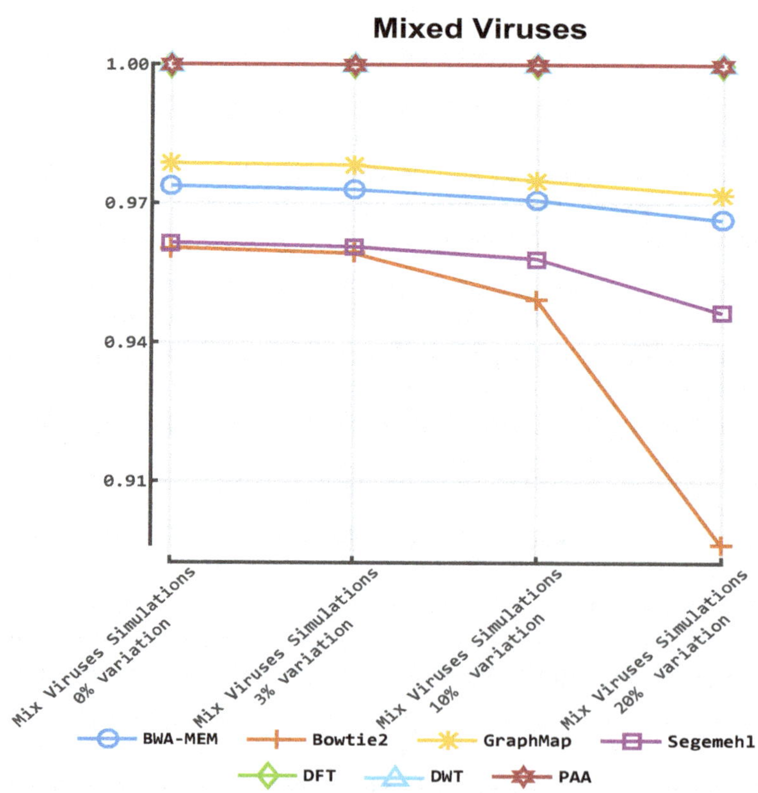

Figure 8. Accuracy of our prototype aligner implementation and four established tools on mixed viruses simulated datasets. The *y*-axis indicates the *F*-measure score, and the *x*-axis depicts the reads data files. The plot depicts the *F*-measures obtained for each aligner on the mixed virus simulations. DFT: discrete Fourier transform; DWT: discrete wavelet transform; PAA: piece-wise aggregate approximation.

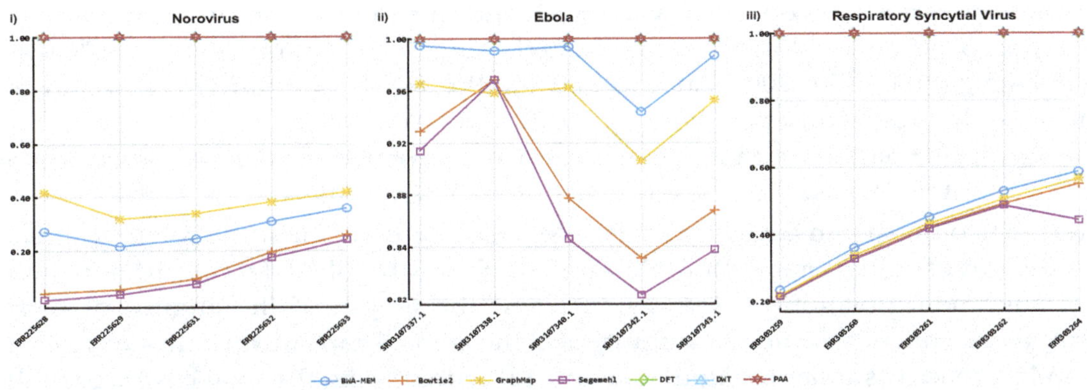

Figure 9. Accuracy of our prototype aligner implementation and four established tools on real sequences datasets. The *y*-axis indicates the *F*-measure score, and the *x*-axis depicts the reads data files. Plot 8-(**i**) depicts the *F*-measures obtained for each aligner on the Norovirus sequences data. Plot 8-(**ii**) illustrates the *F*-measures obtained for each aligner on the Ebola sequences data. Plot 8-(**iii**) illustrates the *F*-measures obtained for each tool on the Respiratory syncytial virus (RSV) sequences data. DFT: discrete Fourier transform; DWT: discrete wavelet transform; PAA: piece-wise aggregate approximation.

3.3. De novo Assembly by Numbers

Lastly, to test the applicability of this approach to the de novo assembly of short reads, we implemented assembly by numbers (ASBN), a prototype algorithm for all-against-all *k*-mer comparison, using data transformations/approximation. Note, preliminary results have been presented as a conference paper [59]. Figure 10 illustrates the main concept of our de novo assembly approach. For the ASBN tool, reads are represented as numerical sequences using an appropriate numerical representation method (Table 1). Here, we used the tetrahedron numerical representation. Every *k*-mer of each numerically represented read was identified and transformed to lower dimensional space using the chosen transformation method. All *k*-mers' transformations were used to build a *VP*-tree, to allow for fast data comparison. Afterwards, all *k*-mers were compared to the rest of the data using the *VP*-tree index. Information from the data comparison was used to construct a weighted graph similar to that shown in Figure 10A. The shortest path on the weighted graph was identified with a breadth-first search (BFS) (Figure 10B). Reads overlaps were used to generate an OLC alignment of short reads (Figure 10C).

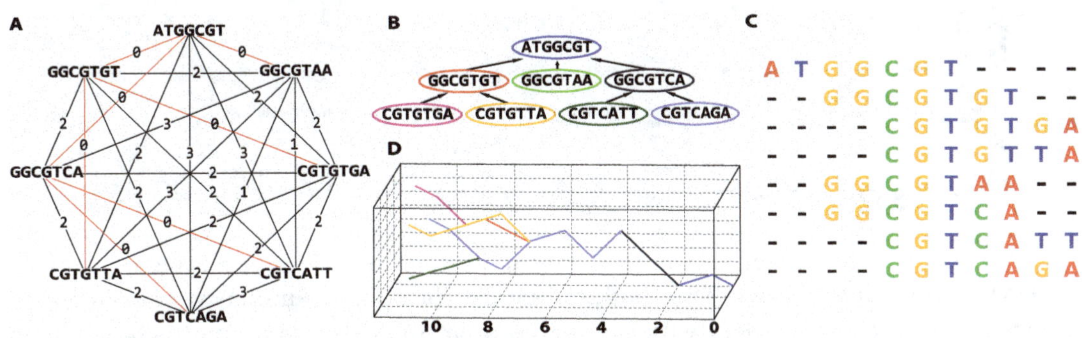

Figure 10. A de novo assembly methodology for numerically represented nucleotide reads. All-against-all sequence comparison (**A**) enables the construction of a read graph with weighted edges. The weight assigned to each edge is the smallest pairwise distance obtained between every possible *k*-mer representation of the two reads. In this example, a 5-mer was used. The smallest distance between every possible *k*-mer can be obtained by either using a sliding window approach or break reads every possible subsequence with length *k*. (**B**) The shortest path in the graph is identified with a breadth-first search algorithm (red coloured edges) thereby (**C**) enabling read alignment. A DNA walk representation of the overlapped reads (**D**) may subsequently be used as a three-dimensional graphical portrayal of the reads, illustrating alignment characteristics.

The ASBN assembler was compared with Megahit (version 1.1.3) [60] and SPAdes (version 3.13.0) [61] de novo assemblers on the HIV-1 HXB2, and mixed virus simulated datasets accordingly. Megahit, SPAdes, BLASTn and Kaiju were evaluated using default parameters. ASBN was evaluated using k-mer lengths 100, 150, 200, 250 and 300 for the HXB2 simulated reads and 50, 100 and 150 for the mixed virus datasets. For the DFT and PAA variants, we evaluated the use of transformation/approximations with 2, 4, 6, 8, 10 and 12 frequencies and PAA coefficients accordingly. For the DWT variant, we tested the cases of 2, 4, 8, 16 and 32 wavelets.

The derived contigs from each assembler were evaluated against the reference genomes used to generate the data simulations with BLASTn [54]. From the BLASTn output, information about the contigs' alignment position on the genome and the length of the alignment were obtained. Subsequently, a measure of assembly contiguity and the sum of gaps/mismatches were calculated and plotted on an X-Y matrix (similar to Figures 11 and 12) with x being the total coverage of the genomes generated and y being the total number of gaps in the coverage. A perfect assembly would have $x =$ full genome length and $y = 0$, indicating that the contig is identical to the genome in terms of length and nucleotide composition. For the HIV-1 HXB2 datasets, the contigs were evaluated against the K03455 genome, and the contigs obtained from the mixed virus datasets were evaluated against the 15 different genomes: KM198529, KM198528, KM198511, KM198500, KM198486, KU296608, KU296553, KU296549, KU296528, KU296416, KP317952, KP317946, KP317934, KP317923 and KP317922.

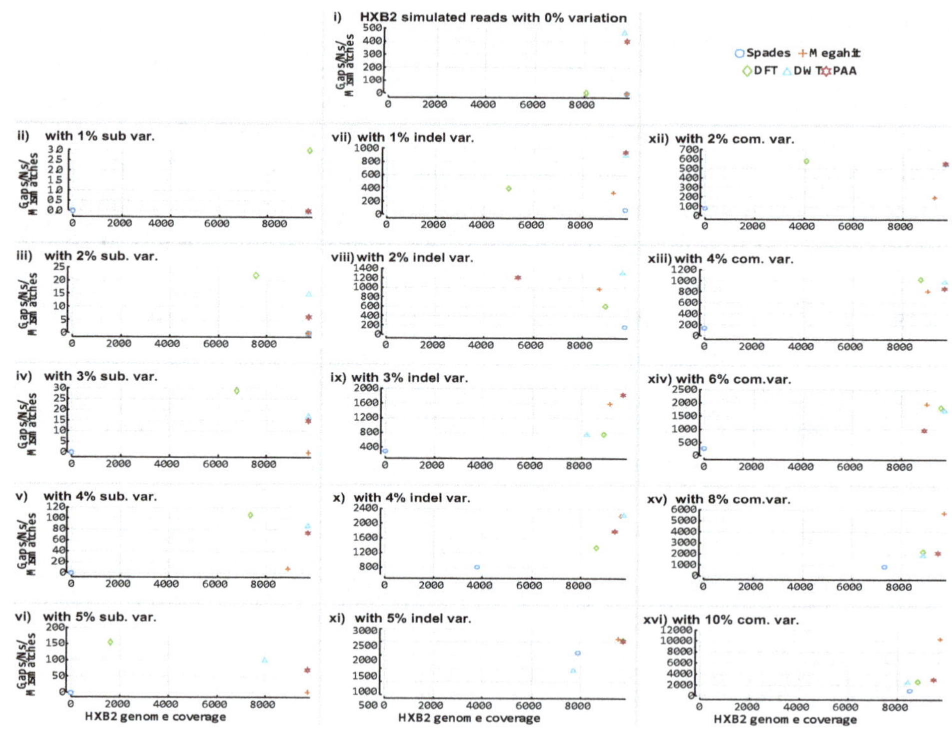

Figure 11. Accuracy of our prototype de novo assembly implementation and two established tools on HIV-1 HXB2 simulated datasets. The contigs obtained for each assembler were evaluated against the reference genome used to generate the simulated data. BLASTn was used to align all contigs to an HIV-1 HXB2 reference genome and determine genome coverage. The y-axis indicates the number of gaps and mismatches that exist in the contigs obtained for each tool, and the x-axis depicts the length of the genome the reported contigs cover. The contigs obtained from the assembly of the HIV-1 HXB2 simulated short read data were evaluated against the K03455 reference genome. Plot 10-**i** illustrates results obtained from all assemblers on variation-free data. Plots 10-**ii** to 10-**vi** illustrate results obtained from all assemblers on data with different levels of substitution variation. Plots 10-**vii** to 10-**xi** illustrate results obtained from all assemblers on data with different levels of insertion/deletion variation. Plots 10-**xii** to 10-**xvi** illustrate results obtained from all assemblers on data with different levels of combined insertion/deletion and substitution variation.

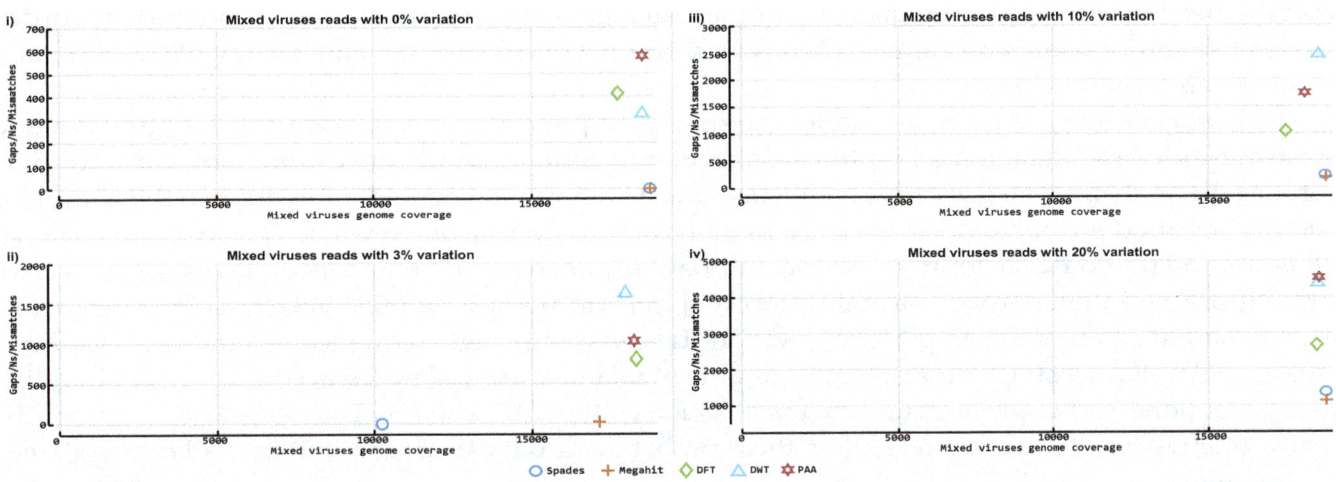

Figure 12. Accuracy of our prototype de novo assembly implementation and two established tools on mixed viruses simulated datasets. The contigs obtained for each assembler were evaluated against the reference genome that was used to generate the simulated data. BLASTn was used to align all contigs to an HIV-1 HXB2 reference genome and determine how much of the particular genome they cover. The y-axis indicates the number of gaps and mismatches that exist in the contigs obtained for each tool, and the x-axis depicts the length of the genome the reported contigs cover. The contigs obtained from the mixed virus simulated dataset were evaluated against the KM198529, KM198528, KM198511, KM198500, KM198486, KU296608, KU296553, KU296549, KU296528, KU296416, KP317952, KP317946, KP317934, KP317923 and KP317922 references genomes. Plots 11-**i** to 11-**iv** illustrate results obtained from all assemblers on data with 0%, 3%, 10% and 20% variation levels accordingly.

Figure 11 illustrates the assembly results of SPAdes, Megahit and all three variants of ASBN on the 16 simulated HIV-1 HXB2 datasets. Figure 12 illustrates the assembly results on the mixed virus simulated databases. Although ASBN processes data and assembles short reads in a lower dimensional space, it nevertheless generated contigs that collectively cover the expected genome length and provided comparable results to both existing state of the art de novo assemblers tested in this experiment (Figures 11 and 12). In all cases, ASBN generated contigs spanning the whole genomes of their respective viral species.

4. Discussion

Although well-established data compression methods for reversible compression of one-dimensional and multivariate signals, images, text and binary exist [62–64], there are very few examples of their application to biological sequence data. We have developed algorithms incorporating signal compression methods for three common biological sequence analysis problems: classification, alignment and de novo assembly of NGS short read virus data. Our results in Figures 3–12 show that this approach permits accurate classification of de novo assembly and reference alignment in spite of high rates of sequence variation or the use of a divergent reference genome. Data approximation/summarisation techniques, such as the DFT, the DWT and the PAA, can be used to extract major features of sequence data and to suppress noise or low-level variation. This allows sequence comparison exploiting the major characteristics of the data, thus enabling the identification of similarities that might otherwise be concealed by minor variation or sequencing error/noise.

Collectively, our results demonstrate that complete nucleotide-level sequence resolution is not a prerequisite of accurate sequence analysis and that analytical performance can be preserved and even enhanced through appropriate dimensionality reduction (compression) of sequences. While our implementations use k-mers, the nature of the transformation/compression methods used shows that optimal k-mer length selection is far less important than the conventional exact k-mer matching methods. The inherent error tolerance of the approach also permits the use of longer k-mers than typically used in conventional sequence comparison algorithms, reducing the computational burden of pairwise comparison, and thus, in de novo assembly specifically, the complexity of building and searching an assembly graph.

Efficient mining of terabase-scale biological sequence datasets requires looking beyond substring-indexing algorithms [65] towards more versatile methods of compression for both data storage and analysis. The use of probabilistic data structures can considerably reduce the computer memory required for in-memory sequence lookups at the expense of a few false positives, and Bloom filters and related data structures have seen broad application in *k*-mer centric tasks, such as error correction [66], in silico read normalisation [67] and de novo assembly [68,69]. However, while these hash-based approaches perform well on datasets with high sequence redundancy, for large datasets with many distinct *k*-mers, large amounts of memory are still necessary [67]. Lower bounding transformations and approximation methods (such as the DFT, the DWT and PAA) can exhibit the same attractive one-sided error offered by these probabilistic data structures, but instead of hash tables, use concrete and reusable sequence representations.

Furthermore, transformations allow compression of standalone sequence composition, enabling flexible reduction of sequence resolution according to analytical requirements, so that redundant sequence precision need not hinder analysis. While the problem of read alignment to a known reference sequence is largely considered solved, assembly of large genomes remains a formidable problem in computing. Moreover, consideration of the metagenomic composition of mixed biological samples, as demonstrated, further extends the scope and scale of the assembly problem beyond what is tractable using conventional sequence comparison approaches. By implementing a reference-based aligner and de novo assembler, we have demonstrated that using compressed numerical representations offers a versatile approach for reconstructing genomes and metagenomes sequenced with short reads.

Emerging long read sequencing technologies bring new challenges for sequence data analysis. Whilst the error rate of Oxford Nanopore sequencing platform, for example, has decreased considerably since the technology's introduction [70,71], the relatively high error rate still limits the scope of downstream analyses [72]. Efficient algorithmic approaches are needed to (1) identify sequence identity/infer homology in spite of abundant insertion/deletion errors associated with the platform, which are problematic for approaches dependent on exact subsequence matching and (2) to overcome issues relating to high data dimensionality and the curse of dimensionality [73]. Both in terms of the raw electric current traces generated by DNA translocation through a nanopore and the corresponding base-called sequences, the resemblance between long reads and time series data from other fields is striking, such that the various transformations/approximations we have implemented will be directly applicable.

In conclusion, nucleotide sequences may be effectively represented as numerical series, enabling the application of existing analytical methods from a variety of mathematical and engineering fields for the purposes of sequence alignment and assembly. By applying established signal decomposition methods, compressed representations of nucleotide sequences can be created, permitting reductions in the spatiotemporal complexity of their analysis, without necessarily compromising analytical accuracy.

Author Contributions: A.T. designed and wrote the methods and software, and performed the data analysis with the help from B.C., M.V.T.P., S.K. and M.C., B.C. and M.C. generated the simulated data, and B.C., M.V.T.P., and M.C. helped perform the data evaluation. A.T. and D.L.R. conceived the study. A.T. and B.C. wrote the manuscript with comments from M.V.T.P., M.C. and D.L.R. All authors read and approved the final manuscript.

References

1. Margulies, M.; Egholm, M.; Altman, W.E.; Attiya, S.; Bader, J.S.; Bemben, L.A.; Berka, J.; Braverman, M.S.; Chen, Y.-J.; Chen, Z. Genome sequencing in microfabricated high-density picolitre reactors. *Nature* **2005**, *437*, 376–380. [CrossRef] [PubMed]
2. Bentley, D.R.; Balasubramanian, S.; Swerdlow, H.P.; Smith, G.P.; Milton, J.; Brown, C.G.; Hall, K.P.; Evers, D.J.; Barnes, C.L.; Bignell, H.R. Accurate whole human genome sequencing using reversible terminator chemistry. *Nature* **2008**, *456*, 53–59. [CrossRef]

3. Rothberg, J.M.; Hinz, W.; Rearick, T.M.; Schultz, J.; Mileski, W.; Davey, M.; Leamon, J.H.; Johnson, K.; Milgrew, M.J.; Edwards, M. An integrated semiconductor device enabling non-optical genome sequencing. *Nature* **2011**, *475*, 348–352. [CrossRef]
4. Eid, J.; Fehr, A.; Gray, J.; Luong, K.; Lyle, J.; Otto, G.; Peluso, P.; Rank, D.; Baybayan, P.; Bettman, B. Real-time DNA sequencing from single polymerase molecules. *Science* **2009**, *323*, 133–138. [CrossRef]
5. Salipante, S.J.; Roach, D.J.; Kitzman, J.O.; Snyder, M.W.; Stackhouse, B.; Butler-Wu, S.M.; Lee, C.; Cookson, B.T.; Shendure, J. Large-scale genomic sequencing of extraintestinal pathogenic Escherichia coli strains. *Genome Res.* **2015**, *25*, 119–128. [CrossRef] [PubMed]
6. Rose, R.; Constantinides, B.; Tapinos, A.; Robertson, D.L.; Prosperi, M. Challenges in the analysis of viral metagenomes. *Virus Evol.* **2016**, *2*. [CrossRef] [PubMed]
7. Li, H.; Durbin, R. Fast and accurate short read alignment with Burrows–Wheeler transform. *Bioinformatics* **2009**, *25*, 1754–1760. [CrossRef] [PubMed]
8. Shrestha, A.M.S.; Frith, M.C.; Horton, P. A bioinformatician's guide to the forefront of suffix array construction algorithms. *Brief. Bioinform.* **2014**, *15*, 138–154. [CrossRef]
9. Myers, E.W. Toward simplifying and accurately formulating fragment assembly. *J. Comput. Biol.* **1995**, *2*, 275–290. [CrossRef]
10. Kececioglu, J.D.; Myers, E.W. Combinatorial algorithms for DNA sequence assembly. *Algorithmica* **1995**, *13*, 7–51. [CrossRef]
11. Earl, D.; Bradnam, K.; John, J.S.; Darling, A.; Lin, D.; Fass, J.; Yu, H.O.K.; Buffalo, V.; Zerbino, D.R.; Diekhans, M. Assemblathon 1: A competitive assessment of de novo short read assembly methods. *Genome Res.* **2011**, *21*, 2224–2241. [CrossRef] [PubMed]
12. Iqbal, Z.; Caccamo, M.; Turner, I.; Flicek, P.; McVean, G. De novo assembly and genotyping of variants using colored de Bruijn graphs. *Nat. Genet.* **2012**, *44*, 226–232. [CrossRef]
13. Pevzner, P.A.; Tang, H.; Waterman, M.S. An Eulerian path approach to DNA fragment assembly. *Proc. Natl. Acad. Sci. USA* **2001**, *98*, 9748–9753. [CrossRef]
14. Bradnam, K.R.; Fass, J.N.; Alexandrov, A.; Baranay, P.; Bechner, M.; Birol, I.; Boisvert, S.; Chapman, J.A.; Chapuis, G.; Chikhi, R. Assemblathon 2: Evaluating de novo methods of genome assembly in three vertebrate species. *Gigascience* **2013**, *2*, 1–31. [CrossRef] [PubMed]
15. Archer, J.; Rambaut, A.; Taillon, B.E.; Harrigan, P.R.; Lewis, M.; Robertson, D.L. The evolutionary analysis of emerging low frequency HIV-1 CXCR4 using variants through time—An ultra-deep approach. *PLoS Comput. Biol.* **2010**, *6*, e1001022. [CrossRef] [PubMed]
16. Clement, N.L.; Thompson, L.P.; Miranker, D.P. ADaM: Augmenting existing approximate fast matching algorithms with efficient and exact range queries. *BMC Bioinform.* **2014**, *15*, S1. [CrossRef] [PubMed]
17. Agrawal, R.; Faloutsos, C.; Swami, A. Efficient similarity search in sequence databases. In Proceedings of the 4th International Conference on Foundations of Data Organization and Algorithms, Chicago, IL, USA, 13–15 October 1993.
18. Chan, K.-P.; Fu, A.-C. Efficient time series matching by wavelets. In Proceedings of the 15th International Conference on Data Engineering, Sydney, Australia, 23–26 March 1999; pp. 126–133.
19. Woodward, A.M.; Rowland, J.J.; Kell, D.B. Fast automatic registration of images using the phase of a complex wavelet transform: Application to proteome gels. *Analyst* **2004**, *129*, 542–552. [CrossRef] [PubMed]
20. Geurts, P. Pattern extraction for time series classification. In Proceedings of the 5th European Conference on Principles of Data Mining and Knowledge Discovery, Freiburg, Germany, 3–7 September 2001; pp. 115–127.
21. Keogh, E.; Chakrabarti, K.; Pazzani, M.; Mehrotra, S. Locally adaptive dimensionality reduction for indexing large time series databases. *ACM SIGMOD Record* **2001**, *30*, 151–162. [CrossRef]
22. Shumway, R.H.; Stoffer, D.S.; Stoffer, D.S. *Time Series Analysis and Its Applications with R examples*, 2nd ed.; Springer: New York, NY, USA, 2006.
23. Silverman, B.; Linsker, R. A measure of DNA periodicity. *J. Theor. Biol.* **1986**, *118*, 295–300. [CrossRef]
24. Cheever, E.; Searls, D.; Karunaratne, W.; Overton, G. Using signal processing techniques for DNA sequence comparison. In Proceedings of the Fifteenth Annual Northeast Bioengineering Conference, Boston, MA, USA, 27–28 March 1989; pp. 173–174.
25. Katoh, K.; Misawa, K.; Kuma, K.i.; Miyata, T. MAFFT: A novel method for rapid multiple sequence alignment based on fast Fourier transform. *Nucleic Acids Res.* **2002**, *30*, 3059–3066. [CrossRef] [PubMed]

26. Kwan, H.K.; Arniker, S.B. Numerical representation of DNA sequences. In Proceedings of the 2009 IEEE International Conference on Electro/Information Technology, Windsor, ON, Canada, 7–9 June 2009; pp. 307–310.
27. Yi, B.-K.; Faloutsos, C. Fast time sequence indexing for arbitrary Lp norms. In Proceedings of the 26th roceedings of 26th International Conference on Very Large Data Bases, Cairo, Egypt, 10–14 September 2000; pp. 385–394.
28. Keogh, E.; Ratanamahatana, C.A. Exact indexing of dynamic time warping. *Knowl. Inf. Syst.* **2005**, *7*, 358–386. [CrossRef]
29. Vlachos, M.; Kollios, G.; Gunopulos, D. Discovering similar multidimensional trajectories. In Proceedings of the 18th International Conference on Data Engineering, San Jose, CA, USA, 26 February–1 March 2002; pp. 673–684.
30. Kotsakos, D.; Trajcevski, G.; Gunopulos, D.; Aggarwal, C.C. *In Data Clustering: Algorithms and Applications*; Aggarwal, C.C., Reddy, C., Eds.; CRC Press: Boca Raton, FL, USA, 2013; Chapter 15; pp. 357–379.
31. Chávez, E.; Navarro, G.; Baeza-Yates, R.; Marroquín, J.L. Searching in metric spaces. *ACM Comput. Surv. (CSUR)* **2001**, *33*, 273–321. [CrossRef]
32. Beckmann, N.; Kriegel, H.-P.; Schneider, R.; Seeger, B. The R*-tree: An efficient and robust access method for points and rectangles. *SIGMOD Rec.* **1990**, *19*, 322–331. [CrossRef]
33. Agrawal, R.; Lin, K.; Sawhney, H.S.; Shim, K. Fast similarity search in the presence of noise, scaling, and translation in time-series databases. In Proceedings of the 21th International Conference on Very Large Data Bases, Zurich, Switzerland, 11–15 September 1995; pp. 490–501.
34. Bingham, S.; Kot, M. Multidimensional trees, range searching, and a correlation dimension algorithm of reduced complexity. *Phys. Lett. A* **1989**, *140*, 327–330. [CrossRef]
35. Bellman, R. *Adaptive Control Processes: A Guided Tour*; Princeton University Press: London, UK, 1961; Volume 4.
36. Verleysen, M.; François, D. The Curse of Dimensionality in Data Mining and Time Series Prediction. In Proceedings of the 8th International Work-Conference on Artificial Neural Networks, Barcelona, Spain, 8–10 June 2005; pp. 758–770.
37. Yianilos, P.N. Data structures and algorithms for nearest neighbor search in general metric spaces. In Proceedings of the 4th annual ACM-SIAM Symposium on Discrete Algorithms, Austin, TX, USA; 1993; pp. 311–321.
38. Bozkaya, T.; Ozsoyoglu, M. Indexing large metric spaces for similarity search queries. *ACM Trans. Database Syst. (TODS)* **1999**, *24*, 361–404. [CrossRef]
39. Uhlmann, J.K. Satisfying general proximity/similarity queries with metric trees. *Inf. Process. Lett.* **1991**, *40*, 175–179. [CrossRef]
40. Nair, A.S.; Sreenadhan, S.P. A coding measure scheme employing electron-ion interaction pseudopotential (EIIP). *Bioinformation* **2006**, *1*, 197. [PubMed]
41. Holden, T.; Subramaniam, R.; Sullivan, R.; Cheung, E.; Schneider, C.; Tremberger, G.; Flamholz, A.; Lieberman, D.H.; Cheung, T.D. ATCG nucleotide fluctuation of Deinococcus radiodurans radiation genes. In Proceedings of the Instruments, Methods, and Missions for Astrobiology X, San Diego, CA, USA, 1 October 2007.
42. Voss, R.F. Evolution of long-range fractal correlations and 1/f noise in DNA base sequences. *Phys. Rev. Lett.* **1992**, *68*, 3805. [CrossRef] [PubMed]
43. Faloutsos, C.; Ranganathan, M.; Manolopoulos, Y. Fast subsequence matching in time-series databases. In Proceedings of the 1994 ACM SIGMOD International Conference on Management of data, Minneapolis, MN, USA, 24–27 May 1994.
44. Mitsa, T. *Temporal Data Mining*; CRC Press: New York, NY, USA, 2010.
45. Mörchen, F. *Time Series Feature Extraction for Data Mining Using DWT and DFT*; Technical Report 3; Departement of Mathematics and Computer Science Philipps-University Marburg: Marburg, Germany, 2003; pp. 735–739.
46. Jensen, A.; la Cour-Harbo, A. *Ripples in Mathematics: The Discrete Wavelet Transform*; Springer: Berlin, Germany, 2001.
47. Wu, Y.-L.; Agrawal, D.; El Abbadi, A. A comparison of DFT and DWT based similarity search in time-series databases. In Proceedings of the 9th International Conference on Information and Knowledge Management, Washington, DC, USA, 6–11 November 2000; pp. 488–495.

48. Caboche, S.; Audebert, C.; Lemoine, Y.; Hot, D. Comparison of mapping algorithms used in high-throughput sequencing: Application to Ion Torrent data. *BMC Genom.* **2014**, *15*, 264. [CrossRef] [PubMed]
49. Cotten, M.; Petrova, V.; Phan, M.V.; Rabaa, M.A.; Watson, S.J.; Ong, S.H.; Kellam, P.; Baker, S. Deep sequencing of norovirus genomes defines evolutionary patterns in an urban tropical setting. *J. Virol.* **2014**, *88*, 11056–11069. [CrossRef] [PubMed]
50. Phan, M.V.; Anh, P.H.; Cuong, N.V.; Munnink, B.B.O.; van der Hoek, L.; My, P.T.; Tri, T.N.; Bryant, J.E.; Baker, S.; Thwaites, G. Unbiased whole-genome deep sequencing of human and porcine stool samples reveals circulation of multiple groups of rotaviruses and a putative zoonotic infection. *Virus Evol.* **2016**, *2*. [CrossRef]
51. Kiyuka, P.K.; Agoti, C.N.; Munywoki, P.K.; Njeru, R.; Bett, A.; Otieno, J.R.; Otieno, G.P.; Kamau, E.; Clark, T.G.; van der Hoek, L. Human Coronavirus NL63 Molecular Epidemiology and Evolutionary Patterns in Rural Coastal Kenya. *J. Infect. Dis.* **2018**, *217*, 1728–1739. [CrossRef] [PubMed]
52. Arias, A.; Watson, S.J.; Asogun, D.; Tobin, E.A.; Lu, J.; Phan, M.V.; Jah, U.; Wadoum, R.E.G.; Meredith, L.; Thorne, L. Rapid outbreak sequencing of Ebola virus in Sierra Leone identifies transmission chains linked to sporadic cases. *Virus Evol.* **2016**, *2*. [CrossRef] [PubMed]
53. Agoti, C.N.; Otieno, J.R.; Munywoki, P.K.; Mwihuri, A.G.; Cane, P.A.; Nokes, D.J.; Kellam, P.; Cotten, M. Local evolutionary patterns of human respiratory syncytial virus derived from whole-genome sequencing. *J. Virol.* **2015**, *89*, 3444–3454. [CrossRef] [PubMed]
54. Altschul, S.F.; Gish, W.; Miller, W.; Myers, E.W.; Lipman, D.J. Basic local alignment search tool. *J. Mol. Biol.* **1990**, *215*, 403–410. [CrossRef]
55. Menzel, P.; Ng, K.L.; Krogh, A. Fast and sensitive taxonomic classification for metagenomics with Kaiju. *Nat. Commun.* **2016**, *7*, 11257. [CrossRef] [PubMed]
56. Langmead, B.; Salzberg, S.L. Fast gapped-read alignment with Bowtie 2. *Nat. Methods* **2012**, *9*, 357. [CrossRef] [PubMed]
57. Sović, I.; Šikić, M.; Wilm, A.; Fenlon, S.N.; Chen, S.; Nagarajan, N. Fast and sensitive mapping of nanopore sequencing reads with GraphMap. *Nat. Commun.* **2016**, *7*, 11307. [CrossRef] [PubMed]
58. Otto, C.; Stadler, P.F.; Hoffmann, S. Lacking alignments? The next-generation sequencing mapper segemehl revisited. *Bioinform.* **2014**, *30*, 1837–1843. [CrossRef] [PubMed]
59. Tapinos, A.; Robertson, D.L. De novo assembly of nucleotide sequences in a compressed feature space. In Proceedings of the 2017 IEEE Conference on Computational Intelligence in Bioinformatics and Computational Biology (CIBCB), Manchester, UK, 23–25 August 2017; pp. 1–7.
60. Li, D.; Liu, C.-M.; Luo, R.; Sadakane, K.; Lam, T.-W. MEGAHIT: An ultra-fast single-node solution for large and complex metagenomics assembly via succinct de Bruijn graph. *Bioinformatics* **2015**, *31*, 1674–1676. [CrossRef]
61. Anton, B.; Sergey, N.; Dmitry, A.; Alexey, A.; Mikhail, D.; Alexander, S.; Valery, M.; Sergey, I.; Son, P.; Andrey, D. SPAdes: A new genome assembly algorithm and its applications to single-cell sequencing. *J. Comput. Biol.* **2012**, *19*, 455.
62. Tapinos, A.; Mendes, P. A method for comparing multivariate time series with different dimensions. *PloS ONE* **2013**, *8*, e54201. [CrossRef]
63. Sheybani, E.O. An Algorithm for Real-Time Blind Image Quality Comparison and Assessment. *Int. J. Electr. Comput. Eng. (IJECE)* **2011**, *2*, 120–129. [CrossRef]
64. Hendriks, R.C.; Gerkmann, T.; Jensen, J. DFT-domain based single-microphone noise reduction for speech enhancement: A survey of the state of the art. In *Synthesis Lectures on Speech and Audio Processing*; Morgan & Claypool: San Rafael, CA, USA, 2013; pp. 1–80.
65. Kouchaki, S.; Tapinos, A.; Robertson, D.L. A signal processing method for alignment-free metagenomic binning: Multi-resolution genomic binary patterns. *Sci. Rep.* **2019**, *9*, 2159. [CrossRef]
66. Shi, H.; Schmidt, B.; Liu, W.; Müller-Wittig, W. A Parallel Algorithm for Error Correction in High-Throughput Short-Read Data on CUDA-Enabled Graphics Hardware. *J. Comput. Biol.* **2010**, *17*, 603–615. [CrossRef]
67. Zhang, Q.; Pell, J.; Canino-Koning, R.; Howe, A.C.; Brown, C.T. These Are Not the K-mers You Are Looking For: Efficient Online K-mer Counting Using a Probabilistic Data Structure. *PLoS ONE* **2014**, *9*, e101271. [CrossRef]
68. Salikhov, K.; Sacomoto, G.; Kucherov, G. Using cascading Bloom filters to improve the memory usage for de Brujin graphs. *Algorithms Mol. Biol.* **2014**, *9*, 364–376. [CrossRef]

69. Berlin, K.; Koren, S.; Chin, C.-S.; Drake, J.P.; Landolin, J.M.; Phillippy, A.M. Assembling large genomes with single-molecule sequencing and locality-sensitive hashing. *Nat. Biotechnol.* **2015**, *33*, 623–630. [CrossRef]
70. Laver, T.; Harrison, J.; O'neill, P.; Moore, K.; Farbos, A.; Paszkiewicz, K.; Studholme, D.J. Assessing the performance of the oxford nanopore technologies minion. *Biomol. Detect. Quantif.* **2015**, *3*, 1–8. [CrossRef]
71. Fu, S.; Wang, A.; Au, K.F. A comparative evaluation of hybrid error correction methods for error-prone long reads. *Genome Biol.* **2019**, *20*, 26. [CrossRef]
72. Watson, M.; Warr, A. Errors in long-read assemblies can critically affect protein prediction. *Nature Biotechnol.* **2019**, *37*, 124. [CrossRef]
73. Radovanović, M.; Nanopoulos, A.; Ivanović, M. Time-series classification in many intrinsic dimensions. In Proceedings of the 2010 SIAM International Conference on Data Mining, Columbus, OH, USA, 29 April–1 May 2010; pp. 677–688.

14

Base-By-Base Version 3: New Comparative Tools for Large Virus Genomes

Shin-Lin Tu [1], Jeannette P. Staheli [2], Colum McClay [1], Kathleen McLeod [1], Timothy M. Rose [2,3] and Chris Upton [1,*]

1 Biochemistry and Microbiology, University of Victoria, Victoria, BC V8W 2Y2, Canada; cindytu@uvic.ca (S.-L.T.); colummcclay@hotmail.com (C.M.); kathleen_mcleod14@hotmail.com (K.M.)
2 Center for Global Infectious Disease Research, Seattle Children's Research Institute, Seattle, WA 98101, USA; jeannette.staheli@seattlechildrens.org (J.P.S.); trose@u.washington.edu (T.M.R.)
3 Department of Pediatrics, University of Washington, Seattle, WA 98195, USA
* Correspondence: cupton@uvic.ca

Abstract: Base-By-Base is a comprehensive tool for the creation and editing of multiple sequence alignments that is coded in Java and runs on multiple platforms. It can be used with gene and protein sequences as well as with large viral genomes, which themselves can contain gene annotations. This report describes new features added to Base-By-Base over the last 7 years. The two most significant additions are: (1) The recoding and inclusion of "consensus-degenerate hybrid oligonucleotide primers" (CODEHOP), a popular tool for the design of degenerate primers from a multiple sequence alignment of proteins; and (2) the ability to perform fuzzy searches within the columns of sequence data in multiple sequence alignments to determine the distribution of sequence variants among the sequences. The intuitive interface focuses on the presentation of results in easily understood visualizations and providing the ability to annotate the sequences in a multiple alignment with analytic and user data.

Keywords: bioinformatics; virus; comparative genomics; software; Base-By-Base; BBB; poxvirus; ASFV; MSA

1. Introduction

Base-By-Base (BBB) [1,2], a multiple sequence alignment (MSA) editor, has been under development for more than 15 years and forms an integral component in the Viral Bioinformatics Resource Center (VBRC) platform (www.4virology.net) that supports comparative genomics of large DNA viruses. Although the viruses supported by VBRC are primarily poxviruses and African swine fever virus (ASFV) because of our research interests, BBB is equally valuable for other viral genomes and nucleic acid and protein sequences, which can be imported into BBB from FASTA or GenBank files. The consistent theme running throughout the development of VBRC's tools has been to provide the virologist/biologist with easy-to-use graphical tools that let the user visualize and interact with the raw sequence data. The simplest example of this is providing the ability to quickly and visually scan a viral genome MSA for alignment errors and make manual corrections or use a second alignment algorithm to realign a section of a larger MSA. Without visually reviewing the output of a bioinformatics analysis, users may unknowingly use algorithms and parameters that are not appropriate for their analysis. For example, different tools may include or ignore gaps when calculating percent identity between sequence pairs; for example, counting a single gap of 100 nucleotides as 100 mismatches in one of a pair of sequences (1 kb alignment) that are otherwise 99% identical would create the illusion of 89% identity. The performance of a visual assessment of MSAs should be a routine "reality check" for researchers, rather than relying solely on the numerical output of alignment tools.

Previously described features of BBB include:

- Java code; uses Java Web Start to launch and automate updating of BBB for users;
- Integration with the VBRC's viral genome database (viral orthologous clusters; VOCs);
- Alignment of sequences or subsequences (MUSCLE, ClustalW, MAFFT);
- MSA editing, with intuitive highlighting of differences between sequences;
- Applicability to gene, protein, and virus genome sequences;
- Ability to edit individual sequences;
- Intuitive graphical user interface (GUI) with ability to view sequence residues or full sequence summaries;
- Display of 6-reading frames; understands and links to gene location (data from GenBank files);
- Multiple methods to annotate sequences and MSAs, treating the BBB file as a "results notebook".

Over the years, as we have needed novel functions in our investigations of viral genomes, we have used BBB as our standard platform for analyses that required manipulation and visual presentation of DNA and protein sequences. Thus, what otherwise might have been a solitary Perl or Python script became a feature within BBB. This enhances the functionality of BBB and provides open user access to new functions as new scripts get added. As a result, BBB has become an integrated platform with multiple features that provides a common user-friendly interface for both input of sequences and output of results.

This communication describes new features that have been incorporated into BBB: New "Advanced/Experimental tools" include j-CODEHOP, **Find Differences**, **SNIP**, and **MAFFT-add**; new "Reports" include **Get Counts**, **Get Unique Positions**, and **Get SNP Counts** (of top 2 sequences); new "Tools" that export alignment after deletion of **Specified Columns** or **Columns Containing Gap(s)**. It should be noted that these single nucleotide differences between viral genomes are not "polymorphisms" in the strictest sense, but we have used the term SNP (single nucleotide polymorphism) as it is a recognizable and understandable term among virologists.

2. Materials and Methods

When the BBB project began more than 15 years ago, Java was chosen as the coding language because (1) web browsers were not as capable as today's JavaScript powered interfaces; (2) Java was the primary language taught to undergraduates at the University of Victoria; and (3) it was relatively platform-independent, promising "code once, deploy everywhere" capability, thereby eliminating the compilation and installation obstacles that hinder users that want to try out new tools.

This updated software application is available from the www.4virology.net website (previously virology.uvic.ca) and code is made available upon request under the GNU General Public License version 3. In addition, the BBB and "consensus-degenerate hybrid oligonucleotide primers" (CODEHOP) source code described below has recently been submitted to the GitHub repository (https://github.com/vbrclab/basebybase; https://github.com/vbrclab/Codehop).

j-CODEHOP can be launched from within BBB (menu: **Advanced**) or from its own web page on the www.4virology.net website. In each case, the initial step is the download of the BBB alignment editor configuration file (*.jnlp), which is started by Java Web Start on the user's own computer by default. Java Web Start requires at least Java 7, but less than Java 11 to run. The Java Runtime Environment (JRE) can be downloaded for free, if needed.

Multiple "help" files for BBB and CODEHOP are available, including a "quick start page", "how to doc" and "help book", which get progressively more detailed. A j-CODEHOP tutorial is also provided. Although users are requested to register their email for use of the VBRC, this is only used to allow the resource to email users occasionally to make them aware of important new features; many users choose to use nonidentifying email addresses.

The cowpox viruses (CPXV) used as examples are: BR (AF482758.2), Norway 1994 MAN (HQ420899.1), Germany 1998 2 (HQ420897.1), Germany 1980 EP4 (HQ420895.1), Germany 2002

MKY (HQ420898.1), EleGri07/1 (KC813507.1), BeaBer04/1 (KC813491.1), RatHei09/1 (KC813504.1), GRI-90 (X94355.2), and HumGra07 (KC813510.1). The core conserved nucleotide alignment (60 kb) was used to generate a maximum-likelihood phylogenetic tree using the GTRGAMMA model in RAxML v.8.2.10 [3].

Currently, BBB has an upper limit of about 500 protein sequences (300 aa each) due to the memory assigned to the tool.

3. Results

3.1. CODEHOP Integration

Even though vast amounts of genomic sequences have been obtained recently, it is unlikely that the complete genome sequence will have been determined for all living species that might provide valuable scientific and medical insights. In order to obtain sequence information for specific genes in unsequenced organisms or pathogens, a primer design strategy for PCR amplification of novel genes using "consensus-degenerate hybrid oligonucleotide primers" (CODEHOPs) was previously developed [4]. CODEHOPs are designed from amino acid sequence motifs that are highly conserved within a gene family, and are used in PCR amplification to identify unknown related family members. Each CODEHOP consists of a pool of primers containing all possible nucleotide sequences within a 3′ degenerate core encoding a stretch of 3–4 highly conserved amino acids (Figure 1). A longer 5′ nondegenerate clamp region in the primers contains the most probable nucleotide predicted for each flanking codon. The degenerate core allows primer binding to all existing target variations in the initial PCR cycles, while the clamp region, once integrated into early PCR products, leads to efficient amplification of the PCR products in later PCR cycles. CODEHOPs designed from two adjacent conserved motifs are used to amplify the gene sequences between these motifs.

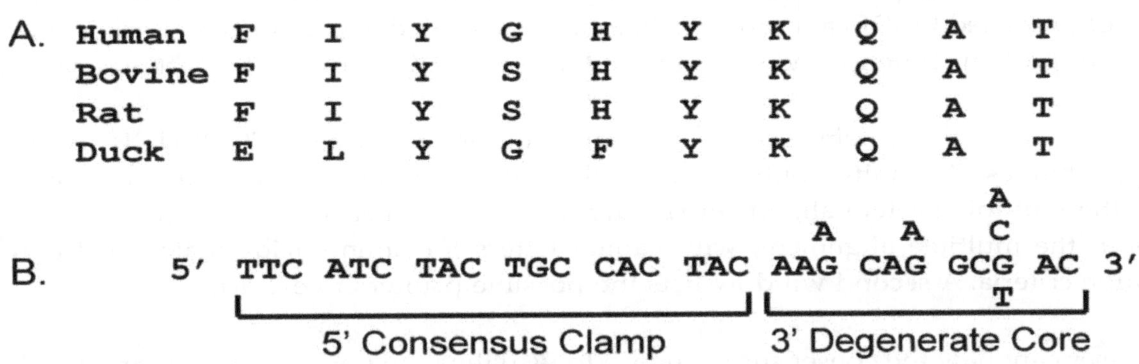

Figure 1. Anatomy of a "consensus-degenerate hybrid oligonucleotide primers" (CODEHOP) PCR primer. A CODEHOP is a pool of related primers containing all possible nucleotide sequences encoding 3 to 4 highly conserved amino acids within a 3′ degenerate core and a 5′ consensus clamp containing the most probable nucleotide at each position for the flanking codons. (**A**) multiple alignment of protein sequences; (**B**) predicted CODEHOP primer pool.

Other methods to identify unknown genes have used degenerate primers, containing most or all of the possible nucleotide sequences encoding amino acid motifs, or a consensus primer containing the most common nucleotide at each codon position within the motifs. However, unlike strictly degenerate or consensus approaches, the CODEHOP PCR approach has proven to be highly successful in amplifying distantly related genes containing significant sequence variations at low copy numbers. The primer design software and the CODEHOP PCR strategy have been utilized for the identification and characterization of new gene orthologs and paralogs in different plant, animal, and bacterial species, as well as for virus typing (e.g., enteroviruses [5]); consequently, the original publication has been cited in more than 800 subsequent publications. In addition, this approach has been successful in identifying new pathogen species and genes, as we have previously published [6–12].

A computer strategy to predict CODEHOP PCR primers from multiply aligned sets of related protein sequences was previously developed, which has been continuously accessible over the internet since 1998 as an integral part of the BLOCKS database developed by Steven Henikoff and hosted by the Fred Hutchinson Cancer Research Center [4]. A description of the CODEHOP program and its uses was published in the 2003 NAR Web services edition [13]. Subsequently, we developed iCODEHOP, an interactive web application independent of the BLOCKS database, to simplify and automate the process of designing CODEHOP PCR primers. The iCODEHOP program added new features, including interactive visualization of predicted CODEHOPs, phylogenetic plots for multiple aligned sequences, and user sessions on the server that allowed data to be stored during the design process [14]. However, due to advances in web browser technology, problems with the stored server sessions, and resource limitations, the iCODEHOP web application could no longer be supported and we have now developed a Java-based iteration called j-CODEHOP for integration into BBB (menu: **Advanced**).

j-CODEHOP guides users through the CODEHOP PCR primer design process, including uploading sequences, creating a multiple alignment, and identifying and visualizing primer pools that match the specified design criteria. A linked tutorial provides a step-by-step guide to demonstrate how to create CODEHOPs, using a sample FASTA file containing related sequences within the uracil DNA glycosylase family. The input to j-CODEHOP can be a set of nonaligned protein sequences or a set of aligned protein sequences. Protein sequence files may be formatted as GenBank (*.gb, *.gbk), EMBL (*.embl), BBB (*.bbb), FASTA (*.fasta, *.fas, *.fa) or CLUSTAL (*.clustal, *.clustalw).

The program's output includes a graphic showing predicted CODEHOP primers at their locations along a consensus protein sequence, a graphical representation of the region of the multiple alignment from which they are derived, and a set of metadata about each primer pool (length, degeneracy, and annealing temperature range). j-CODEHOP enables the user to visually scan the entire set of predicted CODEHOP primers to assess their relative positions and orientations within the consensus protein sequence and select individual CODEHOP primers for further analysis.

For the aligned protein sequences and chosen criteria, j-CODEHOP computes all primer possibilities. The initial output shows the consensus amino acid sequence for conserved blocks of the multiple protein alignment (Figure 2A). This sequence is numbered according to the positions in the multiple alignment, with capital letters for amino acids matching the minimum conservation criteria. A second window lists the possible primers to export.

The user can view the consensus sequence to visualize the positions of the predicted primers, which are shown as arrows, forward or reverse. The amino acid motif targeted by the 3′ degenerate core of the primer is aligned with the primer arrow, as are the flanking amino acids specifying the 5′ non-degenerate clamp. A specific primer can be selected, which will open a third window to show the CODEHOP sequence, the block of aligned sequences used for primer design, and the primer design criteria (Figure 2B). Both forward and reverse CODEHOPs in the correct orientation need to be identified. If an insufficient number of CODEHOPs are predicted, the program can be rerun using more relaxed design criteria or the distance between the group of sequences can be reduced. Detailed methodologies have been previously published that describe the design of CODEHOP PCR primers and their use in identifying novel sequences [4,11,13,14].

Figure 2. j-CODEHOP primer design output. The uracil DNA glycosylase test data set was used as input for the j-CODEHOP program and the following options were used for primer design: (1) Block making alignment tool—"MUSCLE"; (2) Codon table—"Homo sapiens"; (3) Clamp (nondegenerate 5′ region) length—"25"; (4) Core (degenerate 3′ region)—max degeneracy "16", length of degenerate core in aa "4", strictness (%) "0", min aa conservation (%) "80". Default values were used for the Advanced options: (1) 3′ nucleotide—"Invariant 3′ nt"; (2) Min block length—"5"; Primer concentration (nM)—"50"; (3) Restrict 3′ nucleotide to G or C—"unchecked"; (4) Exclude Leu, Ser, and Arg from 3′ region—"unchecked". (**A**) Initial graphical output showing the consensus amino acid sequence for the ordered blocks of multiply aligned protein sequences. Amino acids showing conservation above the chosen minimum value are capitalized. The positions of predicted CODEHOP PCR primers are indicated showing the extent of the amino acid sequence used for primer design, the direction of the primer (forward (F) or reverse (R)), and the degeneracy of the 3′ degenerate region, i.e., 4×, 8× or 16×. A CODEHOP with 4× degeneracy is composed of a pool of 4 different primers that provide all possible sequences encoding the 3–4 highly conserved amino acid motif targeted by the CODEHOP primer. The amino acid sequence of the motif is used to name the primer, ex. "PWNY". (**B**) The output obtained by clicking on a primer of interest in the initial graphical output, in this case primer "PWNY-F 8×". This output shows the primer sequence (5′ to 3′), with the 5′ nondegenerate consensus region in capital letters and the 3′ degenerate region in small letters, using the international code for ambiguous nucleotides, i.e., "Y" (C,T), "R" (A,G), "N" (A,C,G,T), etc. The codons in the primer sequence are aligned with the block of multiply aligned protein sequences. Amino acid positions showing conservation above the chosen minimum are indicated with an asterisk. Amino acids within the multiple alignment which are identical to the consensus sequence are indicated with a dot. The metadata for the chosen primer design criteria are indicated, as is the primer location in amino acids and base pairs. A third panel (not shown) provides a list of primers predicted from the current amino acid block to export. The primer sequence and metadata can be exported in a "comma-separated values" (CSV) spreadsheet format. The panels shown are high-resolution representations of program output.

3.2. Sequence Characteristics

As the -omics revolution progresses, more and more researchers make use of sequence data from the various databases and, increasingly, the data behind publication claims are not presented. For example, a phylogenetic tree may be published without the MSA that was used to generate it. Given that errors in sequence naming and annotation are common in the databases, it is important that researchers check results if they are going to rely on them. Figure 3 shows a BBB **Visual Summary** of 2 virus genomes (menu: **Reports**), which are almost identical except for four large indels and a block of very poorly matching sequence. This type of visualization is a powerful tool for highlighting inconsistencies in alignments. When we further investigated these sequences (BLAST [15] searches and dotplots [16]), we found that the differences between the two genomes were entirely the result of genome assembly errors.

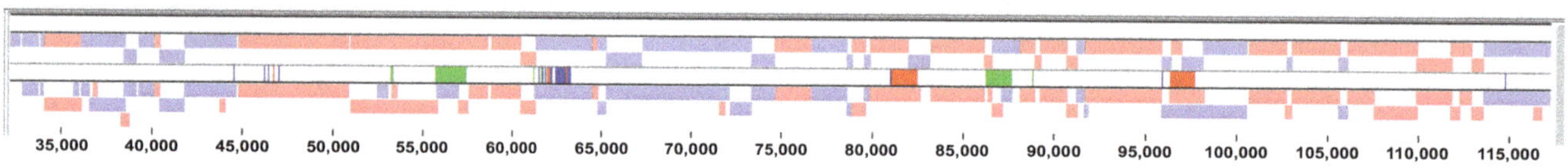

Figure 3. Visual summary from BBB. Pink and pale blue boxes represent genes transcribed to the right and left, respectively, for the genomes of two poxviruses. The centre tract indicates differences between the two sequences: Dark blue lines are SNPs (the abruptly dense SNPs turns out to be falsely assembled out-of-frame sequence from another virus) and green and red blocks show insertions and deletions (erroneously transposed sequences).

Additionally, under the BBB **Reports** menu, the ability to display **Sequence Similarity** and **Sequence Difference** graphs (useful for detecting recombination; not shown) has been supplemented by the plotting of a **Nucleotide Content Graph**. The user has control over which nucleotides are included in the analysis, as well as the size of the sliding window of nucleotides and the number of nucleotides that is used to "step" across the sequence. The tool also allows the user to choose which sequences from an MSA are included in these analyses. Importantly, the option to ignore gapped columns in an MSA has been included.

Additional new **Reports** features that summarize characteristics of an MSA include: (1) **Get Counts**, which counts the number of columns in the MSA with particular features, reporting the number that have a gap, a single nucleotide, two nucleotides (consensus and second type), three nucleotides, and four nucleotides; (2) **Get Unique Positions**, which lists the number of unique positions that are not gaps for each sequence; and (3) **Get SNP Counts**, which examines the top two sequences (sequences can be moved up or down within the MSA to enable sequence selection) and reports the total number of SNPs and the number of each possible substitution.

3.3. Counting Nucleotides Associated with Specific Sequences in MSAs

As noted above, the data supporting a phylogenetic tree are not often provided in manuscripts. Often, it would be useful to know the percent identity between sequences and the numbers of SNPs that distinguish one branch on a tree from another. The ability to generate a nucleotide identity matrix from an MSA is an older feature of BBB. However, now, from within the **Advanced/Experimental Tools** menu, BBB also allows a researcher to query the MSA data that support (or don't support) a phylogenetic tree. The **Find Differences** tool can be used to count the number of SNPs that support a particular branch; e.g., "find nucleotides that are identical in sequences A, B, and C, but different in all other sequences". Figure 4 shows the phylogenetic tree for the central relatively-conserved core (60 kb) of 10 cowpox viruses. For these sequences, the viruses in the DNA sequence identity range from 98.2–99.4%. Counting the number of SNPs unique to each sequence (red numbers in Figure 4) shows that for these cowpox sequences, the branch lengths created may not truly reflect the evolutionary distances. Instead, the lengths were likely compressed due to evidence of recombination shown in Table 1, which artificially reduced distances between distant strains.

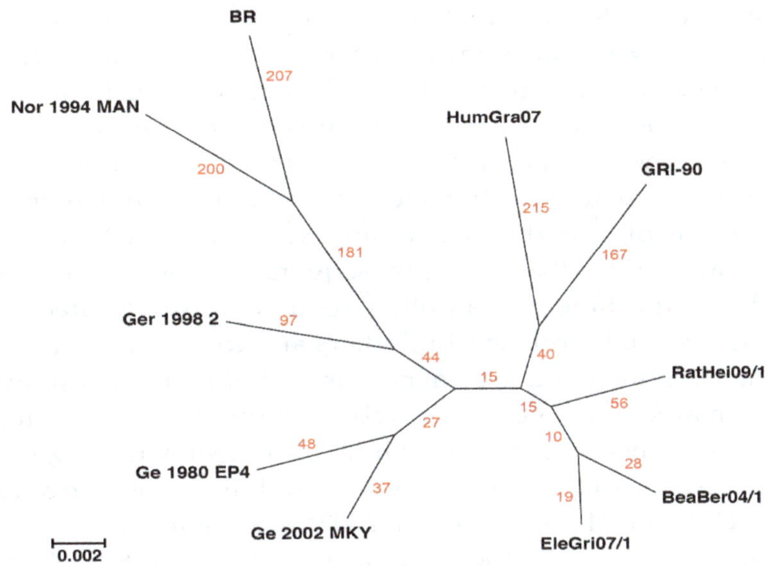

Figure 4. The contrast of unique SNPs found in the genomic core of 10 cowpox viruses (using the BBB **Find Differences** feature) with that of a maximum-likelihood phylogenetic tree. Red numbers denote the number of unique SNPs found for the virus that are not shared with any of the others. The phylogenetic tree branch scale denotes the average number of nucleotide substitution per site.

Table 1. SNPs shared by CPXV-BR, CPXV-Nor1994MAN, and strain noted in the table; all other viruses in Figure 4 have a different nucleotide. SNPs close together are grouped on a single line in the table.

+BeaBer04/1
22,518, 22,519, 22,583
31,870
+RatHei09/1
4677, 4679
4886, 4896, 4899, 4917
9401
16,480
19,731
31,573
35,003
40,781
+Ge 1980 EP4
1204
10,615, 10,618
10,731, 10,747
14,442, 14,457, 14,460, 14,553, 14,574, 14,664, 14,667
15,071, 15,072, 15,076, 15,138, 15,161, 15,163, 15,171
19,409
25,381
30,528, 30,534, 30,547, 30,549, 30,556
32,797, 32,799
35,713, 35,758, 35,812
36,217
41,120
+Ge 2002 MKY,
19,510
47,392

An important feature of the **Find Differences** tool (menu: **Advanced/Experimental Tools**) is that it can allow the matching to be fuzzy. We have termed this feature "tolerance" and it can be viewed as the search "tolerating" one or more (specified by the user) sequences that do not fulfill the query. For example, the query "find nucleotides that are identical in sequences A and B but different in all other sequences, with tolerance = 1" allows any one of the sequences that should be different from A+B to be the same; different sequences are "tolerated" at different positions in the alignment. The software also: (1) Creates a list of all the positions in the alignment that satisfy the query and displays the "tolerated sequence" name if there is one, and (2) displays the distribution of SNPs in the MSA.

These BBB features were created to characterize recombination events among the poxviruses by highlighting the positions of shared SNPs. In any MSA, there will always be coincident SNPs from random events. However, for these cowpox sequences, when "nucleotides that are identical in sequences A, B, and C, but different in all other sequences" are located, some are, as expected, associated with the closest related sequence, but others are from more distant relatives. In addition, many of these coincident SNPs are found to be in nonrandom blocks, suggesting that the arrangements result from recombination among the genomes. Table 1 shows SNPs present only in CPXV-BR and CPXV-Nor1994MAN and one other sequence taken from the tree shown in Figure 4. In several instances, the common SNPs are unexpectedly clustered (Table 1) and likely result from recombination events. The results with CPXV-Ge1980EP4 and CPXV-Ge2002MKY (which are very similar (Figure 4)) as the extra sequence are dramatic; despite their similarity, CPXV-Ge1980EP4 has many more SNPs in common with the other two sequences (Table 1; 33 SNPs) than with CPXV-Ge2002MKY (Figure 4; 3 SNPs).

3.4. Manipulation of Sequences

As previously reported, BBB allows the addition or removal of sequences to an alignment and the removal of columns in an MSA that contains all gap characters which are often generated when removing sequences from an alignment. However, when visually inspecting the relationships between the sequences, it can also be useful to simplify the variation by removing any column that contains a gap character (menu: **Tools/Delete Columns Containing Gap(s) and Export**). Since this action will modify the sequences in use, by deleting residues from some sequences, the program will export the resulting sequence into a new BBB window and prompt the user to enter a new filename. If the sequences in an MSA are very diverged, each will have a relatively large number of unique SNPs. Since these can obscure patterns present among the SNPs shared by subsets of sequences, we also created the **SNIP** feature (menu: **Advanced/Experimental Tools**) that modifies the sequences such that the SNPs that are present only in a single sequence are changed to the consensus nucleotide. Again, because this procedure modifies the actual sequences, users are asked to save the result in a new alignment file.

When using large viral genomes and closely related viruses, SNPs may be relatively infrequent. Therefore, we incorporated a feature into BBB that allows the user to remove any specified column of nucleotides within an MSA. By removing the columns that only contain a single nucleotide, the variation is compressed into a much smaller sequence space and is more easily visualized by the user. First, the **Find Differences** tool (menu: **Advanced/Experimental Tools**) is used to find the columns that are identical (i.e., have no SNPs), then the "Search Log" is used to "List SNP Positions" only. Subsequently, these position values can then be used to delete specified columns (menu: **Tools/Delete Specified Columns and Export**).

3.5. Alignment of Sequences

The options for aligning complete or selected regions of sequences have been updated. Clustal Omega [17] has replaced the option to use ClustalW. Clustal Omega and MUSCLE [18] serve as options to align protein and gene length nucleotide sequences. For the alignment of large viral genomes, MAFFT [19] is the tool of choice. However, the growing number of complete genomes sequenced has translated into a more frequent need to generate larger MSAs, often to update phylogenetic trees. Although MAFFT is available at various web resources and can be easily installed on desktop computers, most users prefer to use MAFFT within BBB (menu: **Tools/Align Selection**). Therefore, we have incorporated the **MAFFT-add** option into the BBB (menu: **Advanced/Experimental Tools**) [20]. This feature allows users to align one or more new sequences to an existing alignment, which significantly reduces the compute time. For example, the alignment of 10 cowpox virus genomes takes approximately 8 min, whereas aligning one new sequence to an alignment of 9 takes a little over 1 min. The **MAFFT-add** function is also useful for scaffolding new contig sequences against a close reference sequence in the process of genome assembly.

4. Discussion

BBB is a foundational program of the Viral Bioinformatics Resource Centre that allows both the viewing and editing of MSAs. It has been developed over many years and serves as a platform for the comparison of large viral genomes, but it is equally useful for small DNA and RNA viral genomes, as well as gene and protein sequence alignments. The data visualization features that it provides are key to its value. These include highlighting differences between sequences in an MSA, plotting graphs of sequence similarity, adding user-comments or primers to sequence regions, and displaying forward and reverse reading frames as well as results of various sequence searches. Importantly, while BBB has provided a common interface to multiple analyses for users, it has also given programmers a standardized data input process and a single visualization canvas. One of the key features of BBB is the ability to read GenBank files so that it becomes aware of the complete set of annotations for the genomes of large viruses, allowing it to display the complete set of annotations for any large virus genome which is extremely useful for comparisons of gene features and single nucleotides.

Here, we have described a significant number of upgrades to BBB that increase the utility of the tool when working with genes, proteins or genomes. The inclusion of j-CODEHOP maintains a functioning version of the CODEHOP algorithm, which otherwise would have been lost to the research community. CODEHOP is a natural fit for BBB, since it is often used to discover novel members of viral families, but also since the generation of the protein MSAs that are used as input by j-CODEHOP to generate CODEHOP primer sequences are already an integral part of BBB. The new plots for similarity are useful for a user looking for recombination events and comparing how different genes are conserved to different degrees. Fuzzy searches for MSA columns that support or fail to support particular phylogenetic relationships bring a new process to the screening of viral genomes for small regions that have been exchanged among viruses. Following the identification of particular MSA nucleotide columns, new editing features in BBB now allow the user to manipulate these MSA columns, thereby simplifying the visualization for the user.

A variety of other tools exist to manipulate and characterize MSAs, including Jalview [21] (primarily for proteins), and IVisTMSA [22] and AliView [23] (primarily for large sequence sets). However, BBB has multiple unique features, including those described here, which make it a valuable multipurpose bioinformatics tool, especially for the comparison and characterization of viral genomes and more.

Author Contributions: Conceptualization, T.M.R. and C.U.; Methodology, C.M., J.P.S., and K.M.; Software, C.M.; Validation, K.M. and S.-L.T.; Writing—Original Draft Preparation, T.M.R., J.P.S., and S.-L.T.; Writing—Review and Editing, S.-L.T. and C.U.; Supervision, C.U.; Funding Acquisition, C.U.

Acknowledgments: The authors would like to thank Richard Boyce for the JavaScript CODEHOP code and the many University of Victoria Co-op students that have contributed to the software available at Viral Bioinformatics Resource Centre.

References

1. Brodie, R.; Smith, A.J.; Roper, R.L.; Tcherepanov, V.; Upton, C. Base-By-Base: Single nucleotide-level analysis of whole viral genome alignments. *BMC Bioinform.* **2004**, *5*, 96–99. [CrossRef] [PubMed]

2. Hillary, W.; Lin, S.-H.; Upton, C. Base-By-Base version 2: Single nucleotide-level analysis of whole viral genome alignments. *Microb. Inform. Exp.* **2011**, *1*, 2. [CrossRef] [PubMed]
3. Stamatakis, A. RAxML version 8: A tool for phylogenetic analysis and post-analysis of large phylogenies. *Bioinformatics* **2014**, *30*, 1312–1313.[CrossRef] [PubMed]
4. Rose, T.M.; Schultz, E.R.; Henikoff, J.G.; Pietrokovski, S.; McCallum, C.M.; Henikoff, S. Consensus-degenerate hybrid oligonucleotide primers for amplification of distantly related sequences. *Nucleic Acids Res.* **1998**, *26*, 1628–1635. [CrossRef] [PubMed]
5. Nix, W.A.; Oberste, M.S.; Pallansch, M.A. Sensitive, Seminested PCR Amplification of VP1 Sequences for Direct Identification of All Enterovirus Serotypes from Original Clinical Specimens. *J. Clin. Microbiol.* **2006**, *44*, 2698–2704. [CrossRef] [PubMed]
6. Van Devanter, D.R.; Warrener, P.; Bennett, L.; Schultz, E.R.; Coulter, S.; Garber, R.L.; Rose, T.M. Detection and analysis of diverse herpesviral species by consensus primer PCR. *J. Clin. Microbiol.* **1996**, *34*, 1666–1671.
7. Børsting, C.; Hummel, R.; Schultz, E.R.; Rose, T.M.; Pedersen, M.B.; Knudsen, J.; Kristiansen, K. *Saccharomyces carlsbergensis* contains two functional genes encoding the acyl-CoA binding protein, one similar to the ACB1 gene from *S. cerevisiae* and one identical to the ACB1 gene from *S. monacensis*. *Yeast* **1997**, *13*, 1409–1421. [CrossRef]
8. Wilson, C.A.; Wong, S.; Muller, J.; Davidson, C.E.; Rose, T.M.; Burd, P. Type C retrovirus released from porcine primary peripheral blood mononuclear cells infects human cells. *J. Virol.* **1998**, *72*, 3082–3087. [PubMed]
9. Rose, T.M. CODEHOP-mediated PCR—A powerful technique for the identification and characterization of viral genomes. *Virol. J.* **2005**, *2*, 20. [CrossRef] [PubMed]
10. Philipp-Staheli, J.; Marquardt, T.; Thouless, M.E.; Bruce, A.G.; Grant, R.F.; Tsai, C.-C.; Rose, T.M. Genetic variability of the envelope gene of Type D simian retrovirus-2 (SRV-2) subtypes associated with SAIDS-related retroperitoneal fibromatosis in different macaque species. *Virol. J.* **2006**, *3*, 11. [CrossRef] [PubMed]
11. Staheli, J.P.; Boyce, R.; Kovarik, D.; Rose, T.M. CODEHOP PCR and CODEHOP PCR primer design. *Methods Mol. Biol.* **2011**, *687*, 57–73. [PubMed]
12. Chouhy, D.; Kocjan, B.J.; Staheli, J.P.; Bolatti, E.M.; Hošnjak, L.; Sagadin, M.; Giri, A.A.; Rose, T.M.; Poljak, M. Detection of novel Betapapillomaviruses and Gammapapillomaviruses in eyebrow hair follicles using a single-tube "hanging droplet" PCR assay with modified pan-PV CODEHOP primers. *J. Gen. Virol.* **2018**, *99*, 109–118. [CrossRef] [PubMed]
13. Rose, T.M.; Henikoff, J.G.; Henikoff, S. CODEHOP (COnsensus-DEgenerate Hybrid Oligonucleotide Primer) PCR primer design. *Nucleic Acids Res.* **2003**, *31*, 3763–3766. [CrossRef] [PubMed]
14. Boyce, R.; Chilana, P.; Rose, T.M. iCODEHOP: A new interactive program for designing COnsensus-DEgenerate Hybrid Oligonucleotide Primers from multiply aligned protein sequences. *Nucleic Acids Res.* **2009**, *37*, W222–W228. [CrossRef] [PubMed]
15. Altschul, S.F.; Gish, W.; Miller, W.; Myers, E.W.; Lipman, D.J. Basic local alignment search tool. *J. Mol. Biol.* **1990**, *215*, 403–410. [CrossRef]
16. Brodie, R.; Roper, R.L.; Upton, C. JDotter: A Java interface to multiple dotplots generated by dotter. *Bioinformatics* **2004**, *20*, 279–281. [CrossRef] [PubMed]
17. Sievers, F.; Higgins, D.G. Clustal Omega, Accurate Alignment of Very Large Numbers of Sequences. In *Multiple Sequence Alignment Methods*; Methods in Molecular Biology; Humana Press: Totowa, NJ, USA, 2013; Volume 1079, pp. 105–116.
18. Edgar, R.C. MUSCLE: Multiple sequence alignment with high accuracy and high throughput. *Nucleic Acids Res.* **2004**, *32*, 1792–1797. [CrossRef] [PubMed]
19. Katoh, K.; Standley, D.M. MAFFT Multiple Sequence Alignment Software Version 7: Improvements in Performance and Usability. *Mol. Biol. Evol.* **2013**, *30*, 772–780. [CrossRef] [PubMed]
20. Kazutaka Katoh, M.C.F. Adding unaligned sequences into an existing alignment using MAFFT and LAST. *Bioinformatics* **2012**, *28*, 3144–3146. [CrossRef] [PubMed]
21. Waterhouse, A.M.; Procter, J.B.; Martin, D.M.A.; Clamp, M.; Barton, G.J. Jalview Version 2—A multiple sequence alignment editor and analysis workbench. *Bioinformatics* **2009**, *25*, 1189–1191. [CrossRef] [PubMed]
22. Pervez, M.T.; Babar, M.E.; Nadeem, A.; Aslam, N.; Naveed, N.; Ahmad, S.; Muhammad, S.; Qadri, S.; Shahid, M.; Hussain, T.; et al. IVisTMSA: Interactive Visual Tools for Multiple Sequence Alignments. *Evol. Bioinform. Online* **2015**, *11*, 35–42. [CrossRef] [PubMed]
23. Larsson, A. AliView: A fast and lightweight alignment viewer and editor for large datasets. *Bioinformatics* **2014**, *30*, 3276–3278. [CrossRef] [PubMed]

PERMISSIONS

All chapters in this book were first published by MDPI; hereby published with permission under the Creative Commons Attribution License or equivalent. Every chapter published in this book has been scrutinized by our experts. Their significance has been extensively debated. The topics covered herein carry significant findings which will fuel the growth of the discipline. They may even be implemented as practical applications or may be referred to as a beginning point for another development.

The contributors of this book come from diverse backgrounds, making this book a truly international effort. This book will bring forth new frontiers with its revolutionizing research information and detailed analysis of the nascent developments around the world.

We would like to thank all the contributing authors for lending their expertise to make the book truly unique. They have played a crucial role in the development of this book. Without their invaluable contributions this book wouldn't have been possible. They have made vital efforts to compile up to date information on the varied aspects of this subject to make this book a valuable addition to the collection of many professionals and students.

This book was conceptualized with the vision of imparting up-to-date information and advanced data in this field. To ensure the same, a matchless editorial board was set up. Every individual on the board went through rigorous rounds of assessment to prove their worth. After which they invested a large part of their time researching and compiling the most relevant data for our readers.

The editorial board has been involved in producing this book since its inception. They have spent rigorous hours researching and exploring the diverse topics which have resulted in the successful publishing of this book. They have passed on their knowledge of decades through this book. To expedite this challenging task, the publisher supported the team at every step. A small team of assistant editors was also appointed to further simplify the editing procedure and attain best results for the readers.

Apart from the editorial board, the designing team has also invested a significant amount of their time in understanding the subject and creating the most relevant covers. They scrutinized every image to scout for the most suitable representation of the subject and create an appropriate cover for the book.

The publishing team has been an ardent support to the editorial, designing and production team. Their endless efforts to recruit the best for this project, has resulted in the accomplishment of this book. They are a veteran in the field of academics and their pool of knowledge is as vast as their experience in printing. Their expertise and guidance has proved useful at every step. Their uncompromising quality standards have made this book an exceptional effort. Their encouragement from time to time has been an inspiration for everyone.

The publisher and the editorial board hope that this book will prove to be a valuable piece of knowledge for researchers, students, practitioners and scholars across the globe.

LIST OF CONTRIBUTORS

Jong-Sik Moon
National Core Research Center for Hybrid Materials Solution, Pusan National University, Busan 46241, Korea

Eun Jung Choi
Department of BIT Fusion Technology Center, Pusan National University, Busan 46241, Korea

Na-Na Jeong and Jong-Ryeul Sohn
Department of Public Health Sciences, Graduate School, Korea University, Seoul 02708, Korea

Dong-Wook Han
Department of Cogno-Mechatronics Engineering, College of Nanoscience & Nanotechnology, Pusan National University, Busan 46241, Korea

Jin-Woo Oh
Department of BIT Fusion Technology Center, Pusan National University, Busan 46241, Korea
Department of Nanofusion Technology, College of Nanoscience & Nanotechnology, Pusan National University, Busan 46241, Korea

Stephan Peter
Ernst-Abbe University of Applied Sciences Jena, Carl-Zeiss-Promenade 2, 07745 Jena, Germany
Bio Systems Analysis Group, Department of Mathematics and Computer Science, University of Jena, Ernst-Abbe-Platz 2, 07743 Jena, Germany

Martin Hölzer, Kevin Lamkiewicz and Manja Marz
RNA Bioinformatics and High-Throughput Analysis, Friedrich Schiller University Jena, Leutragraben 1, 07743 Jena, Germany
European Virus Bioinformatics Center, Leutragraben 1, 07743 Jena, Germany

Pietro Speroni di Fenizio and Peter Dittrich
Bio Systems Analysis Group, Department of Mathematics and Computer Science, University of Jena, Ernst-Abbe-Platz 2, 07743 Jena, Germany

Hassan Al Hwaeer
Mathematics and Computer Applications Department, Al-Nahrain University, Al-Jadriya, Baghdad 10072, Iraq

Stefan Schuster
Chair of Bioinformatics, Matthias-Schleiden-Institute, University of Jena, Ernst-Abbe-Platz 2, 07743 Jena, Germany

Bashar Ibrahim
European Virus Bioinformatics Center, Leutragraben 1, 07743 Jena, Germany
Chair of Bioinformatics, Matthias-Schleiden-Institute, University of Jena, Ernst-Abbe-Platz 2, 07743 Jena, Germany

Franziska Hufsky
European Virus Bioinformatics Center, 07743 Jena, Germany
RNA Bioinformatics and High-Throughput Analysis, Friedrich Schiller University Jena, 07743 Jena, Germany

Sejal Modha
MRC-University of Glasgow Centre for Virus Research, Glasgow G61 1QH, UK

Martha R. J. Clokie
Department of Genetics and Genome Biology, University of Leicester, Leicester LE1 7RH, UK

Stefanie Deinhardt-Emmer
European Virus Bioinformatics Center, 07743 Jena, Germany
Institute of Medical Microbiology, Jena University Hospital, Am Klinikum 1, D-07747 Jena, Germany
Section for Experimental Virology, Jena University Hospital, Hans-Knöll-Straße 2, D-07745 Jena, Germany
Center for Sepsis Control and Care, Jena University Hospital, D-07747 Jena, Germany

Bas E. Dutilh
European Virus Bioinformatics Center, 07743 Jena, Germany
Theoretical Biology and Bioinformatics, Science4Life, Utrecht University, Padualaan 8, Utrecht 3584 CH, The Netherlands
Centre for Molecular and Biomolecular Informatics, Radboud Institute for Molecular Life Sciences, Radboud University Medical Centre, Geert Grooteplein 26, Nijmegen 6525 GA, The Netherlands

Samantha Lycett
Infection & Immunity Division, Roslin Institute, University of Edinburgh, Midlothian EH25 9RG, UK

Peter Simmonds
Nuffield Department of Medicine, University of Oxford, Peter Medawar Building, South Parks Road, Oxford OX1 3SY, UK

Volker Thiel
European Virus Bioinformatics Center, 07743 Jena, Germany
Institute of Virology and Immunology, 3012 Bern, Switzerland
Department of Infectious Diseases and Pathobiology, Vetsuisse Facility, University of Bern, 3012 Bern, Switzerland

Aare Abroi
University of Tartu, Institute of Technology, 50411 Tartu, Estonia

Evelien M. Adriaenssens
European Virus Bioinformatics Center, 07743 Jena, Germany
Quadram Institute Bioscience, Norwich Research Park, Norwich NR4 7UQ, UK

Marina Escalera-Zamudio
Department of Zoology, University of Oxford, Parks Rd, Oxford OX1 3PS, UK

Jenna Nicole Kelly
Institute of Virology and Immunology, 3012 Bern, Switzerland
Department of Infectious Diseases and Pathobiology, Vetsuisse Facility, University of Bern, 3012 Bern, Switzerland

Lu Lu and Julian Susat
Institute of Clinical Molecular Biology, Kiel University, 24118 Kiel, Germany

Thomas Sicheritz
Natural History Museum of Denmark, University of Copenhagen, DK-1123 Copenhagen, Denmark

Luca D. Bertzbach and Benedikt B. Kaufer
Institute of Virology, Freie Universität Berlin, Robert von Ostertag-Straße 7-13, 14163 Berlin, Germany

Florian Pfaff and Dirk Höper
Institute of Diagnostic Virology, Friedrich-Loeffler-Institut, Federal Research Institute for Animal Health, Südufer 10, 17493 Greifswald-Insel Riems, Germany

Viktoria I. Pauker and Axel Karger
Institute of Molecular Virology and Cell Biology, Friedrich-Loeffler-Institut, Federal Research Institute for Animal Health, Südufer 10, 17493 Greifswald-Insel Riems, Germany

Ahmed M. Kheimar
Institute of Virology, Freie Universität Berlin, Robert von Ostertag-Straße 7-13, 14163 Berlin, Germany
Department of Poultry Diseases, Faculty of Veterinary Medicine, Sohag University, Sohag 82424, Egypt

Sonja Härtle
Department of Veterinary Sciences, Institute for Animal Physiology, Ludwig-Maximilians-Universität München, 80539 Munich, Germany

Dong-Myeong Shin
Department of Mechanical Engineering, The University of Hong Kong, Hong Kong 999077, China

Suck Won Hong
Department of Cogno-Mechatronics Engineering, Department of Optics and Mechatronics Engineering, Pusan National University (PNU), Busan 46241, Korea

Yoon-Hwae Hwang
Department of Nanoenergy Engineering & BK21 PLUS Nanoconvergence Technology Division, Pusan National University (PNU), Busan 46241, Korea

Lynda Handala, Tony Fiore and Francois Helle
EA4294, Agents Infectieux, Résistance et Chimiothérapie, Centre Universitaire de Recherche en Santé, Centre Hospitalier Universitaire et Université de Picardie Jules Verne, 80054 Amiens, France

Yves Rouillé
University of Lille, CNRS, INSERM, CHU Lille, Pasteur Institute of Lille, U1019-UMR8204-CIIL-Center for Infection and Immunity of Lille, 59019 Lille, France

Petia Atanasova, Eunjin Choi and Joachim Bill
Institute for Materials Science, University of Stuttgart, Heisenbergstr. 3, 70569 Stuttgart, Germany

Vladimir Atanasov and Jochen Kerres
Institute of Chemical Process Engineering, University of Stuttgart, Böblinger Straße 78, 70199 Stuttgart, Germany

Lisa Wittum, Christina Wege and Sabine Eiben
Institute of Biomaterials and Biological Systems, University of Stuttgart, Pfaffenwaldring 57, 70569 Stuttgart, Germany

Alexander Southan
Institute of Interfacial Process Engineering and Plasma Technology, University of Stuttgart, Pfaffenwaldring 31, 70569 Stuttgart, Germany

Michael Kiening
Department of Bioinformatics, Wissenschaftszentrum Weihenstephan, Technische Universität München, Maximus-von-Imhof-Forum 3, D-85354 Freising, Germany

Hans-Jörg Hellinger and Thomas Rattei
Division of Computational Systems Biology, Department of Microbiology and Ecosystem Science, University of Vienna, Althanstraße 14, 1090 Vienna, Austria

Ivo Hofacker
Research Group Bioinformatics and Computational Biology, Faculty of Computer Science, University of Vienna, Währingerstr. 29, 1090 Vienna, Austria
Department of Theoretical Chemistry, Faculty of Chemistry, University of Vienna, Währingerstrasse 17, 1090 Vienna, Austria

Dmitrij Frishman
Department of Bioinformatics, Wissenschaftszentrum Weihenstephan, Technische Universität München, Maximus-von-Imhof-Forum 3, D-85354 Freising, Germany
St. Petersburg State Polytechnic University, St. Petersburg 195251, Russia

Martina Zoli, Susanne Koethe, Martin Beer and Michael Eschbaumer
Institute of Diagnostic Virology, Friedrich-Loeffler-Institut, Federal Research Institute for Animal Health, 17493 Greifswald, Germany

Sara Hägglund and Jean-François Valarcher
Swedish University of Agricultural Sciences, Host-pathogen interaction group, Division of Ruminant Medicine, 75007 Uppsala, Sweden

Sandra Blaise-Boisseau, Stephan Zientara and Labib Bakkali-Kassimi
Laboratoire de Santé Animale de Maisons-Alfort, UMR 1161 virologie, INRA, Ecole Nationale Vétérinaire d'Alfort, ANSES, Université Paris-Est, 94700 Maisons-Alfort, France

Eve Laloy
Laboratoire de Santé Animale de Maisons-Alfort, UMR 1161 virologie, INRA, Ecole Nationale Vétérinaire d'Alfort, ANSES, Université Paris-Est, 94700 Maisons-Alfort, France
Biopôle EnvA, Ecole Nationale Vétérinaire d'Alfort, Université Paris-Est, 94700 Maisons-Alfort, France

Daniela Zühlke and Katharina Riedel
Institute of Microbiology, Department for Microbial Physiology and Molecular Biology, University of Greifswald, 17489 Greifswald, Germany

Joshua B. Singer, Emma C. Thomson, Joseph Hughes, Elihu Aranday-Cortes, John McLauchlan, Ana da Silva Filipe, Lily Tong and Robert J. Gifford
MRC-University of Glasgow Centre for Virus Research, Glasgow G61 1QH, UK

Eleanor Barnes, M. Azim Ansari and David Smith
Peter Medawar Building for Pathogen Research, Nuffield Department of Medicine, University of Oxford, Oxford OX1 3SY, UK

Jean L. Mbisa, David F. Bibby, Daniel Bradshaw and Carmen F. Manso
Virus Reference Department, National Infection Service, Public Health England, Colindale, London NW9 5EQ, UK

Iruthayapandi Selestin Raja
Monocrystalline Bank Research Institute, Pusan National University, Busan 46241, Korea

Chuntae Kim and Jin-Woo Oh
Department of Nanofusion Technology, College of Nanoscience & Nanotechnology, Pusan National University, Busan 46241, Korea

Su-Jin Song, Moon Sung Kang and Dong-Wook Han
Department of Cogno-Mechatronics Engineering, College of Nanoscience & Nanotechnology, Pusan National University, Busan 46241, Korea

Yong Cheol Shin
Department of Medical Engineering, Yonsei University, College of Medicine, Seoul 03722, Korea

Suong-Hyu Hyon
Joint Faculty of Veterinary Medicine, Kagoshima University, Kagoshima 890-8580, Japan

Roman Ochsenreiter
Department of Theoretical Chemistry, University of Vienna, Währingerstraße 17, 1090 Vienna, Austria

Ivo L. Hofacker and Michael T. Wolfinger
Department of Theoretical Chemistry, University of Vienna, Währingerstraße 17, 1090 Vienna, Austria
Research Group BCB, Faculty of Computer Science, University of Vienna, Währingerstraße 29, 1090 Vienna, Austria

Avraam Tapinos
School of Biological Sciences, The University of Manchester, Manchester M13 9PT, UK

Bede Constantinides
School of Biological Sciences, The University of Manchester, Manchester M13 9PT, UK
Modernising Medical Microbiology Consortium, Nuffield Department of Clinical Medicine, John Radcliffe Hospital, University of Oxford, Oxford OX3 9DU, UK

My V. T. Phan
Department of Viroscience, Erasmus Medical Centre, Doctor Molewaterplein 40, 3015 GD Rotterdam, The Netherlands

Samaneh Kouchaki
School of Biological Sciences, The University of Manchester, Manchester M13 9PT, UK
Department of Engineering Science, Institute of Biomedical Engineering, University of Oxford, Oxford OX3 7DQ, UK

Matthew Cotten
Department of Viroscience, Erasmus Medical Centre, Doctor Molewaterplein 40, 3015 GD Rotterdam, The Netherlands
MRC-University of Glasgow Centre for Virus Research, Glasgow G61 1QH, UK
MRC/UVRI & LSHTM Uganda Research Unit Entebbe, P.O. Box 49 Entebbe, Uganda

David L. Robertson
School of Biological Sciences, The University of Manchester, Manchester M13 9PT, UK
MRC-University of Glasgow Centre for Virus Research, Glasgow G61 1QH, UK

Shin-Lin Tu, Colum McClay, Kathleen McLeod and Chris Upton
Biochemistry and Microbiology, University of Victoria, Victoria, BC V8W 2Y2, Canada

Jeannette P. Staheli
Center for Global Infectious Disease Research, Seattle Children's Research Institute, Seattle, WA 98101, USA

Timothy M. Rose
Center for Global Infectious Disease Research, Seattle Children's Research Institute, Seattle, WA 98101, USA
Department of Pediatrics, University of Washington, Seattle, WA 98195, USA

Index

www.ingramcontent.com/pod-product-compliance
Lightning Source LLC
LaVergne TN
LVHW060513210926
848485LV00005B/1077